Advances in International Environmental Politics

WITHDRAWN
UTSA LIBRARIES

Advances in International Environmental Politics

2nd edition

Edited by

Michele M. Betsill
Professor, Colorado State University, USA

Kathryn Hochstetler
Professor, University of Waterloo, Canada

Dimitris Stevis
Professor, Colorado State University, USA

Editorial matter, selection, introduction and conclusion © Michele M. Betsill, Kathryn Hochstetler, and Dimitris Stevis 2014
Individual chapters © Respective authors 2014

All rights reserved. No reproduction, copy or transmission of this publication may be made without written permission.

No portion of this publication may be reproduced, copied or transmitted save with written permission or in accordance with the provisions of the Copyright, Designs and Patents Act 1988, or under the terms of any licence permitting limited copying issued by the Copyright Licensing Agency, Saffron House, 6–10 Kirby Street, London EC1N 8TS.

Any person who does any unauthorized act in relation to this publication may be liable to criminal prosecution and civil claims for damages.

The authors have asserted their rights to be identified as the authors of this work in accordance with the Copyright, Designs and Patents Act 1988.

First published 2014 by
PALGRAVE MACMILLAN

Palgrave Macmillan in the UK is an imprint of Macmillan Publishers Limited, registered in England, company number 785998, of Houndmills, Basingstoke, Hampshire RG21 6XS.

Palgrave Macmillan in the US is a division of St Martin's Press LLC, 175 Fifth Avenue, New York, NY 10010.

Palgrave Macmillan is the global academic imprint of the above companies and has companies and representatives throughout the world.

Palgrave® and Macmillan® are registered trademarks in the United States, the United Kingdom, Europe and other countries.

ISBN 978–1–137–33896–9 hardback
ISBN 978–1–137–33899–0 paperback

This book is printed on paper suitable for recycling and made from fully managed and sustained forest sources. Logging, pulping and manufacturing processes are expected to conform to the environmental regulations of the country of origin.

A catalogue record for this book is available from the British Library.

A catalog record for this book is available from the Library of Congress.

Contents

List of Figures and Tables	vii
Notes on Contributors	viii
List of Acronyms	xiv

1. General Introduction   1
   Michele M. Betsill, Kathryn Hochstetler, and Dimitris Stevis

Part I The Context of International Environmental Politics

2. The Trajectory of International Environmental Politics   13
   Dimitris Stevis

3. Theoretical Perspectives on International Environmental Politics   45
   Matthew Paterson

4. Methods in International Environmental Politics   78
   Kathryn Hochstetler and Melinda Laituri

Part II Major Research Areas

5. International Political Economy and the Environment   107
   Jennifer Clapp

6. Gender and International Environmental Politics   137
   Nicole Detraz

7. Knowledge and the Environment   161
   Eva Lövbrand

8. Transnational Actors in International Environmental Politics   185
   Michele M. Betsill

9. Environmental Security   211
   Larry A. Swatuk

10 Global Governance and the Environment 245
    Frank Biermann

Part III Frameworks for Evaluating Global Environmental Politics

11 The Effectiveness of International Environmental Regimes: Existing Knowledge, Cutting-edge Themes, and Research Strategies 273
    Oran R. Young

12 Sustainable Development: The Institutionalization of a Contested Policy Concept 300
    Sander Happaerts and Hans Bruyninckx

13 Environmental and Ecological Justice 328
    Chukwumerije Okereke and Mark Charlesworth

14 Transparency and International Environmental Politics 356
    Aarti Gupta and Michael Mason

15 General Conclusion 381
    Michele M. Betsill, Kathryn Hochstetler, and Dimitris Stevis

Index 391

Figures and Tables

Figures

3.1 Relationship between theoretical starting points and theoretical perspectives in IEP 47

Tables

13.1 Headcount of delegates from various countries who attended annual UN climate change meetings 342

Contributors

Michele M. Betsill is Professor of Political Science as well as founder and co-leader of the Environmental Governance Working Group at Colorado State University, USA. Her research investigates the multiple ways in which global environmental issues are governed from the global to the local level across the public and private spheres, with particular interest in questions of politics and authority in global climate governance. She is the author (with Harriet Bulkeley) of *Cities and Climate Change: Urban Sustainability and Global Environmental Governance* (2003) and co-editor (with Elisabeth Corell) of *NGO Diplomacy: The Influence of Non-governmental Organizations in International Environmental Negotiations* (2008). Her current projects focus on the governance and legitimacy of carbon markets, transnational environmental governance, and the governance of low-carbon transitions. She is also the co-editor of *Global Environmental Change* and a member of the scientific steering committee of the Earth System Governance network.

Frank Biermann is Professor of Political Science and of Environmental Policy Sciences at the VU University Amsterdam, The Netherlands, and Visiting Professor of Earth System Governance at Lund University, Sweden. He specializes in the study of global environmental politics, with emphasis on climate negotiations, UN reform, global adaptation governance, public–private governance mechanisms, the role of science, and North–South relations. Biermann has authored, co-authored, or edited 16 books and published about 150 articles in peer-reviewed journals and chapters in academic books, along with more than 100 papers, reports, and contributions to policy-oriented journals. His most recent book is *Earth System Governance: World Politics in the Anthropocene*, forthcoming. Biermann serves as General Director of the Netherlands Research School for Socio-Economic and Natural Sciences of the Environment and chairs the Earth System Governance Project, a global research network under the new alliance Future Earth.

Hans Bruyninckx is Executive Director of the European Environment Agency (EEA), having taken office on June 1, 2013. He completed his PhD in 1996 at Colorado State University, USA, on the topic of international environmental politics, subsequently teaching at several other universities in the United States and Europe. From 2010 until his appointment at the European Environment Agency, he was head of the HIVA Research Institute in Leuven, Belgium, a policy-oriented research institute associated with the Katholieke Universiteit Leuven, Belgium, where he was also head of the Political Science department from 2007 to 2010. Dr Bruyninckx has also been a senior member of the interdisciplinary Leuven Centre for Global Governance Studies and promoter-coordinator of the Flemish Policy Research Centre on Transitions for Sustainable Development (TRADO). Over the last 20 years, he has conducted and managed policy-oriented research in the areas of environmental politics, climate change, and sustainable development. He was responsible for research in the domains of policy evaluation, monitoring and reporting, methodology development, environmental policy integration, and more recently also on long-term transition policies.

Mark Charlesworth is an Honorary Research Fellow in Social Sciences at Keele University, UK, where he completed his PhD in Environmental Politics and International Relations. He is an expert reviewer for all three working groups of the Intergovernmental Panel on Climate Change in the Fifth Assessment Report. He works as the Qualification Quality Manager for a leading UK center for training in building energy efficiency assessment and has previously worked as an engineer in a range of disciplines.

Jennifer Clapp holds a Canada Research Chair in Global Food Security and Sustainability in the Faculty of Environment at the University of Waterloo, Canada, and is a recent recipient of a Trudeau Fellowship. Her research focuses on the global governance of problems that arise at the intersection of the global economy, the environment, and food security. Recent books include *Food* (2012); *Hunger in the Balance: The New Politics of International Food Aid* (2012); *Paths to a Green World: The Political Economy of the Global Environment* (2011, 2nd edn); *The Global Food Crisis: Governance Challenges and Opportunities* (2009); and *Corporate Power in Global Agrifood Governance* (2009).

Nicole Detraz is Assistant Professor of Political Science at the University of Memphis, USA. Her research focuses on the intersections of security, gender, and the environment. Her work has recently appeared in *Security Studies, Global Environmental Politics, International Feminist Journal of*

Politics, and *International Studies Perspectives*, as well as multiple edited volumes. She is the author of *International Security and Gender* (2012).

Aarti Gupta is Associate Professor in the Environmental Policy Group of the Department of Social Sciences, Wageningen University, The Netherlands. Her research focuses on global environmental governance, including the role of science, knowledge, and transparency therein. Additional areas of interest include the challenges of securing accountability and managing fragmentation in global environmental governance. She has published extensively on these topics, with an empirical focus on biosafety, climate, and forest governance. Her recent publications include a special issue of *Ecological Economics* on Accountability and Legitimacy in Earth System Governance, co-edited with Frank Biermann; a special issue on the interdisciplinary challenges of REDD+ governance in *Current Opinion on Environmental Sustainability* (co-edited); and *Transparency in Global Environmental Governance: Critical Perspectives* (2014), co-edited with Michael Mason. She is an associate editor of the journal *Global Environmental Politics*.

Sander Happaerts is a research manager at HIVA (Research Institute for Work and Society, KU Leuven) and a lecturer of Global Environmental Politics (Faculty of Social Sciences, KU Leuven). He has expertise in the fields of sustainable development, transitions, climate change, and sustainable materials management. Sander's PhD (KU Leuven, 2011) won an award in the Thesis Competition of the EU Committee of the Regions. He published extensively on the sustainable development policies of subnational governments and on their embeddedness in national and international governance. In 2010, he held a visiting fellowship at the Université du Québec à Montréal (UQÀM, Canada), where he joined the Social Responsibility and Sustainable Development Research Chair. Before joining KU Leuven, Sander worked as an intern for the Permanent Representation of Belgium to the United Nations in New York.

Kathryn Hochstetler is CIGI Chair of Governance of the Environment in the Balsillie School of International Affairs, Canada, and Professor of Political Science at the University of Waterloo, Canada. She has published widely on environmental politics and movements in Brazil and other Latin American countries as well as on international environmental negotiations. Her most recent book is the prize-winning *Greening Brazil: Environmental Activism in State and Society* (2007, with Margaret

E. Keck). Her current research focuses on the role of the emerging powers in global climate negotiations as well as Brazilian and South African initiatives to implement their international commitments at home in the energy sector. Other work examines South–South economic relations in both trade and finance and their implications for environmental outcomes.

Melinda Laituri is Professor in the Department of Ecosystem Science and Sustainability in the Warner College of Natural Resources at Colorado State University, USA. She researches water resources and disasters using Geographic Information Systems (GIS). She works with indigenous cultures to develop cultural data layers for analysis within GIS.

Eva Lövbrand is Associate Professor at the Department of Thematic Studies: Water and Environment and affiliated with the Centre for Climate Science and Policy Research, both at Linköping University in Sweden. Much of Eva's work has focused on the ideas, knowledge claims, and expert practices that inform and legitimize global environmental politics and governance. Global climate governance has served as Eva's prime empirical example, but she has also critically interrogated scholarly efforts to 'democratize' expertise across other environmental issue areas. Eva's work has been extensively published in journals such as *Global Environmental Politics; Science, Technology & Human Values; Science; Critical Policy Studies; Review of International Relations;* and *Global Environmental Change*. She is also co-editor of the volume *Environmental Politics and Deliberative Democracy: Examining the Promise of New Modes of Environmental Governance* (2010).

Michael Mason is Associate Professor in the Department of Geography and Environment, the London School of Economics and Political Science, UK. His research interests encompass environmental politics and governance, notably issues of accountability, security, and climate change adaptation. He is currently Principal Investigator in an international research project examining climate vulnerability and security in the Jordan River Basin. Alongside articles in a wide range of academic journals, he is the author of *Environmental Democracy* (1999) and *The New Accountability: Environmental Responsibility across Borders* (2005). He is also co-editor of (with Amit Mor) *Renewable Energy in the Middle East* (2009) and (with Aarti Gupta) *Transparency in Global Environmental Governance: Critical Perspectives* (2014).

Chukwumerije Okereke is Reader (Associate Professor) in Environment and Development at the Department of Geography and Environmental Science, University of Reading, UK. He is also a Senior Visiting Fellow at the Environmental Change Institute, University of Oxford, UK. He is a Lead Author in Working Group three for the Fifth Assessment Report of the Intergovernmental Panel on Climate Change.

Matthew Paterson is Professor of Political Science at the University of Ottawa, Canada. His research focuses on the political economy of global environmental change, both as a general theoretical approach and in relation to global climate change and the politics of the automobile. His publications include *Global Warming and Global Politics* (1996), *Understanding Global Environmental Politics: Domination, Accumulation, Resistance* (2000), *Automobile Politics: Ecology and Cultural Political Economy* (2007), and most recently *Climate Capitalism: Global Warming and the Transformation of the Global Economy* (with Peter Newell, 2010). He has recently been working on carbon market politics, as well as a Lead Author by the Intergovernmental Panel on Climate Change, working on the chapter on international cooperation for the Fifth Assessment Report.

Dimitris Stevis is Professor of Political Science at Colorado State University, USA. In broad terms his research examines the social governance of the world political economy in the areas of labor and the environment. He is currently involved in a number of research projects. One project investigates Global Framework Agreements between multinational companies and labor unions. A second project investigates the environmental politics of labor unions and the labor politics of environmental organizations. A third project examines Colorado's New Energy Economy within global production networks. His recent publications include *Globalization and Labor: Democratizing Global Governance* (2008, with Terry Boswell) and *Global Framework Agreements in a Union-Hostile Environment: The Case of the USA* (2013, with Michael Fichter).

Larry A. Swatuk is Associate Professor and Director of the International Development Programme at the University of Waterloo, Canada. He is also Adjunct Professor of International Development at St Mary's University and External Research Fellow, Centre for Foreign Policy Studies, Dalhousie University, both in Halifax, Nova Scotia, Canada. Prior to coming to Waterloo, Dr Swatuk spent 11 years at the University of Botswana, first in the Department of Political and Administrative Studies in Gaborone, and later at the Okavango Research Institute in Maun, where he was Associate Professor and Director of the Research Program on Natural Resources Governance. Among his recent publications is a co-edited collection with Matthew A. Schnurr, entitled *Environmental Change, Natural Resources and Social Conflict: Towards Critical Environmental Security*.

Oran R. Young is Research Professor at the Bren School of Environmental Science and Management, University of California (Santa Barbara). His theoretical work deals with the nature and effectiveness of governance systems in stateless societies (for example, international society). At the applied level, he works on efforts to solve problems of governance relating to the atmosphere (for example, climate change), the oceans, and the polar regions. His most recent book is *On Environmental Governance: Sustainability, Efficiency, and Equity*. Professor Young is past chair of the Board of Governors of the University of the Arctic and of the Scientific Committee of the International Human Dimensions Programme on Global Environmental Change.

Acronyms

AP	Actual Performance
BASD	Business Action for Sustainable Development
CCP	Cities for Climate Protection
CDM	Clean Development Mechanism
CDP	Carbon Disclosure Project
CEC	Commission on Environmental Cooperation
CERES	Coalition for Environmentally Responsible Economies
CERs	Certified Emission Reductions
CSD	Commission on Sustainable Development
ECSP	Environmental Change and Security Project
EEA	European Environment Agency
EIR	Extractive Industries Review
EKC	Environmental Kuznets Curve
ENGO	Environmental Non-Governmental Organization
ESS	Environmental Security Studies
EU	European Union
EUSDS	European Union Sustainable Development Strategy
FAO	Food and Agriculture Organization
FDI	Foreign Direct Investment
FoE	Friends of the Earth
FSC	Forest Stewardship Council
G77	Group of 77
G8	Group of 8
GATT	General Agreement on Tariffs and Trade
GEF	Global Environment Facility
GGCA	Global Gender and Climate Alliance
GHGs	Greenhouse Gases
GIS	Geographic Information Science/Systems
GIT	Geographic Information Technologies
GMOs	Genetically Modified Organisms
GRI	Global Reporting Initiative
HUGE	Human, Gender and Environmental Security
ICSU	International Council for Science
IDGEC	Institutional Dimensions of Global Environmental Change
IEJ	International Environmental Justice

IEP	International Environmental Politics
ILO	International Labor Organization
IMF	International Monetary Fund
INGO	International Non-Governmental Organization
IOM	International Organization for Migration
IPBES	Intergovernmental Platform on Biodiversity and Ecosystem Services
IPCC	Intergovernmental Panel on Climate Change
IPE	International Political Economy
IR	International Relations
ISA	International Studies Association
ISO	International Organization for Standardization
IUCN	International Union for Conservation of Nature
KZN	KwaZulu Natal
LDC	Least Developed Countries
MA	Millennium Ecosystem Assessment
MEAs	Multilateral Environmental Agreements
MIA	Micro Insurance Academy
MNC	Multinational Corporation
NAACP	National Association for the Advancement of Colored People
NAAEC	North American Agreement on Environmental Cooperation
NAFTA	North American Free Trade Agreement
NAPAs	National Adaptation Programmes of Action
NGO	Non-Governmental Organization
Nrg4SD	Network of Regional Governments for Sustainable Development
OECD	Organization for Economic Cooperation and Development
PCBs	Polychlorinated Biphenyls
PD	Prisoner's Dilemma
PPMs	Process and Production Methods
PRI	Principles for Responsible Investment
SADC	Southern African Development Community
STS	Science and Technology Studies
TMN	Transnational Municipal Networks
TNC	Transnational Corporation
UN	United Nations
UNCED	United Nations Conference on Environment and Development

UNCHE	United Nations Conference on the Human Environment
UNCSD	United Nations Conference on Sustainable Development
UNDP	United Nations Development Program
UNEP	United Nations Environment Program
UNESCO	United Nations Educational, Scientific and Cultural Organization
WEDO	Women's Environment and Development Organization
WEF	World Economic Forum
WEO	World Environment Organization
WMO	World Meteorological Organization
WRI	World Resources Institute
WSSD	World Summit on Sustainable Development
WTO	World Trade Organization
WWF	World Wide Fund for Nature

1
General Introduction

Michele M. Betsill, Kathryn Hochstetler, and Dimitris Stevis

The study of global environmental politics has grown in both quantity and quality over the last 40 years, and international relations (IR) scholars have been increasingly more involved, particularly since the late 1980s. The goal of this book is to provide a state-of-the art review of how IR scholars approach the study of global environmental politics, a field we refer to as 'international environmental politics' (IEP).[1] The term *politics* in the phrase 'international environmental politics' signals that we are interested in the ways humans organize themselves to relate to their physical *environments*. Our chapter authors adopt a broad range of more specific understandings of both of these terms. To say that we are focusing on *global* environmental politics means we are concentrating on environmental issues whose causes or consequences cross national boundaries. From a sub-discipline that attracted mostly American scholars, IR scholarship on IEP has now spread throughout much of the world, although rather unevenly.

Over the years, a number of important volumes, often in multiple editions, have tracked the practice of global environmental politics (Brenton, 1994; McCormick, 1995; Caldwell and Weiland, 1996; Guha, 2000; Chasek et al., 2013). While these volumes also provide important insights into IEP, their primary focus is on the practice of global environmental politics itself. Several other volumes have offered a combination of chapters that examine aspects of IEP, along with particular sectors of the international environment (Hurrell and Kingsbury, 1992; Choucri, 1993; Vogler and Imber, 1996; Laferrière and Stoett, 1999; Chasek, 2000; Elliott, 2004; O'Neill, 2009; Mitchell, 2010; Axelrod et al., 2005), and a number of handbooks offer extensive overviews of the field (Dauvergne, 2011; Falkner, 2013). There have also been a few chapter- and article-length attempts at synthesizing the field of IEP as a whole (Stevis et al., 1989; Jancar, 1991/1992; Alker and Haas, 1993; Jacobsen, 1996, 1999; Mitchell, 2001; Stevis, 2010). Our book complements these efforts with its systematic attempt to identify the field's central research areas and to provide authoritative accounts of the major concepts, research agendas, and debates involved in their study.

Accordingly, this book examines the major theoretical approaches and substantive debates in IEP as reflected in a sample of graduate syllabi and texts.[2] We have asked a number of scholars with active research agendas in these areas to provide an account of the past study of that issue area as well as of the major questions and debates that characterize it presently. We have also asked them to apply their insights to a case study of their choice in order to illuminate the theoretical issues that they have addressed as well as to demonstrate how these insights can be employed to better understand specific questions.

As a result the book is intended to introduce graduate and advanced undergraduate students to IEP, particularly those with some previous exposure to IR. It can also serve as a complement to the types of volumes mentioned above in more introductory courses. IR scholars who embark on the study of IEP will also find this book helpful both as a review of the relevant literature and as a guide to how research is being done. Academics from various disciplines, including those who are interested in learning more about IR scholarship on IEP, either for teaching or in order to initiate a new research project, will find that this book offers authoritative, accessible, and sophisticated accounts of this area of research.

The contributors to this book were chosen with an eye towards representing the increasing globalization and diversity of IR research on IEP.[3] While we collectively provide an authoritative account of English-language literature, most of the contributors are also familiar with literature published in various other languages and have sought to integrate it where relevant. As a result, this book will appeal to the above-mentioned audiences throughout the English-speaking world as well as to anyone who uses English for their research or writing.

The book's chapters discuss a number of themes that are crucial to understanding the theory, method, and substantive content of the field of IEP. Our organizing framework stresses the international politics roots of this field, as the chapters are focused on broad and enduring areas of study in IR more generally. As Stevis' chapter (Chapter 2) on the history of IEP shows, such disciplinary frameworks have been important influences on how the field defines its questions and seeks its answers.

Specific substantive environmental issues such as biodiversity or water are studied quite differently, depending on whether they are framed as, for example, elements of the international political economy or instances of non-state governance.

The chapters are organized into three major parts. The chapters in Part I – The Context of International Environmental Politics – place the later chapters in a theoretical and historical context. They review the historical development of IEP as well as the theoretical and methodological approaches used by IR scholars in its study. All three of these chapters stress the diverse perspectives and tools that have been developed over several decades. This is a field with few orthodoxies and many debates, not unlike the field of IR as a whole (Reus-Smit and Snidal, 2009; Viotti and Kauppi, 2012; Dunne et al., 2013; Burchill and Linklater, 2013). The chapters in Part II – Major Research Areas – introduce a variety of actors, institutions, and structures that have influenced global environmental politics. Each chapter provides an overview of how a particular topic has risen to prominence, discusses the major theoretical views of that topic, and identifies lines of future research. In addition, each chapter includes original arguments and evidence in a case study to help illustrate some of the theoretical concepts and debates raised in the chapter. A similar framework is used in Part III – Frameworks for Evaluating Global Environmental Politics. The chapters in this final part discuss four key frameworks or standards that have been proposed for evaluating the quality and outcomes of global environmental politics: sustainability, effectiveness, justice, and transparency.

The chapters also address several cross-cutting themes that we believe are central to IEP and the practice of global environmental politics, regardless of issue area, theoretical perspective, or methodological approach. The North–South dimension is one such prominent theme, emerging in nearly every chapter. The interface between local and higher levels of politics is also central in many of the chapters, providing links to the comparative environmental politics field within political science (Steinberg and VanDeveer, 2012). In the conclusion, we discuss how the relatively straightforward treatment of domestic–international linkages in concepts such as 'two-level games' has evolved into discussions of complex interactions across scales captured in ideas like 'multi-level governance'. Such discussions also challenge the state-centrism of many IR theories by tracking the emergence of other types of actors and new forms of governance in IEP, giving rise to debates about the role of the state.

Each of the contributors is an accomplished scholar in her or his own right, and all authors have been encouraged to summarize existing research as well as to stake out their own positions. Most of the authors explicitly position themselves within particular perspectives, illustrating the multi-vocal nature of the field. While individual chapters may reflect some perspectives more heavily than others, across the book as a whole we have attempted to achieve balance, providing readers with a picture of the rich diversity of approaches used by IR scholars in IEP.

With one exception Parts II and III include original chapters with specific cases.[4] The cases are meant to illuminate the theoretical debates and concepts identified in each of the chapters and to provide readers with examples of empirical research conducted by scholars across IEP. The case studies cover a variety of issues, including climate change, biodiversity, sustainable development goals, the green economy, trade in hazardous waste, transboundary resource management, and the establishment of a World Environment Organization. The goal of the case studies is to show how the authors engage some of the theoretical issues they discuss rather than to prove or support a specific argument, as would be the case in a research chapter. The various authors employ a range of methods and approach their subject matter from a diversity of theoretical perspectives. As a result, the case studies reinforce the book's central aim – to introduce readers to the major approaches and debates that characterize IEP.

The book begins with a presentation of the historical trajectory of IR scholarship on global environmental politics. In his chapter, Dimitris Stevis draws on an extensive review of IEP publications, research organizations, and programs, as well as interviews with several senior IEP scholars, to outline the trajectory of IEP since World War II (WWII). He divides the field's history into four distinct periods and traces the genealogy of worldviews on IEP and of the research topics examined in the remainder of the book. He concludes that IR scholarship on global environmental politics has broadened and deepened in terms of both what is being studied and how it is being studied.

Matthew Paterson's chapter (Chapter 3) introduces the major theoretical approaches used in IEP. He organizes the chapter according to what he sees as six fundamental starting points for inquiry that guide most analyses. In the process, he examines an array of IR theories, including realism, liberal institutionalism, constructivism, pluralism, Marxism, feminism, and dependency theory, as well as perspectives developed specifically to understand environmental politics, including ecoauthoritarianism and Green political theory. This chapter stresses the importance of recognizing the strengths and weaknesses of different theories for addressing particular types of IEP research questions.

In their chapter on methods (Chapter 4), Kathryn Hochstetler and Melinda Laituri note that IR scholars have devoted little attention to the methods they use in IEP. Their aim is thus to outline a number of different approaches, discuss how they are used, and identify their potential pitfalls. The chapter is oriented around two major categories of methods: positivist (including qualitative, quantitative, rational choice, and geospatial approaches) and critical (including qualitative and structural approaches). Given the diversity of the field, they conclude that methodological pluralism is desirable but encourage scholars to pay more attention to their methodological choices in order to avoid unnecessary and unintended weaknesses in their studies.

Leading off the second part of the book, Jennifer Clapp orients her chapter (Chapter 5) on international political economy and the environment around three competing evaluations of the relationship: that growth in the global economy is positive for the environment, that the environment is harmed by growth in the global economy, and the third view that either outcome is possible and depends on the presence or absence of global rules that support the possible positive outcomes. These three positions reappear in her discussions of the more specific impacts of global trade, finance, and investment flows on the environment and their governance. All of these flows occur in Clapp's case study of the international transfer of hazardous wastes from rich to poor countries.

Nicole Detraz's chapter (Chapter 6) explores the links between gender and the environment in IEP, noting that people often experience environmental problems differently because of socially constructed ideas about the appropriate roles and responsibilities of men and women. She applies a gender lens to a number of central issues in IEP, including sustainability/sustainable development, population and consumption, environmental justice, and environmental security, and argues that this perspective challenges some of the traditional assumptions of IEP scholars and raises new questions that have been overlooked by the research community. The case of biodiversity protection in South America provides an interesting space to explore real-world examples of the interplay between gender, environmental degradation, and policymaking.

In the next chapter, Eva Lövbrand (Chapter 7) reviews scholarship on the role of knowledge in IEP. Environmental policy debates, whether on local or global scales, are often permeated by claims to particular types of knowledge and expertise, perceptions of environmental risk and scientific uncertainty. However, the role attributed to science and other forms of knowledge in global environmental affairs differs according to theoretical tradition. This chapter contrasts how rationalist and constructivist IR theories portray the interaction between knowledge and power and the subsequent effect on global environmental politics. The chapter also reviews the contemporary critique of the privileged status of scientific knowledge in the governance of the environment and examines normative calls for more legitimate forms of knowledge and expertise in environmental affairs. The contrasting perspectives are exemplified through a brief case study of the Intergovernmental Panel on Climate Change (IPCC).

The chapter on transnational actors in IEP (Chapter 8), by Michele Betsill, begins by highlighting the contribution of IEP scholars to broader debates about the role of non-state actors in IR. Betsill then presents findings on how transnational actors engage in the practice of global environmental politics, in terms of both their involvement in traditional multilateral processes dominated by states and the emergence of a distinct transnational political sphere. She also discusses some of the methodological challenges encountered by IEP scholars in assessing the impact of transnational actors. A brief case study of the Cities for Climate Protection program, a transnational network of municipal governments involved in the governance of climate change, illustrates these points and concepts.

Larry Swatuk's chapter (Chapter 9) links IEP to one of the central concerns of mainstream IR scholarship – security. Following a discussion of how environmental concerns have reshaped understandings of security in IR, Swatuk distinguishes between two types of environmental security scholars: those concerned primarily with problem-solving, particularly within a society of self-regarding states, and those taking a more critical and holistic approach to issues of security. He further elaborates the critical perspective in his case study of transboundary natural resource management practices in Southern Africa.

Frank Biermann addresses the question of global environmental governance (Chapter 10). He starts by clarifying the main uses of the term and suggests a more empirical approach that distinguishes global governance from IR at large. He then proceeds to discuss various aspects of global environmental governance, particularly participation by categories of actors other than states, the emergence of private governance, and the segmentation of global environmental governance. Drawing upon these insights, he elaborates on how Southern participation can be enhanced and advances a proposal to turn the United Nations Environment Programme into a World Environment Organization, a move that would address segmentation as well as participation.

In the first chapter on frameworks for evaluating global environmental politics in Part III (Chapter 11), Oran Young provides a state-of-the-art review of current knowledge about the effectiveness of international environmental regimes. The chapter begins with a discussion of conceptual and definitional issues before identifying a number of key findings about the determinants of institutional effectiveness. The chapter then introduces several cutting-edge themes or areas for future research and reviews some of the methodological tools available for tackling these themes, along with recommendations on which strategies are likely to produce policy-relevant results. Young is cautiously optimistic that IEP scholars can make valuable contributions to efforts to strengthen existing regimes or to create new ones.

The next chapter by Sander Happaerts and Hans Bruyninckx (Chapter 12) examines the emergence of sustainable development as a central discourse in global environmental politics and its study. In the first part, they trace the emergence of the concept from the early 1970s to the Brundtland Report (1987), the United Nations Conference on Environment and Development (1992), and the World Summit on Sustainable Development (2002). They then examine various debates about the meaning of the concept in policy and academic debates, which are further illustrated by an account of research on the institutionalization of sustainable development at various levels, from the global to the local. The chapter concludes by examining the downturn of the sustainable development discourse as reflected in recent debates at the 2012 United Nations Conference on Sustainable Development.

Chukwumerije Okereke and Mark Charlesworth trace the evolution of key debates on justice in IEP (Chapter 13). They consider social ecology, deep ecology, empty-belly environmentalism, and intergenerational justice as aspects of distributive justice and highlight procedural justice as a growing component of IEP scholarship. A brief case study of climate change illustrates how these concepts and perspectives inform the practice of global environmental politics. Although practical actions are yet to catch up with rhetoric, they argue that questions of ethics and justice can no longer be regarded as external or marginal concerns in global environmental governance.

The chapter on transparency in IEP by Aarti Gupta and Michael Mason (Chapter 14) examines the changing procedural quality of global environmental governance. Specifically, they focus on increasing expectations for openness in decision-making and disclosure of information as an antidote to various deficiencies in global environmental governance arrangements. The chapter analyzes the promises and pathologies of institutionalizing transparency in diverse areas of global environmental governance, showing how the pathologies go beyond failings of institutional design to reflect the contested nature of global environmental governance. In their case study, Gupta and Mason advance an analytical framework for comparative assessment of transparency in global environmental politics which is applied to a diverse set of initiatives that rely on environmental governance through disclosure.

In the final chapter (Chapter 15), the editors briefly reflect on the status of IR scholarship on global environmental politics as a whole based on the individual chapters in the book. We conclude that the field of IEP has become broader and deeper over time in terms of research agendas, substantive concerns, theoretical approaches, and the geographical and disciplinary origins of researchers. Consistent with this finding, we note that the field lacks a single normative core. We then make several observations related to the three cross-cutting themes – North–South relations, domestic–international linkages, and the role of the state. Looking ahead, we speculate on the future trajectory of substantive, methodological, and theoretical debates in IEP. Finally, we discuss the role of IR in IEP and consider how IEP scholars might create bridges to a number of other disciplines.

Notes

1. We have decided to label the object of study 'global environmental politics' and its study 'international environmental politics'.
2. We fully recognize that other scholars might make different choices about the theoretical approaches and substantive debates to include in such a book. Some readers may find gaps in the issues presented and/or prefer that a topic addressed within one or more chapters be treated separately. We acknowledge these potential critiques and can only say that the organization of the book reflects conscious decisions based on our own experiences teaching and researching in the field of IEP, constraints dictated by the publisher, and/or the usual challenges of coordinating an edited book.
3. Despite our best efforts, the book does not include contributions from Southern scholars to the extent we would have liked.
4. The exception is Oran Young's chapter, which is a revised version of an article that appeared in the *Proceedings of the National Academies of Sciences* and contains many rich empirical examples throughout the text.

Works cited

Alker, H. J. and P. M. Haas (1993) 'The Rise of Global Ecopolitics', in N. Choucri (ed.) *Global Accord: Environmental Challenges and International Responses* (Cambridge, MA: MIT Press), 133–71.

Axelrod, R. S., D. Downie and N. J. Vig (eds) (2005) *The Global Environment: Institutions, Law and Policy, 2nd edition* (Washington, DC: CQ Press).

Brenton, T. (1994) *The Greening of Machiavelli* (London: Earthscan Publications).

Burchill, S. and A. Linklater (eds) (2013) *Theories of International Relations,* 5th edition (Basingstoke: Palgrave Macmillan).

Caldwell, L. K. and P. S. Weiland (1996) *International Environmental Policy: from the Twentieth to the Twenty-first Century* (Durham: Duke University Press).

Chasek, P. S. (ed.) (1999) *The Global Environment in the Twenty-first Century: Prospects for International Cooperation* (Tokyo: United Nations University).

Chasek, P., D. Downie and J. Brown (2013) *Global Environmental Politics,* 6th edition (Boulder: Westview Press).

Choucri, N. (ed.) (1993) *Global Accord: Environmental Challenges and International Responses* (Cambridge, MA: MIT Press).

Dauvergne, P. (ed.) (2012) *Handbook of Global Environmental Politics,* 2nd edition (Cheltenham, UK: Edward Elgar).

Dunne, T., M. Kurki and S. Smith (eds) (2013) *International Relations: Discipline and Diversity,* 3rd edition (Oxford: Oxford University Press).

Elliott, L. (2004) *The Global Politics of the Environment,* 2nd edition (New York: New York University Press).

Falkner, R. (ed.) (2013) *The Handbook of Global Climate and Environmental Policy* (Chichester, UK: Wiley-Blackwell).

Guha, R. (2000) *Environmentalism: A Global History* (New York: Longman).

Hurrell, A. and B. Kingsbury (eds) (1992) *The International Politics of the Environment* (Oxford: Oxford University Press).

Jacobsen, S. (1996) *North–South Relations and Global Environmental Issues: A Review of the Literature* (Copenhagen: Centre for Development Research).

Jacobsen, S. (1999) 'International Relations and Global Environmental Change: Review of the Burgeoning Literature on the Environment', *Cooperation and Conflict* 34, 2, 205–36.

Jancar, B. (1991/92) 'Environmental Studies: State of the Discipline', *International Studies Notes* 16/17, 25–31.

Laferrière, E. and P. Stoett (1999) *International Relations Theory and Ecological Thought: Towards a Synthesis* (London: Routledge).

McCormick, J. (1995) *The Global Environmental Movement* (New York: Wiley).

Mitchell, R. (2001) 'International Environment', in W. Carlsnaes, T. Risse and B. A. Simmons (eds) *Handbook of International Relations* (London: Sage), 500–16.

Mitchell, R. (2010) *International Politics and the Environment* (Los Angeles: Sage Publications).

O'Neill, K. (2009) *The Environment and International Relations* (Cambridge: Cambridge University Press).

Reus-Smit, C. and D. Snidal (eds) (2009) *The Oxford Handbook of International Relations* (Oxford: Oxford University Press).

Steinberg, P. and S. VanDeveer (2012) *Comparative Environmental Politics: Theory, Practice, and Prospects* (Cambridge, MA: The MIT Press).

Stevis, D. (2010) 'International Relations and the Study of Global Environmental Politics: Past and Present', in B. Denemark (ed.) *The International Studies Encyclopedia,* volume vii (Hoboken: Wiley-Blackwell), 4476–4507.

Stevis, D., V. J. Assetto and S. P. Mumme (1989) 'International Environmental Politics: A Theoretical Review of the Literature', in J. P. Lester (ed.) *Environmental Politics and Policy: Theories and Evidence* (Durham: Duke University Press), 289–313.

Viotti, P. and M. Kauppi (2012) *International Relations Theory*, 5th edition (Boston: Longman).

Vogler, J. and M. F. Imber (eds) (1996) *The Environment and International Relations* (London: Routledge).

Part I
The Context of International Environmental Politics

2
The Trajectory of International Environmental Politics

Dimitris Stevis

Introduction

This chapter traces the study of global environmental politics by international relations (IR) scholars (hereafter referred to as IEP) since World War II (WWII).[1] The overarching question is whether IEP is a cohesive subfield articulated around a core set of concepts and debates, as seems to be the case with the study of the environment in some other social sciences such as Economics or Sociology. A related second question is whether one or more IR perspectives dominate IEP. A third question, strongly related to the second one, asks whether IEP addresses questions of social and ecological purpose in addition to the architecture of global environmental politics. I have chosen to address these questions by offering a genealogy, rather than a simple chronology of IEP, across four periods since WWII. In order to do so I pay close attention to three dimensions. The first is that of the broader political dynamics that influenced IEP during each period. The second is the deployment of IR perspectives within IEP and, relatedly, whether IEP has addressed global inequalities and power asymmetries. Finally, I ask whether and how IEP has treated the relations between nature and people over time.

The periodization employed is based on benchmark periods, rather than dates, derived from a close examination of the quantity and focus of IEP research over time. The first period extends from the mid-1940s to the mid-1960s. While explicit IEP scholarship was very limited, this period's framing of debates over population and resources has cast a much longer shadow on IEP than we often acknowledge. During the second period, from the mid-1960s to the late 1970s, the main lines of research emerged in the shadow of non-IR research on the global environment. From the late 1970s to the mid-1990s, IR engaged IEP decisively, if unevenly. In fact there was a lull in direct IEP research, partly due to the 'construction' of sustainable development and partly due to the political shifts to the right in the United States and the United Kingdom – underscoring the predominance of IEP research in those countries. Yet, this is also a period during which many of the scholars that have shaped the subsequent period were trained. The launching of journals such as *Global Environmental Change* (1990–), *International Environmental Affairs* (1991–1998), and *Environmental Politics* (1992–) can be offered as support to drive the point home that the 1980s were a period of gestation. Finally, the period from the mid-1990s to the present is characterized by an explosion in IEP with two more explicitly IEP journals launched during the early 2000s – *Global Environmental Politics* and *International Environmental Agreements*.

The discussion of each period follows a similar organization. I start by briefly outlining the political context, geographic origins, and substantive focus of IEP research (for more detail and references, see Stevis, 2010) and then comment on the grand narratives that colored each period. I then examine each period in terms of research areas, with particular attention to continuities, changes, and internal debates. In order to facilitate reading I start with the more prominent ones during each period. In discussing these research areas I pay close attention to the use of IR theories, whether research addresses global inequities, and whether and how IEP frames the relations between humanity and nature.

By 'research areas' I refer to the kinds of common research concerns like those used to organize this book. A genealogy based on research areas can tell us a great deal about what IEP scholars consider promising ways to think about environmental problems. Quite often particular research agendas tend to dominate a research area, but the two are not synonymous. Most of IEP research can be placed within political economy, governance, and security. Sustainable development, for instance, can be viewed as a political economy issue while effectiveness and transparency as governance issues. However, one can validly argue that the research areas of theories, methodology, actors, knowledge, gender, transparency, effectiveness, sustainable development, justice, and others also have their own dynamics and institutional mechanisms – such as specialists, journals, associations, and conferences – and can address problems across the range of IEP. Knowledge can not only inform governance but also helps us identify security tensions, judge effectiveness, or clarify the causes and degree of environmental harms and benefits. The study of environmental justice draws extensively from the wide body of literature on justice and can very well speak to governance, political economy, or security.

The choices of research areas are frequently associated with particular theoretical perspectives. Liberal institutionalists, for instance, are more likely to study governance than structural realists who are more likely to study conflict. Historical materialists are more likely to focus on political economy, but they are also likely to take such a broad approach to it that it includes governance. Paterson (chapter 3 in this book) provides a broad range of theoretical perspectives that have been prominent in IEP. Building on his efforts I will comment on the relative influence of various perspectives with respect to particular issue areas as well as during the whole period.

IR views do not vary solely along ontological and epistemological lines; they also involve core normative views about how the world is and can be. Some IR perspectives tend to privilege social change more than others. However, there is no *a priori* reason why radical views would consider environmental policies part of the solution rather than the problem. Dependency theorists, for instance, did not address the environment and may have been somewhat skeptical of its rise on the global agenda. While most liberal institutionalist research does not privilege social change or equity, there are compelling arguments that there is a liberal approach to environmental justice. The environment–development issue is fraught with debates on their balance. This brings me to the last theme.

Environmental politics is both about nature and about people. As one examines IEP it is possible to place research into four sets of worldviews based on the relationship between nature and people. Instrumental/resource IEP research sees the environment as a resource to be allocated or a problem that needs to be solved largely in order to avoid or mitigate tensions among people. Environmentalist IEP focuses on harms and benefits, in the process recognizing that the environment does have its own dynamics and value, albeit from an anthropocentric point of view. What Clapp and Dauvergne (2011: chapter 1) call 'market environmentalists' and 'institutionalists' may fall in either category. Bioenvironmentalists bring in nature, whether because they feel that there are natural limits to what it can absorb or because they extend some 'standing' to it. In the case of bioenvironmentalists, in Clapp's and Dauvergne's scheme, the emphasis is more on nature, while in the case of social greens – social environmentalists here – the goal is to fuse society and nature (for an overview, see Dryzek, 2013). In any event, the goal of this chapter is to explore whether, over time, IR perspectives have internalized the environment or whether non-IR perspectives have been employed to enrich the universe of IEP theory, thus greening IR theory (on this question, see Laferrière and Stoett, 1999 and 2006; Gale and M'Gonigle, 2000; Chistoff and Eckersley, 2013).

From the mid-1940s to the mid-1960s: Hidden origins

The 1960s is generally considered as a turning point in environmental politics. As the growing field of environmental history shows, turning points are the culminations of older processes (for examples, see Aage, 2008; Robertson, 2012). While transboundary and other transnational issues were the most prominent in practical terms, joined by nuclear testing, it is worth noting that the pre-1960s IEP cast the environment in decisively global terms (Osborn, 1948; Vogt, 1948). Two factors seem to account for this: the global ends and means of American politics and the resource and naturalist legacies of colonial empires (for historical accounts, see Nicholson, 1970; Boardman, 1981; McCormick, 1989; Brenton, 1994; Caldwell and Weiland, 1996; for a brief overview, see Stevis, 2005).

The substantive foci of IEP during this period were on the extraction or use of resources (including ocean resources), the implications of population increases on these resources, and the promise of technology. Most of the IEP research came from non-IR scholars who worked in the natural or physical sciences (Vogt, 1948; Osborn, 1948; Brown, 1954; Thomas, 1956), natural resource economists (Spengler, 1949; Scott, 1954; Crutchfield, 1956), and geopolitical analysts (Sprout and Sprout, 1957; Kristof, 1960; Konigsberg, 1960). Ecological worldviews, such as those articulated by Leopold (1949), did not play a role in this proto-IEP, further suggesting that its focus was more on the geopolitics of resources and less on nature. Little of the research was published in flagship IR journals.

The grand narrative of the era was that of scarcities due to increasing human population. From the very beginning there were three versions of 'scarcity' employed. Analysts such as Vogt and Osborn framed scarcity in absolute terms, while resource economists dealt largely with conjunctural scarcities. Geopoliticians such as the Sprouts paid more attention to relative scarcities – appropriation by some meant scarcity for others. The logic of absolute scarcity morphed into the 'limits to growth' narrative of the subsequent period and continues to be an important assumption of views associated with global environmental change (Turner, 2008) and bioenvironmentalism. We can also see during this period the emergent criticism of the absolute scarcity line of thought by what have been called 'cornucopians' (Kaplan, 1949). The logic of relative scarcity – developed further by Hardin – underlies much of the environmental conflict literature today. The logic of conjunctural scarcity continues to be central to environmental, if not ecological, economics as well as among various neoliberal proponents of environmental policy (for example, the Copenhagen Consensus).

Despite the paucity of explicit IEP work, there are some identifiable linkages to the present. The most identifiable IR perspective was that of classical realist geopolitics that also influenced non-IR scholars (Osborn, 1948, 1953; Vogt, 1948). The concern over the relations between nature and people was geopolitical, despite the use of ecological principles such as 'carrying capacity' (Robertson, 2012). This approach is evident in the tentative introduction of the environment into IR by Sprout and Sprout (1957), who treated the resource environment as one component of the 'milieu' within which actors operate. The geopolitical interest in nature and resources is further underlined by research published in the *Journal of Conflict Resolution* during this period (Konigsberg, 1960; Kristof, 1960). A great deal of environmental conflict literature stands on these earlier foundations.

Debates about common property resources also received attention (Gordon, 1954; Crutchfield, 1956). The fundamental debate here centered on the most appropriate management of these resources (private or public, domestic or international). These debates are important for understanding the public choice institutionalism undergirding the common pool resources research agenda that has become so prominent over time (Keohane and Ostrom, 1995; Barkin and Shambaugh, 1999) as well as Oran Young's work on regimes. In short, here we see some of the origins of the study of governance.

We also see during this period some explicit references to governance by international organizations (Kotok, 1945; Goodrich, 1951; Pollock, 1956), but there was no work on non-governmental organizations despite the prominent role of the International Union for Conservation of Nature (IUCN) and the International Council for Science (ICSU) (Boardman, 1981). The promise, more so than limits or sociology, of science and technology was central to the early accounts (Osborn 1948: Chapter 5; Brown et al., 1957). Starting in the early 1960s, however, social and natural scientists, such as Bookchin (1974[1962]), Carson (1962), Commoner (1966), and Dubos (1968), started addressing the perils of unexamined and unregulated technology, focusing on the impacts of nuclear tests and of chemicals, and their work shaped the subsequent period and beyond.

On balance, and in hindsight, we can identify the origins of some of the major research areas and approaches that have come to characterize IEP. Most of the intellectual impetus came from outside IR and from people whose training was in the natural or physical sciences or resource economics, but commonsensical geopolitical thinking was a central element of their approach.

From the mid-1960s to late 1970s: The origins of IEP

Those who think of the 1960s and 1970s as a period during which environmental politics, including global environmental politics, rose higher on the agenda are right. Yet the concerns and policy priorities of the 1960s did not come out of thin air. Many of the same people who played a key role during this period were prominent in the previous one, as well. The framing of the environment as a global issue accelerated during this period, growing more complex over time as a result of the assertion of the global South (see Ward, 1966). Latin American countries were joined by the growing number of newly independent countries in efforts to forge a common Southern agenda, largely through the United Nations. Global conferences, such as UNESCO's Biosphere Conference (1968) and the 1972 Stockholm Conference on the Human Environment, and the formation of international environmental non-governmental organizations (ENGO)s and new think tanks provide strong evidence of the increasing globalization of environmental politics as well as the more vocal presence of the global South (for background, see Nicholson, 1970; Wilson, 1971; Rowland, 1973; Boardman, 1981; McCormick, 1989; Brenton, 1994; Caldwell and Weiland, 1996; MacDonald, 2003; Stevis, 2010).

Most of IEP came from non-IR scholars, but IR scholars now entered the picture very clearly (for example, International Journal, 1972/1973; Kay and Skolnikoff, 1972; Shields and Ott, 1974). A great deal of IEP came from the United States, the United Kingdom, and Canada, but it was also growing rapidly in Scandinavia and continental Europe. Some of that work was *about* the South (for example, Woodhouse, 1972), but almost none came from the South, even as the South became more important in global environmental politics. In substantive terms transboundary pollution and resources issues remained prominent, but global issues, such as the climate, rose on the agenda.

The most prominent narratives of this period were those of limits and scarcities, increasingly challenged by ideas that were later called 'sustainable development'. The relative scarcity approach, exemplified in the extreme by Hardin (1968; Hardin and Baden, 1977), employs a causal logic between people and resources that goes directly through institutional arrangements. Hardin's 'tragedy of the commons' is as much about more people claiming scarce resources as it is about the absence of an appropriate governance mechanism. Before and after Hardin there have been various framings and solutions of the problem of the 'commons', and the overall approach continues to be a vibrant strand of thinking about resources and the environment. Over the decades the commons has come to denote spaces – whether the oceans or the atmosphere – that are not within jurisdictions or which cross jurisdictions and which demand appropriate kinds of governance (Ostrom, 1990; Goldman, 1998; Barkin and Shambaugh, 1999; Vogler, 2000).

For those that emphasized absolute scarcity (for example, Ehrlich, 1968; Meadows et al., 1972), the natural limits of industrial growth and wasteful practices, aggravated by population, required not only new governance but different civilizational priorities (for a review, see Onuf, 1983). The limits approach was subjected to strong criticism from various angles. Some argued that it sought to legitimate the existing global inequities (Galtung, 1973; Cole, 1973; Herrera, 1976). Yet, where Hardin and the resource economists were most interested in questions of resource allocation and management – and very little about equity – some of those who focused on limits during this period were more likely to engage distributive issues, albeit with a lag (see Meadows et al., 1992; Ehrlich et al., 1995).

While bioenvironmentalists set the tone, social environmentalists, including critics of the impact of industrial practices on health and nature (Carson, 1962; Commoner, 1966 and 1971; Dubos, 1968; Bookchin, 1974[1962]), analysts interested in development (Ward, 1966; Falk, 1971), and, increasingly, those concerned about nuclear weapons (for a history of the people and the debates, see Egan, 2007), were able to broaden the horizons of environmentalism in the North toward social environmentalism.

Influenced by the strategy of the New International Economic Order, there emerged an alternative view of globality that sought to connect environment and development (see Rowland, 1973). The key elements here were the argument that the South needed to industrialize through a resetting of global rules and that the North had important responsibilities, with respect to both the causes and the solutions of the environmental crisis. In short, this approach introduced global equity issues into the global environmental debate, albeit not in terms of environmental ethics or justice. Starting with the release of the Founex Report (1972) on Development and Environment, the aggregate approaches from the North entered into a long contestation with the country-level distributive approaches of the South. The Stockholm conference placed the challenge of fusing environment and development on the global policy agenda.

The proponents of marrying environment and development agreed that environmental problems were largely the result of Northern practices. Beyond that there was serious divergence. Developmentalists (de Araujo Castro, 1972; *Economic and Political Weekly*, 1972) and their pro-growth neoliberal supporters in the North (Beckerman, 1974) largely privileged growth, whether on developmental or liberal grounds. But, very early on, there emerged voices that advanced ecological concatenations of environmental and economic policies, whether in the South or the North. The work of Ignacy Sachs (1970, 1974, 1980) best characterizes a proactive effort toward integrating environmental and ecological priorities ('ecodevelopment') into economic policies.

It is fair to say that geopolitical, environmentalist, bioenvironmentalist, and social environmentalist worldviews increasingly influenced IEP and contested for political primacy during this period (see Clemens, 1972/1973). Geopolitical voices gravitated toward security (for example, Sprout and Sprout, 1965, 1971) and environmentalist voices gravitated toward governance (for example, Kay and Skolnikoff, 1972). Bioenvironmentalism motivated the work of scholars who sought to provide an ecological foundation for IEP (for example, Soroos, 1977; Pirages, 1978, 1983; Orr and Soroos, 1979) while social environmentalism influenced those concerned about North–South relations (Woodhouse, 1972; Dahlberg, 1973, 1979; Dahlberg et al., 1984). During the mid-1970s a number of these scholars, many from outside IR, formed the Environmental Studies Section of the International Studies Association (ISA). From the point of view of IEP the formation of the Section is one pivotal development demarcating the rise of IEP as a recognized subfield of IR.

Influenced by the debates on environment and development there was also some research on particular processes of the world economy, such as the impacts of the Green Revolution on the South (Farvar and Milton, 1972; Dahlberg, 1973), where we also see a more reflexive approach to science and technology developing to complement the strong criticisms emanating from social environmentalists. It took a while, however, for IEP to interrogate knowledge, science, and technology (see Ruggie, 1975, for the state of affairs). Economic processes also received some attention, especially from economists (for example, Baumel, 1971; Kneese et al., 1971).

Geopolitical thought continued to engage the environment, largely spearheaded by Harold and Margaret Sprout, who came to IEP from political geography (Sprout and Sprout, 1957, 1965, 1971). Their approach to the environment was colored by concern about human conflict rather than by any intrinsic value assigned to nature or any consideration of the impacts of human activity on nature and human well-being, if such activities did not cause conflict. This geopolitical line of thought was further amplified by the early work of Choucri and North (1975), which has cast an important shadow onto subsequent research on environment and conflict. But during these same years Scandinavian analysts sought to connect ecology and peacemaking – *The Journal of Peace Research* formed in 1964 – presaging another line of thought on environmental security (Cjessing, 1967).

Organizational thinking was an equally important theoretical entrance for IR scholars, as evident in the various articles and special issues of *International Organization*, which moved increasingly toward addressing the environment directly, rather than through resources and technology (Kay and Skolnikoff, 1972). Questions of governance were examined, often in more breadth and depth by international lawyers (Hargrove, 1972; Contini and Sand, 1972; Brownlie, 1972). Along with political economists (for example, Poleszynksi, 1977), these scholars called for global ecological standards, initiating the long road toward the measurement of environmental effectiveness. It is possible to subsume much of that research under liberal institutionalism, but the work of that time paid more attention to organizations rather than social institutions, such as regimes. It was only toward the end of this period that Ruggie (1975) and Oran Young (1977) started applying institutional thinking to the environment. Finally, Ophuls (1977) offered what remains one of the strongest statements on eco-authoritarianism.

Specific actors also received some limited attention. The impacts of corporations on the international environment were not addressed in any sustained fashion, despite the fact that MNCs rose on the agenda of IR scholars, starting in the late 1960s (Gladwin, 1977). Intergovenmental Organizations (IGOs) were the subject of more research (Kay and Skolnikoff, 1972) while ENGOs were identified as an important development but were not researched extensively (Smith, 1972; Rowland, 1973; Feraru, 1974).

Questions of equity and social power were placed on the agenda by social environmentalists and received additional impetus from the emergence of a Southern agenda centered around the relations of environment and development. Yet, while questions of power and equity were central to the environment–development debate and environmental ethics was emerging (for example, Stone, 1973), IEP scholars did not explicitly engage questions of environmental ethics and justice during this period, with the exception of the World Order Models Project (Falk, 1971, 1975). The dependency and historical materialist views that were prominent at the time did not address the environment.

On balance, the theoretical direction of IEP was contested between geopolitics, environmentalism, bioenvironmentalism, and social environmentalism. Some of IEP sought to establish ecopolitical theories in IEP, but most of the literature came from liberal analysts that focused on organizations more than institutions and realists who focused on conflict over resources. With respect to specific research areas there are clearer lineages to the present concerning governance, political economy, environmental conflict and peace, sustainable development, technology more than knowledge, and the role of non-state actors.

From the late 1970s to the mid-1990s: Institutions to the rescue?

The 1970s are considered a turning point in post-WWII politics from embedded liberalism to global neoliberalism. During the 1980s neoliberals emerged victorious in the United States and the United Kingdom, leading to deeper challenges to welfare capitalism across the world. At the same time the rise of major southern economies modified the North–South contours of the world political economy and has influenced global environmental politics, among other issues, ever since. Global transactions, particularly trade, reached their highest levels since before WWI, increasingly joined by investment, infrastructure, and communications – hence the emergence of the debates over globalization. For many, the dissolution of the Soviet Union and the end of the Cold War, along with economic liberalization in both China and Russia, signaled the deepening of not only globalization but of liberal globalization.

The contested nature of global environmental politics was evident along the whole range of specific environmental and resource issues that characterized the period, such as the global commons, species protection, deforestation, hazardous technologies and toxics, ozone depletion, and, increasingly, climate change. Research in each one of those areas grew, as did research on environment and development. Not surprisingly, given the move toward sustainable development, there was increasing research *about* the South (Dahlberg, 1979; I. Sachs, 1980; Brookfield, 1980–1981). But, surprisingly, there was very little on the environment in development and area studies journals such as *World Development* and *Third World Quarterly* (see James, 1978, for review and references; Gordon, 1975). It was not until the very late 1980s that some work on sustainable development started appearing, to pick up steam in the very early 1990s (for example, Harris, 1991; Schor, 1991; Brandon and Brandon, 1992). Voices *from* the South also rose in prominence, driven by the growing debate over environment and development (Cardoso, 1980; Balasubramaniam, 1984; Biswas,1984; Ghosh, 1984; Agarwal and Narain, 1991; Shiva, 1991; Banuri and Holmberg, 1992; Jasanoff, 1993; various chapters in W. Sachs, 1993). Much of that research was published in the North where many of those scholars were also trained or worked.

The grand narratives over the environment that continue to the present matured during this period. On the one hand, there was the debate over the fusion of environment and development which had started with the Founex Report and was placed centrally on the global agenda with the United Nations Commission on Environment and Development and its *Our Common Future* (1987) report, 'negotiated' in preparation for the 1992 Rio Conference on Environment and Development (see Harrison, 2000; Clark and Munn, 1986). Despite the title of the report, sustainable development could hardly center around an aggregate view of the world, as had been the case with the limits to growth and scarcity frames. From the very beginning it was about distributive issues and the reconciliation of environment and development in an unequal world.

On the other hand, the discourses of limits and scarcity had lost significant ground for a variety of reasons, including criticisms from the right regarding their alarmist tenor and from the left because of their inattention to their distributive implications. Increasingly, the global environmental change discourse emerged on the foundations of enhanced technological capabilities to monitor the globe from space and through other monitoring technologies and offered itself as an alternative to sustainable development. As Buttel et al. (1990) pointed out – in the inaugural issue of the journal *Global Environmental Change* – the global change discourse offered a place for those social forces, such as corporations, which felt excluded from the limits to growth and scarcities approaches as well as from the distributive implications of sustainable development (for the practice of global environmental politics during this period, see Thomas, 1992; Brenton, 1994; Caldwell and Weiland, 1996; Bernstein, 2001; MacDonald 2003).

IEP took off starting in the late 1980s (see Stevis et al., 1989; for collections that capture discussions and developments within IEP, see Jancar, 1991–1992; Soroos, 1991; Hurrell and Kingsbury, 1992; Choucri, 1993; Kamieniecki, 1993). The main research areas that were emergent during the previous period continued to mature and diversify with the study of governance becoming most prominent. The role of international organizations continued to receive attention (M'Gonigle and Zacher, 1979; Boardman, 1981; Kay and Jacobson, 1983) as did international law (Schneider, 1979; Springer, 1983), knowledge and science (Andresen and Østreng, 1989), and international negotiations (Carroll, 1988; Benedick, 1991), the latter receiving a great deal of impetus as a result of the success of the ozone negotiations.

The move from the administrative or organizational approach to IGOs toward a more institutional perspective that can be traced back to the late 1970s clearly influenced IEP scholars during this period (see Jacobsen, 1996). Building on this shift toward international institutions and his own public choice institutionalism, Oran Young turned regime analysis into an important research agenda within IEP (Young, 1977, 1989). Others suggested taking a global policy approach rather than the more ontologically institutional regime strategy (Soroos, 1986). 'Global governance', as a term, also entered IEP at the end of this period (Sand, 1992), but the late 1980s and early 1990s are characterized by theorizing of inter-state governance regimes, despite the fact that regime analysis had lost some ground in IR since its heydays of the early 1980s.

The main trends within the environmental security research area – both its conflict and peace dimensions – were even more apparent by the end of this period. With respect to environmental conflict there is a clear lineage between the scarcity arguments and Choucri and North's (1975) work on lateral pressures and conflict which was expanded to account for more explicitly environmental factors (North, 1977; Orr and Soroos, 1979; Gurr, 1985). This kind of geopolitical environmental conflict research received increasing attention (Homer-Dixon, 1991), while there was some debate over the wisdom of legitimating environmental concerns by appropriating the language of conventional security (Deudney, 1990).

A second research stream, arguably older than its better-known cousin, addressed the impacts of the military on the environment, both in terms of military activities and in terms of the military's use of resources (Westing, 1977, 1988; Hveem, 1979; Galtung, 1982). This approach sought to embed the institutions of conflict and war in the broader political order. Building on these foundations there matured during the 1980s an important expansion of the meaning of security beyond its traditional geopolitical meaning and toward human security (Brown, 1977; Myers, 1989; for overviews, see Levy, 1995; Rønnfeldt, 1997).

The move toward placing the environment within the political economy continued to grow because of increasing global economic integration and North–South debates. At the more structural level, the move from ecodevelopment (Riddell, 1981; Balasubramaniam, 1984; Glaeser, 1984) toward sustainable development involved debates over the relations between North and South but, also, an attempt at integrating ecology and society (Biswas, 1984; Redclift, 1984 and 1992; Bartelmus, 1986).

In addition to addressing broad questions of political economy, associated with sustainable development, IEP also started to address particular processes of the global political economy. Thus, there was important work on the relations between the internationalization of the economy and the environment (for example, Rubin and Graham, 1982), MNC strategies and pollution havens (Gladwin, 1977; Pearson, 1987; Leonard, 1988), global economic organizations and the environment (Stein and Johnson, 1979; Rich, 1985), and the export of hazardous technology and toxics (Shrivastava, 1987). During this period there also emerged arguments promoting the greening of the corporation (Elkington and Burke, 1987; Cairncross, 1992) that have exerted such influence ever since. It is fair to say, however, that by the end of this period the study of the international political economy of the environment was still limited, compared to the prominence of international political economy debates in IR.

In terms of actors this is a period when societal entities, including hybrid networks, started receiving more attention (Boardman, 1981; Willetts, 1982; Haas, 1990; Peterson, 1992) while the outlines of the move toward seeing ENGOs as more than pressure organizations were emergent (Lipschutz with Mayer, 1996; Wapner, 1996). The state, in fact, received the least systematic attention, other than in some discussions of environmental foreign policy formation (for example, Benedick, 1991; but see Lipschutz and Conca, 1993), despite the raging debates on the nature and role of the state throughout the social sciences.

Questions of power and equity had been implicit in the politics of IEP at least since the environment–development relationship was placed on the agenda. Environmental ethics also has a long history, going back as far as the 1970s (Stone, 1974; Nash, 1989), and for some, further back. The journal *Environmental Ethics* was launched in 1979. But it was not until the late 1980s that IEP addressed questions of international environmental justice explicitly (Weiss, 1988 and 1990; d'Amato, 1990). As in other areas, international lawyers were ahead of the curve in this case. Reflecting a move toward liberal environmentalism and geopolitics, IEP scholarship first addressed intergenerational issues rather than the more apparent and immediate intra-generational ones (Agarwal and Narrain, 1991; Schor, 1991). Finally, the social role of knowledge, science, and technology science received more attention (Haas, 1990; Taylor and Buttel, 1992; Litfin, 1994; Beck, 1999), but not as much as one would have anticipated given some of the early 1970s research (Dahlberg, 1973; Ruggie, 1975) and the increasing engagement between social and natural scientists in various national and global interdisciplinary programs.

By the mid-1990s IEP had come into its own as graduate schools produced more scholars trained in the subject, the Environmental Studies Section started growing more rapidly, and other related organizations welcomed international environmental scholars. While IEP had clearly broadened during this period to address a wider variety of substantive areas and moved into additional research areas, it is not clear that IEP had deepened theoretically. To my knowledge there were no sustained efforts to rethink international relations in the light of nature similar to the ecological economics, or sociological debates about risk society, ecological modernization, or the treadmill of production (but see Vogler and Imber, 1996; Laferrière and Stoett, 1999; Gale and M'Gonigle, 2000; Kutting, 2000). The efforts toward some kind of ecopolitics that had emerged in the 1970s had largely disappeared. Rather, much of the work that took place during these years applied some insights from liberal institutionalism and realism to nature. The influence of liberal institutionalism and realism continued into the subsequent period and remains potent to the present but, increasingly, IEP has again become broader and deeper.

From the mid-1990s to the present: (re)contesting IEP?

The dynamics of the previous period became more apparent during the 1990s and the beginning of the third millennium. IEP has grown exponentially during this period, hence the number of edited volumes (for example, Chasek, 1999; Laferrière and Stoett, 2006), monographs (for example, Laferrière and Stoett, 1999; Elliott, 2004; O'Neil, 2009; Mitchell, 2010), articles and chapters (Jacobsen, 1996, 1999; Stevis, 2010), and handbooks (for example, Haas, 2003; Dauvergne, 2011; Falkner, 2013; Harrris, 2013) that seek to provide overviews of IEP.

While there is research in a variety of issue areas, climate change has colonized most corners of environmental research, raising some concerns among scholars about the political and research implications of its hegemony (Swyngedouw, 2010; Methman, 2013). The geographical diffusion of IEP has continued and accelerated. European and Japanese institutions and scholars are now as prominent as North American or British ones while other English-speaking countries have contributed fundamentally to our thinking about IEP. Analysts from Australia and New Zealand, for instance, have often played a leading role (for an overview of Australian work, see Elliott, 2009; also Christoff and Eckersley, 2013). Research *about* the global South has proliferated, but there is also more research *from* the global South (see Jacobsen, 1996; Miller, 1996; Guha and Martinez-Alier, 1997; Dwivedi, 1997; Guha, 2000; Najam et al. 2007; Stevis, 2010 for a brief account). While much of that research is still taking place within Northern-led networks, the role of Southern scholars is more visible compared to the previous periods, particularly with respect to questions of sustainable development and equity. Global research networks – a trend that accelerated around 1990 – made it easier for some organizations and researchers from some Southern countries to now participate in the study of global environmental politics. This diffusion is evident in terms of research organizations, professional associations, journals, and publishers. But despite this ongoing diffusion, IEP remains North-centered largely because of resources (for a historical account of the practice of global environmental politics, see Elliott, 2004; Chasek et al., 2013).

A fuller understanding of IR/IEP during this period must address the grand narratives of our era, that is, global environmental change and sustainable development and the emergent social-ecological approach, the latter largely situated within global environmental change (Folke, 2006; Young et al., 2006; Nelson et al., 2007). It is worth noting that bibliographic searches using the terms 'environment' or 'environmental' and searches using 'sustainable development' and 'sustainability' produce noticeably different results in terms of both titles and venues of publication. This divergence is also evident in the titles of journals. The global environmental change narrative shares the aggregate globality evident in 'limits to growth', with policies often directed *at* the South or the weaker sectors of the population while questions of social power and equity are a secondary element among many.

Like limits and scarcities during the 1970s and global change now, the social-ecological narrative recognizes that ecosystems can collapse and with them social-ecological systems. Like sustainable development, it recognizes that humans and nature are elements of the whole and that it is possible to enhance the resilience of social-ecological systems through mitigation and adaptive strategies. What remains to be seen is whether adaptation and resilience will stay within the parameters of nature or will legitimate techno-social engineering with all its implications (for example, Keith, 2013), particularly since social-ecological thought has been slower or uncomfortable engaging questions of social power and equity (Hornborg, 2009; but also Adger et al., 2006; Phelan et al., 2012).

The sustainable development narrative pays more attention to questions of power and equity and is more likely to place the South and the weak at the center of analysis. However, in the hands of developmentalists (from both North and South) there is no guarantee that it will produce more environmentally sound syntheses. Moreover, the narrowly economic cost–benefit analysis often adopted ensures that policies are directed at the weak, rather than empowering them (Copenhagen Consensus; Beckerman and Pasek, 2001).

Theorizing is not the most prominent research area in IEP, but there is more evidence of IEP challenging its theoretical horizons with respect to both IR theories (evidence of which follows) and reaching out to ecology and other disciplines (Laferrière and Stoett, 1999, 2006; Princen, 2005; Wapner, 2010; Christoff and Eckersley, 2013; Patterson, chapter 3 in this book).

Governance probably remains the most prominent research area, but attention to political economy has risen. Within global environmental governance one can identify a variety of research agendas (Biermann and Pattberg, 2008 and 2012; Young and Schroder, 2008; O'Neill, 2009; Biermann, chapter 10 in this book; Andonova and Mitchell, 2012). It is fair to say that the term 'governance' has come to subsume various other terms used previously, such as effectiveness, agreements, regimes, law, policy, and so on, but focused research on those topics does continue. In addition, the interfaces of governance with political economy, justice, and other major themes often blur their boundaries. Important questions permeating the study of governance include public and private governance (Betsill, chapter 8 and Biermann, chapter 10 in this book), the scales of governance and their interplay, the relative focus on explicit and implicit environmental politics, legitimacy of governance, and the implications of governance, including with respect to democracy, transparency (Gupta and Mason, chapter 14 in this book), and equity (Okereke and Charlesworth, chapter 13 in this book).

The emergence of private governance, whether self-regulatory or civic-regulatory, has been one of the most important developments of the second part of this period (Betsill, chapter 8 in this book; Biermann, chapter 10 in this book). From the earlier work that focused on the significance of civil society we have now moved toward more detailed accounts of how corporations govern their environmental practices (Prakash, 2000; Angle et al., 2007; Falkner, 2008), the implications of self-regulation, and the promise and limits of civil regulation (Lipschutz with Mayer, 1996; Cashore et al., 2004; Biermann and Pattberg, 2012). Supporters of societal governance range from those that see a great deal of promise to those that treat it as an unavoidable necessity that, if done properly with some state intervention, can re-embed capital. Skeptics, on the other hand, consider private environmental governance run by corporations as a challenge to democracy and an exercise in public relations.

Partly due to the emergence of private governance and partly due to the significance of the issues themselves there is now important research on the processes of governance. Transparency (see Gupta and Mason, chapter 14 in this book) has received a great deal of attention, strongly motivated by the emergence of various schemes of private governance. Environmental democracy has long been an issue in environmental politics and now has received attention from IEP scholars (Bäckstrand, 2006) as has legitimacy (Bernstein, 2011).

Two areas that, in my view, require additional attention are environmental foreign policy (de Sombre, 2000; for a review, see Barkdull and Harris, 2002) and the implementation of environmental agreements and policies at the national level (Weiss and Jacobson, 1998). The fusion of domestic and international politics carries a great deal of promise because international-level explanations would benefit from more nuanced research on national and subnational processes (Steinberg and VanDeveer, 2012).

The issue that has truly exploded since the 1990s, however, is that of societal politics (see Betsill, chapter 8 in this book; Princen and Finger, 1994; Wapner, 1996; Betsill and Correll, 2008). Many analysts examine ENGOs without broader claims about the implications for the structure of world politics. A perspective that has been prominent in IR/IEP, however, has argued that there was evidence of an emerging global civil society that was reconfiguring the state and was distinct from states and capital (Lipschutz with Mayer, 1996; Wapner, 1996). The emergence of global societal politics does seem related to the end of the Cold War and the spread of US-type interest-group politics around the world, at the same time that the United States and the global financial institutions sought to devolve social policy power from the state. The fact that transnational politics emerged in the 1970s but had to be brought 'back in' in the 1990s lends some plausibility to this hypothesis.

It is worth noting, in that vein, that there is still limited sustained research on the state as an actor (for a review, see Stevis and Bruyninckx, 2006), but students of foreign policy (Barkdull and Harris, 2002) and the green state (Litfin, 1998; Barry and Eckersley, 2005) are adding to the discussion. With the state's role growing again in response to the early 21st century's economic and security crises, it seems that the real developments with respect to societal politics are obscuring fundamental developments at the level of the state and, more significantly, the historical relations between state and society over time. Since much of the literature on non-state actors is based on assumptions about the state, a more sustained focus on the state remains important.

While interest in societal politics has proliferated, IR scholars have not paid commensurate attention to social categories, whether gender, ethnicity, or class (for a baseline overview, see Elliott 2004: chapter 6; Detraz, chapter 6 in this book). Some important research has addressed questions of gender (Bretherton, 1998, 2003), but searches in key IR and IEP journals suggest that gender does not receive extensive attention. The gap is even more serious with respect to ethnic groups and class (for example, Newell, 2005) topics that have received a great deal of attention in environmental sociology. While some relevant research appears in more specialized journals, the limited treatment of social categories by IR/IEP scholars is cause for discussion. In the author's view the limited emphasis on the state and subaltern social categories reflects our subdiscipline's inattention to social power and its organization. Hopefully, this gap will be addressed as part of our increasing engagement with international environmental justice (see Parks and Roberts, 2007; Ehresman and Stevis, 2010; Okereke and Charlesworth, chapter 13 in this book).

IEP scholars have also fully engaged political economy (see Haas, 2003, for a baseline, although many of the chosen readings could well fit into governance or security; Stevis and Assetto, 2001; Clapp and Dauvergne, 2011; Newell, 2012; Clapp and Helleiner, 2012). Theoretically research on the political economy of the global environment comes from a variety of IR perspectives.

Research on sustainable development has also grown (Harrison, 2000; Kates et al., 2005; Happaerts and Bruyninckx, chapter 12 in this book), but there is some evidence of it slowing down or mutating into research on greening the economy (Natural Resources Forum, 2011). Increasingly, IEP scholars address the interface between environment and economy from additional broad angles. A number of authors, for instance, have focused on consumption (see Princen et al., 2002; Iles, 2006) while others on greening the economy (Kutting, 2004; Natural Resources Forum, 2011).

Specific sectors and processes of the world political economy have also received growing attention (Clapp, chapter 5 in this book). During the 1990s, for example, the relations between trade/investment and the environment were one of the most important issues due to the formation of the World Trade Organization (WTO)and the negotiation of the North American Free Trade Agreement (NAFTA) (Esty, 1994; Vogel, 1995; Audley, 1997; Kirton and Maclaren, 2002; McCarthy, 2004). Increasingly, analysts have also focused on the role of economic IGOs (Najam et al., 2007; Park, 2007), MNCs (Prakash, 2000; Levy and Newell, 2005; Falkner, 2008) and, less so, state agencies (Keohane and Levy, 1996; Hicks et al., 2008). The impacts of financial processes and development programs on the environment have also become an important but still less developed line of research (Clapp, chapter 5 in this book; Clapp and Helleiner, 2012).

This period also witnessed an explosion of research on the relations between environment and security divisible into various broad streams (see Deudney and Matthew, 1999; Khagram and Ali, 2006; Matthew et al., 2010; Deligianis, 2012; Swatuk, chapter 9 in this book). One approach has focused on the relations between resources and environmental conflict and can be traced directly to the geopolitics of the 1970s. A second line of research, also ranging back to the 1970s, focused on the ecocidal impacts of military establishments, while a third line focuses on the challenges of reaching agreement under conditions of disagreement rather than actual conflict. Finally, there has emerged a strong line of critical environmental security studies (for example, Dalby, 2002; Detraz, 2012).

Conclusions

In closing this chapter I would like to address the questions posed upfront. I interpret the evidence to suggest that IEP is strongly tethered to the various IR subfields and perspectives rather than some foundational ecological themes, concepts, or debates. The prospect of an IEP centered around such central themes emerged during the 1970s when the commons, lifeboat ethics, and limits informed some of the emergent IEP scholarship. While similar core concepts, such as social-ecological systems, are used by networks of researchers, they are one tendency among various.

Some IR perspectives, such as liberalism and liberal constructivism, are well represented in IEP and can be said to be dominant at present. But other IR perspectives, such as realism, ecosocialism, critical and poststructural IR, feminism, and so on, have carved their niches, some quite significant. Here I must also note that some of that research is published in non-environmental journals, for example, conflict in security journals, political economy in IPE journals, critical work in IR/theory journals, and so on. IEP, as a whole, does now address social and ecological purpose in dealing with political economy, governance, peace research, ecological justice, the politics of knowledge, and so on. However, a great deal of the research in IEP tends to focus on architecture and form, reflecting the continued dominance of liberalism, liberal constructivism, and realism in IR. In short, the plurality and inclusiveness of IEP cannot be taken for granted and it may very well be the case that, like in IR as a whole (see Tickner and Weaver, 2009), IEP is splintering more than engaging in dialog.

While governance, political economy, and security are central research areas, the various other research areas can now show robust work that is not readily reducible to governance, security, or political economy. The various chapters in this volume strongly demonstrate that the various research areas are not insular yet have also developed their own internal dynamics and priorities. Despite the fact that certain approaches and research programs within IEP tend to be self-referential, there seems to be more constructive exchange across perspectives than in many other IR subfields.

Where does this research come from? Not surprisingly most of the early research in the English language came from the United States, Canada, and the United Kingdom, with the former soon becoming dominant. During the last several decades the rest of the North has caught up in terms of research producers and research output. Increasingly, Southern research has become more visible. Even though many Southern scholars have been trained in and are employed by Northern institutions, or work within global networks enabled by or dependent on Northern institutions and resources, Southern institutions and research are now more visible. Like IR, as a whole, IEP scholarship is now extensive enough to require a close examination of its own history and sociology (Tickner and Waever, 2009).

Note

1. I use global environmental politics to refer to the practice of politics and international environmental politics to refer to its study, as we do throughout this book.

Annotated bibliography

Orr, D. W. and M. R. Soroos (eds) (1979) *The Global Predicament: Ecological Perspectives on World Order* (Chapel Hill: University of North Carolina Press).

This volume captures the approaches of the first generation of IEP scholars who sought to bridge ecological thought and international politics.

Laferrière, E. and P. Stoett (1999) *International Relations Theory and Ecological Thought: Towards a Synthesis* (London: Routledge).

A very incisive discussion of the challenges that ecological thought presents for IR. The authors offer an account that is well grounded in both ecological theory and IR theory.

Handbooks by: P. Dauvergne (2011) *Handbook of Global Environmental Politics*, 2nd edition (Cheltenham, UK: Edward Elgar); Robert Falknker (2013) *Handbook of Global Climate and Environment Policy* (Chichester: Wiley-Blackwell) and Paul Harris (2013) *Routledge Handbook of Global Environmental Politics* (London: Routledge).

These handbooks cover a broad swath of research areas and issues. Because of its theoretical inclusiveness and sound theoretical foundations, readers of the present volume will be well prepared to turn these handbooks into useful complements in their effort to gain sound foundations on particular research and issue areas.

Works cited

Aage, H. (2008) 'Economic Ideology about the Environment: From Adam Smith to Bjorn Lomborg', *Global Environment* 2, 1, 8–45.

Adger, W. N. et al. (eds) (2006) *Fairness in Adaptation to Climate Change* (Cambridge, MA: The MIT Press).

Agarwal, A. and S. Narain (1991) *Global Warming in an Unequal World: A Case of Environmental Colonialism* (New Delhi, India: Centre for Science and Environment).

Andonova, L. B. and R. B. Mitchell (2012) 'The Rescaling of Global Environmental Politics', *Annual Review of Environment and Resources* 35, 1, 255–282.

Andresen, S. and W. Østreng (1989) *International Resource Management: The Role of Science and Politics* (London: Belhaven Press).

Angle, D., T. Hamilton and M. Huber (2007) 'Global Environmental Standards for Industry', *Annual Review of Environment and Resources* 32, 295–316.

Audley, J. J. (1997) *Green Politics and Global Trade: NAFTA and the Future of Environmental Politics* (Washington, DC: Georgetown University Press).

Bäckstrand, K. (2006) 'Democratizing Global Environmental Governance? Stakeholder Democracy after the World Summit on Sustainable Development', *European Journal of International Relations* 12, 4, 467–498.

Balasubramaniam, A. (1984) *Ecodevelopment: Towards a Philosophy of Environmental Education* (Singapore: Konigswinter Friedrich Naumann Stiftung).

Banuri, T. and J. Holmberg (1992) *Governance for Sustainable Development: A Southern Perspective* (London: International Institute for Environment and Development).

Barkdull, J. and P. Harris (2002) 'Environmental Change and Foreign Policy: A Survey of Theory', *Global Environmental Politics* 2, 2, 63–91.

Barkin, J. S. and G. Shambaugh (eds) (1999) *Anarchy and the Environment: The International Relations of Common Pool Resources* (Albany: State University of New York Press).

Barry, J. and R. Eckersley (eds) (2005) *The State and the Global Ecological Crisis* (Cambridge, MA: MIT Press).

Bartelmus, P. L. (1986) *Environment and Development* (Reading, MA: Allen and Unwin).

Baumel, W. (1971) *Environmental Protection, International Spillovers, and Trade* (Stockholm: Almquist and Wiksell).

Beck, U. (1999) *World Risk Society* (Cambridge: Polity).

Beckerman, W. (1974) *In Defense of Economic Growth* (London: J. Cape).

Beckerman, W. and J. Pasek (2001) *Justice, Posterity and the Environment* (New York: Oxford University Press).

Benedick, R. E. (1991) *Ozone Diplomacy: New Directions in Safeguarding the Planet* (Cambridge, MA: Harvard University Press).

Bernstein, S. (2001) *The Compromise of Liberal Environmentalism* (New York: Columbia University Press).

Bernstein, S. (2011) 'Legitimacy in Intergovernmental and Non-state Global Governance', *Review of International Political Economy* 18, 1, 17–51.

Betsill, M. and E. Correll (eds) (2008) *NGO Diplomacy: The Influence of Non-governmental Organizations in International Environmental Negotiations* (Cambridge, MA: The MIT Press).

Biermann, F. and P. Pattberg (2008) 'Global Environmental Governance: Taking Stock, Moving Forward', *Annual Review of Environment and Resources* 33, 277–294.

Biermann, F. and P. Pattberg (eds) (2012) *Global Environmental Governance Reconsidered* (Cambridge, MA: MIT Press).

Biswas, A. K. (1984) *Climate and Development* (Dublin: Tycooly International).

Boardman, R. (1981) *International Organization and the Conservation of Nature* (Bloomington: Indiana University Press).

Bookchin, M. (1974[1962]) *Our Synthetic Environment* (New York: Harper Colophon Books).

Brandon, K. E. and C. Brandon (eds) (1992) 'Linking Environment to Development: Problems and Possibilities', *World Development* 20, 4, Special Issue.

Brenton, T. (1994) *The Greening of Machiavelli: The Evolution of International Environmental Politics* (London: Royal Institute of International Affairs).

Bretherton, C. (1998) 'Global Environmental Politics: Putting Gender on the Agenda?', *Review of International Studies* 24, 1, 85–100.
Bretherton, C. (2003) 'Movements, Networks, Hierarchies: A Gender Perspective on Global Environmental Governance', *Global Environmental Politics* 3, 2, 103–119.
Brookfield, H. C. (1980–1981) 'Development and the Environment: Politics, Production and Pollution', *Pacific Affairs* 53, 4, 661–681.
Brown, H. (1954) *The Challenge of Man's Future: An Inquiry Concerning the Condition of Man During the Years That Lie Ahead* (New York: Viking Press).
Brown, H., J. Bonner and J. Weir (1957) *The Next Hundred Years: Man's Natural and Technological Resources* (New York: Viking Press).
Brown, L. (1977) *Redefining National Security* (Washington: Worldwatch Institute).
Brownlie, I. (1972) 'The Human Environment: Problems of Standard Setting and Enforcement', *International Relations* 4, 1, 29–37.
Buttel, F. H., A. P. Hawkins and A. G. Power (1990) 'From Limits to Growth to Global Change: Constraints and Contradictions in the Evolution of Environmental Science and Ideology', *Global Environmental Change* 1, 1, 57–66.
Cairncross, F. (1992) *Costing the Earth: The Challenge for Governments, the Opportunities for Business* (Boston, MA: Harvard Business School Press).
Caldwell, L. K. and P. S. Weiland (1996) *International Environmental Policy: From the Twentieth to the Twenty-First Century* (Durham: Duke University Press).
Cardoso, F. F. (1980) 'Development and the Environment: The Brazilian Case', *CEPAL Review* 12, 111–128.
Carroll, J. E. (1988) *International Environmental Diplomacy: The Management and Resolution of Transfrontier Environmental Problems* (Cambridge: Cambridge University Press).
Carson, R. (1962) *Silent Spring* (Boston: Houghton Mifflin).
Cashore, B., G. Auld and D. Newsom (2004) *Governing Through Markets: Forest Certification and the Emergence of Non-State Authority* (New Haven: Yale University Press).
Chasek, P. S. (ed.) (1999) *The Global Environment in the Twenty-First Century: Prospects for International Cooperation* (Tokyo: United Nations University Press).
Chasek, P., D. Downie and J. Brown (2013) *Global Environmental Politics*, 6th edition (Boulder, CO: Westview Press).
Choucri, N. (1993) *Global Accord: Environmental Challenges and International Responses* (Cambridge, MA: MIT Press).
Choucri, N. and R. C. North (1975) *Nations in Conflict: National Growth and International Violence* (San Francisco: W. H. Freeman).
Christoff, P. and R. Eckersley (2013) *Globalization and the Environment* (Lanham, MD: Rowman and Littlefield Publishers).
Cjessing, G. (1967) 'Ecology and Peace Research', *Journal of Peace Research* 4, 2, 125–139.
Clapp, J. and E. Helleiner (2012) 'International Political Economy and the Environment: Back to the Basics?', *International Affairs*, 88, 3, 485–501.
Clapp, J. and P. Dauvergne (2011) *Paths to a Green World: The Political Economy of the Global Environment*, 2nd edition (Cambridge, MA: The MIT Press).
Clark, W. C. and R. E. Munn (1986) *Sustainable Development of the Biosphere* (Cambridge: Cambridge University Press).

Clemens, W., Jr. (1972–1973) 'Ecology and International Relations', *International Journal* 28, 1–27.
Cole, H. S. D. (ed.) (1973) *Models of Doom: A Critique of the Limits to Growth* (New York: Universe Books).
Commoner, B. (1966) *Science and Survival* (New York: Viking Press).
Commoner, B. (1971) *The Closing Circle: Nature, Man and Technology* (New York: Knopf).
Contini, P. and P. Sand (1972) 'Methods to Expedite Environmental Protection: International Ecostandards', *The American Journal of International Law* 66, 1, 37–59.
Crutchfield, J. A. (1956) 'Common Property Resources and Factor Allocation', *The Canadian Journal of Economics and Political Science/Revue Canadienne d'Economique et de Science Politique* 22, 3, 292–300.
Dahlberg, K. A. (1973) 'The Technological Ethic and the Spirit of International Relations', *International Studies Quarterly* 17, 1, 55–88.
Dahlberg, K. A. (1979) *Beyond the Green Revolution: The Ecology and Politics of Global Agricultural Development* (New York: Plenum Press).
Dahlberg, K., M. S. Soroos and A. T. Feraru (1984) *Environment and the Global Arena: Actors, Values, Policies and Futures* (Durham, NC: Duke University Press).
Dalby, S. (2002) *Environmental Security* (Minneapolis: University of Minnesota Press).
D'Amato, A. (1990) 'Do We Owe a Duty to Future Generations to Preserve the Global Environment?', *American Journal of International Law* 84, 1, 190–198.
Dauvergne, P. (ed) (2011) *Handbook of Global Environmental Politics*, 2nd edition (Cheltenham, UK: Edward Elgar).
de Araujo Castro, J. A. (1972) 'Environment and Development: The Case of the Developing Countries', *International Organization* 26, 2 (International Institutions and the Environmental Crisis), 401–416.
Deligianis, T. (2012) 'The Evolution of Environment-Conflict Research: Toward a Livelihood Framework', *Global Environmental Politics* 12, 1, 78–100.
DeSombre, E. (2000) *Domestic Sources of International Environmental Policy* (Cambridge, MA: MIT Press).
Detraz, N. (2012) *International Security and Gender* (Cambridge, UK: Polity Press).
Deudney, D. (1990) 'The Case against Linking Environmental Degradation and National Security', *Millennium* 19, 3, 461–476.
Deudney, D. and R. A. Matthew (eds) (1999) *Contested Grounds: Security and Conflict in the New Environmental Politics* (Albany: State University of New York Press).
Dryzek, J. (2013) *The Politics of the Earth: Environmental Discourses*, 3rd edition (Oxford: Oxford University Press).
Dubos, R. (1968) *Man, Medicine, and Environment* (New York: Praeger).
Dwivedi, O. P. (1997) *India's Environmental Policies, Programmes and Stewardship* (Basingstoke: Macmillan).
Economic and Political Weekly (1972) 'Development and Environment', *Economic and Political Weekly* 7, 25, 1172–1174.
Egan, M. (2007) *Barry Commoner and the Science of Survival: The Remaking of American Environmentalism* (Cambridge, MA: MIT Press).

Ehresman, T. and D. Stevis (2010) 'International Environmental and Ecological Justice', In G. Kütting (ed.) *Global Environmental Politics: Concepts, Theories and Case Studies* (London: Routledge), 87–104.

Ehrlich, P. R. (1968) *The Population Bomb* (New York: Ballantine Books).

Ehrlich, P., A. Ehrlich and H. D. Gretchen (1995) *The Stork and the Plow: The Equity Answer to the Human Dilemma* (New York: Putnam's).

Elliott, L. M. (2004) *The Global Politics of the Environment*, 2nd edition (New York: New York University Press).

Elliott, L. M. (2009) 'Australian Scholarship, International Relations and the Environment: Commitment, Critique and Contestation', *Australian Journal of Politics and History* 55, 3, 394–414.

Elkington, J. and T. Burke (1987) *The Green Capitalists: How Industry Can Make Money – and Protect the Environment* (London: Victor Gollancz).

Esty, D. (1994) *Greening the GATT: Trade, Environment and the Future* (Washington, DC: International Institute for Economics).

Falk, R. A. (1971) *This Endangered Planet: Prospects and Proposals for Human Survival* (New York: Random House).

Falk, R. A. (1975) *A Study of Future Worlds* (New York: Free Press).

Falkner, R. (2008) *Business Power and Conflict in International Environmental Politics* (Basingstoke: Palgrave Macmillan).

Falkner, R. (ed.) (2013) *Handbook of Global Climate and Environment Policy* (Chichester: Wiley-Blackwell).

Farvar, M. T. and J. P. Milton (eds) (1972) *The Careless Technology: Ecology and International Development* (Garden City, NY: Natural History Press).

Feraru, A. T. (1974) 'Transnational Political Interests and the Global Environment', *International Organization* 28, 1, 31–60.

Folke, C. (2006) 'Resilience: The Emergence of a Perspective for Social-Ecological Analyses', *Global Environmental Change* 16, 253–267.

Founex Report (1972) *Development and Environment* (Report and Working Papers of a Panel of Experts Convened by the Secretary-General of the United Nations Conference on the Human Environment, Founex, Switzerland, 4–12 June 1971) (Paris: Mouton).

Gale, F. and R. M. M'Gonigle (eds) (2000) *Nature, Production, Power: Towards an Ecological Political Economy* (Cheltenham, UK: Edward Elgar Publishing).

Galtung, J. (1973) 'The Limits to Growth and Class Politics', *Journal of Peace Research* 10, 1–2, 101–114.

Galtung, J. (1982) *Environment, Development and Military Activity: Towards Alternative Security Doctrines* (Oslo: Universitesforlaget).

Ghosh, P. K. (1984) *Population, Environment and Resources, and Third World Development* (Westport, CT: Greenwood Press).

Gjessing, G. (1967) 'Ecology and Peace Research', *Journal of Peace Research* 4, 2, 125–139.

Gladwin, T. (1977) *Environment, Planning and the Multinational Corporation* (Greenwich, Conn: JAI Press).

Glaeser, B. (1984) *Ecodevelopment: Concepts, Projects, Strategies* (Oxford: Oxford University Press).

Goldman, M. (ed.) (1998) *Privatizing Nature: Political Struggles for the Global Commons* (New Brunswick: Rutgers University Press).

Goodrich, C. (1951) 'The United Nations Conference on Resources', *International Organization* 5, 1, 48–60.
Gordon, H. S. (1954) 'The Economic Theory of a Common-Property Resource: The Fishery', *Journal of Political Economy* 62, 2, 124–142.
Gordon, L. (1975) 'Environment, Resources and Directions of Growth', *World Development* 3, 2–3, 113–121.
Guha, R. (2000) *Environmentalism: A Global History* (New York: Longman).
Guha, R. and J. Martinez-Alier (1997) *Varieties of Environmentalism: Essays North and South* (London: Earthscan).
Gurr, T. (1985) 'On the Political Consequences of Scarcity and Economic Decline', *International Studies Quarterly* 29, 1, 51–75.
Haas, P. M. (1990) *Saving the Mediterranean: The Politics of International Environmental Cooperation* (New York: Columbia University Press).
Haas, P. M. (ed.) (2003) *Environment in the New Global Economy, Vol. I: Analytic Approaches to the IPE of the Environment* (Cheltenham, UK: An Elgar Reference Collection).
Haas, P., R. Keohane and M. Levy (eds) (1993) *Institutions for the Earth: Sources of Effective International Environmental Protection* (Cambridge, MA: MIT Press).
Hardin, G. J. (1968) 'The Tragedy of the Commons', *Science* 162, 1243–1248.
Hardin, G. J. and J. Baden (1977) *Managing the Commons* (San Francisco: W.H. Freeman).
Hargrove, J. L. (ed.) (1972) *Law, Institutions, and the Global Environment: Papers and Analysis of the Proceedings* (Dobbs Ferry, NY: Oceana Publications).
Harris, J. (1991) 'Global Institutions and Ecological Crisis', *World Development* 19, 1, 111–122.
Harris, P. (ed.) (2013) *Routledge Handbook of Global Environmental Politics* (New York, NY: Routledge).
Harrison, N. (2000) *Constructing Sustainable Development* (Albany, NY: SUNY Press).
Herrera, A. O. (1976) *Catastrophe or New Society? A Latin American World Model* (Ottawa: International Development Research Centre).
Hicks, R., B. Parks J. T. Roberts et al. (2008) *Greening Aid? Understanding the Environmental Impact of Development Assistance* (Oxford: Oxford University Press).
Hornborg, A. (2009) 'Zero-Sum World: Challenges in Conceptualizing Environmental Load Displacement and Ecologically Unequal Exchange in the World-System', *International Journal of Comparative Sociology*, 50, 3–4, 237–262.
Homer-Dixon, T. (1991) 'On the Threshold: Environmental Changes as Causes of Acute Conflict', *International Security* 16, 2, 76–116.
Hurrell, A. and B. Kingsbury (eds) (1992) *The International Politics of the Environment: Actors, Interests, and Institutions* (Oxford: Clarendon Press).
Hveem, H. (1979) 'Militarization of Nature: Conflict and Control over Strategic Resources and Some Implications for Peace Policies', *Journal of Peace Research* 16, 1, 1–26.
Iles, A. (2006) 'The International Political Economy of Making Consumption Sustainable', *Review of International Political Economy* 13, 2, 340–358.
International Journal (1972/1973) 'Earth Politics', *International Journal* 28, 1, 1–174.
Jacobsen, S. (1996) *North-South Relations and Global Environmental Issues: A Review of the Literature* (Copenhagen: Centre for Development Research).

Jacobsen, S. (1999) 'International Relations and Global Environmental Change: Review of the Burgeoning Literature on the Environment', *Cooperation and Conflict* 34, 2, 205–236.
James, J. (1978) 'Growth, Technology and the Environment in Less Developed Countries: A Survey', *World Development* 6, 937–965.
Jancar, B. (1991–1992) 'Environmental Studies: State of the Discipline', *International Studies Notes* 16/17, 1/3, 25–31.
Jasanoff, S. (1993) 'India at the Crossroads in Global Environmental Policy', *Global Environmental Change* 3, 1, 32–52.
Kamieniecki, S. (ed.) (1993) *Environmental Politics in the International Arena: Movement, Parties, Organizations and Policy* (Albany: SUNY Press).
Kaplan, J. (1949) 'Resource Conservation and Foreign Policy', *World Politics* 1, 2, 257–265.
Kay, D. and E. Skolnikoff (eds) (1972) *World Eco-Crisis* (Madison: University of Wisconsin Press).
Kay, D. and H. Jacobson (eds) (1983) *Environmental Protection: The International Dimension* (Totowa, NJ: Allanheld, Osmun).
Keith, D. (2013) *A Case for Climate Engineering* (A Boston Review Book, Cambridge, MA: MIT Press).
Keohane, R. O. and E. Ostrom (1995) *Local Commons and Global Interdependence: Heterogeneity and Cooperation in Two Domains* (London: Sage).
Keohane, R. O. and M. Levy (eds) (1996) *Institutions for Environmental Aid: Pitfalls and Promises* (Cambridge, MA: MIT Press).
Khagram, S. and S. Ali (2006) 'Environment and Security', *Annual Review of Environment and Resources* 31, 1, 395–411.
Kirton, J. J. and V. W. Maclaren (2002) *Linking Trade, Environment, and Social Cohesion: NAFTA Experiences, Global Challenges* (Burlington, VT: Ashgate).
Kneese, A., S. Rolfe and J. Harned (1971) *Managing the Environment: International Economic Cooperation for Pollution Control* (New York: Praeger Publishers).
Konigsberg, C. (1960) 'Climate and Society: A Review of the Literature', *Journal of Conflict Resolution (The Geography of Conflict)* 4, 1, 67–82.
Kotok, E. I. (1945) 'International Policy on Renewable Natural Resources', *The American Economic Review* 35, 2, 110–119.
Kristof, L. (1960) 'The Origins and Evolution of Geopolitics', *The Journal of Conflict Resolution* 4, 1, 15–51.
Kütting, G. (2000) *Environment, Society and International Relations: Towards More Effective International Environmental Agreements* (London: Routledge).
Kütting, G. (2004) *Globalization and the Environment: Greening Global Political Economy* (Albany: SUNY Press).
Laferrière, E. and P. J. Stoett (1999) *International Relations Theory and Ecological Thought: Towards a Synthesis* (London: Routledge).
Laferrière, E. and P. J. Stoett (eds) (2006) *International Ecopolitical Theory: Critical Approaches* (Vancouver: University of British Columbia Press).
Leonard, J. (1988) *Production and the Struggle for the World Product: Multinational Corporations, Environment, and International Comparative Advantage* (New York: Cambridge University Press).
Leopold, A. (1949) *A Sand County Almanac and Sketches Here and There* (New York: Oxford University Press).
Levy, M. (1995) 'Is the Environment a National Security Issue?', *International Security* 20, 2, 35–62.

Levy, D. and P. Newell (eds) (2005) *The Business of Global Environmental Governance* (Cambridge, MA: The MIT Press).
Lipschutz, R. and K. Conca (eds) (1993) *The State and Social Power in Global Environmental Politics* (New York: Columbia University Press).
Lipschutz, R., with J. Mayer (1996) *Global Civil Society and Global Environmental Governance: The Politics of Nature from People to Place* (Albany: SUNY Press).
Litfin, K. (1994) *Ozone Discourses: Science and Politics in Global Environmental Cooperation* (New York: Columbia University Press).
Litfin, K. (1998) *The Greening of Sovereignty in World Politics* (Cambridge, MA: MIT Press).
MacDonald, G. (2003) 'Environment: Evolution of a Concept', *The Journal of Environment and Development* 12, 2, 151–176.
Matthew, R. A., J. Barnett, B. McDonald and K. L. O'Brien (eds) (2010) *Global Environmental Change and Human Security* (Cambridge, MA: MIT Press).
McCarthy, J. (2004) 'Privatizing Conditions of Production: Trade Agreements as Neoliberal Environmental Governance', *Geoforum* 35, 327–441.
McCormick, J. (1989) *Reclaiming Paradise: The Global Environmental Movement* (Bloomington: Indiana University Press).
Meadows, D. H., D. L. Meadows, J. Randers and W. B. Behrens, III (1972) *The Limits to Growth: A Report for the Club of Rome's Project on the Predicament of Mankind* (New York: Universe Books).
Meadows, D. H., D. L. Meadows and J. Randers (1992) *Beyond the Limits: Confronting Global Collapse, Envisioning a Sustainable Future* (Post Mills, VT: Chelsea Green).
Methman, C. (2013) 'The Sky is the Limit: Global Warming as Global Governmentality', *European Journal of International Relations* 19, 1, 69–91.
M'Gonigle, R. M. and M. W. Zacher (1979) *Pollution, Politics and International Law: Tankers at Sea* (Berkeley: University of California Press).
Miller, M. (1995) *The Third World in Global Environmental Politics* (Boulder, CO: Lynne Rienner).
Najam, A., M. Halle and R. Melendez-Ortiz (eds) (2007) *Trade and Environment: A Resource Book* (Winnipeg, Canada: International Centre for Trade and Sustainable Development, International Institute for Sustainable Development and ISD, ICTSD and The Ring).
Nash, R. (1989) *The Rights of Nature: A History of Environmental Ethics* (Madison, WI: University of Wisconsin Press).
Natural Resources Forum (2011) *Viewpoints* [Green Economy] 35, 63–72.
Nelson, D., W. N. Adger and K. Brown (2007) 'Adaptation to Environmental Change: Contributions of a Resilience Framework', *Annual Review of Environment and Resources* 32, 395–419
Newell, P. (2005) 'Race, Class and the Global Politics of Environmental Inequality', *Global Environmental Politics* 5, 3, 50–94.
Newell, P. (2012) *Globalization and the Environment: Capitalism, Ecology and Power* (Cambridge, UK: Polity Press).
Nicholson, M. (1970) *The Environmental Revolution: A Guide for the New Masters of the World* (New York: McGraw-Hill).
North, R. C. (1977) 'Toward a Framework for the Analysis of Scarcity and Conflict', *International Studies Quarterly,* Special Issue on International Politics of Scarcity, 21, 4, 569–591.

O'Neill, K. (2009) *The Environment and International Relations* (Cambridge: Cambridge University Press).
Onuf, N. (1983) 'Reports to the Club of Rome', *World Politics* 36, 1, 121–146.
Ophuls, W. (1977) *Ecology and the Politics of Scarcity: Prologue to a Political Theory of the Steady State* (San Francisco: W.H. Freeman).
Orr, D. W. and M. R. Soroos (eds) (1979) *The Global Predicament: Ecological Perspectives on World Order* (Chapel Hill: University of North Carolina Press).
Osborn, F. (1948) *Our Plundered Earth* (Boston: Little, Brown).
Osborn, F. (1953) *The Limits of the Earth* (Boston: Little, Brown).
Ostrom, E. (1990) *Governing the Commons* (Cambridge: Cambridge University Press).
Park, S. (2007) 'The World Bank Group: Championing Sustainable Development Norms?', *Global Governance* 13, 4, 535–556.
Parks, B. and J. Timmons Roberts (2006) 'Environmental and Ecological Justice', In M. Betsill, K. Hochstetler, and D. Stevis (eds.) (2006) *Palgrave Advances in International Environmental Politics* (Basingstoke: Macmillan), 329–360.
Pearson, C. S. (1987) *Multinational Corporations, Environment, and the Third World: Business Matters* (Durham, NC: Duke University Press).
Peterson, M. J. (1992) 'Transnational Activity, International Society and World Politics', *Millennium* 21, 3, 371–388.
Phelan, L., A. Henderson-Sellers and R. Taplin (2012) 'The Political Economy of Addressing the Climate Crisis in the Earth System: Undermining Perverse Resilience', *New Political Economy* 18, 2, 198–226.
Pirages, D. (1978) *Global Ecopolitics: The New Context for International Relations* (North Scituate, MA: Duxbury Press).
Pirages, D. (1983) 'The Ecological Perspective and the Social Sciences', *International Studies Quarterly* 27, 3, 243–255.
Poleszynski, D. (1977) 'Waste Production and Overdevelopment: An Approach to Ecological Indicators', *Journal of Peace Research* 14, 4, 285–298.
Pollock, S. (1956) 'The International Allocation of Resources: New Concepts and Problems of Administration', *The Canadian Journal of Economics and Political Science/Revue Canadienne d'Economique et de Science Politique* 22, 4, 461–466.
Prakash, A. (2000) *Greening the Firm: The Politics of Corporate Environmentalism* (New York: Cambridge University Press).
Princen, T. (2005) *The Logic of Sufficiency* (Cambridge, MA: MIT Press).
Princen, T. and M. Finger (eds) (1994) *Environmental NGOs in World Politics: Linking the Local and the Global* (London: Routledge).
Princen, T., M. Maniates and K. Conca (eds) (2002) *Confronting Consumption* (Cambridge, MA: MIT Press).
Redclift, M. R. (1984) *Development and the Environmental Crisis: Red or Green Alternatives* (London: New York).
Redclift, M. R. (1992) 'Sustainable Development and Global Environmental Change: Implications of a Changing Agenda', *Global Environmental Change: Human and Policy Dimensions* 2, 1, 32–42.
Rich, B. (1985) 'The Multilateral Development Banks, Environmental Policy, and the United States', *Ecology Law Quarterly* 12, 4, 681–746.
Riddell, R. (1981) *Ecodevelopment: Economics, Ecology, and Development: An Alternative to Growth Imperative Models* (New York: St. Martin's Press).

Roberts, J. T. and B. Parks (2007) *A Climate of Injustice: Global Inequality, North – South Politics, and Climate Policy* (Cambridge, MA: MIT Press).
Robertson, T. (2012) 'Total War and the Total Environment: Fairfield Osborn, William Vogt, and the Birth of Global Ecology', *Environmental History* 17, 336–364.
Rønnfeldt, C. (1997) 'Three Generations of Environment and Security Research', *Journal of Peace Research* 34, 4, 473–482.
Rowland, W. (1973) *The Plot to Save the World: The Life and Times of the Stockholm Conference on the Human Environment* (Toronto: Clarke, Irwin).
Rubin, S. J. and T. R. Graham (1982) *Environment and Trade: The Relation of International Trade and Environmental Policy* (Totowa, NJ: Allanheld Osmun).
Ruggie, J. (1975) 'International Responses to Technology – Concepts and Trends', *International Organization* 29, 3, 557–583.
Sachs, I. (1970) 'Environmental Concern and Development Planning', *International Conciliation* 39, 72.
Sachs, I. (1974) 'Environment et planification: quelques pistes de recherches et d'action' [Environment and Planning: Some Areas of Research and Action], *Information sur les Sciences Sociales* 13, 6, 17–29.
Sachs, I. (1980) *Stratégies de l'écodéveloppement [Strategies of Ecodevelopment]* (Paris: Éditions ouvrières).
Sachs, W. (ed.) (1993) *Global Ecology: A New Arena of Political Conflict* (London: Zed Books).
Sand, P. H. (1992) *The Effectiveness of International Environmental Agreements: A Survey of Existing Legal Instruments* (Cambridge, England: Grotius Publications).
Schneider, J. (1979) *World Public Order of the Environment: Towards an International Ecological Law and Organization* (Toronto: University of Toronto Press).
Schor, J. (1991) 'Global Equity and Environmental Crisis: An Argument for Reducing Working Hours in the North', *World Development* 19, 1, 73–84.
Shields, L. and M. Ott (1974) 'The Environmental Crisis: International and Supranational Approaches', *International Relations* 4, 6, 629–648.
Shiva, V. (1991) *The Violence of the Green Revolution: Third World Agriculture, Ecology, and Politics* (London: Zed Books).
Shrivastava, P. (1987) *Bhopal: Anatomy of a Crisis* (Cambridge, MA: Ballinger).
Smith, J. E. (1972) 'The Role of Special Purpose and Nongovernmental Organizations in the Environmental Crisis', *International Organization* (*International Institutions and the Environmental Crisis*) 26, 2, 302–326.
Soroos, M. S. (1977) 'The Commons and Lifeboat as Guides for International Ecological Policy', *International Studies Quarterly* 21, 4, Special Issue on International Politics of Scarcity, 647–674.
Soroos, M. S. (1986) *Beyond Sovereignty: The Challenge of Global Policy* (Columbia: University of South Carolina Press).
Soroos, M. (ed.) (1991) 'The Human Dimensions of Global Environmental Change', *International Studies Notes*, Special Issue, Winter, 16, 1.
Spengler, J. (1949) 'The World's Hunger: Malthus 1948', *Proceedings of the Academy of Political Science* 23, 2, 53–72.
Springer, A. (1983) *The International Law of Pollution: Protecting the Global Environment in a World of Sovereign States* (Westport, CT: Quorum Books).
Sprout, H. and M. Sprout (1957) 'Environmental Factors in the Study of International Politics', *Journal of Conflict Resolution* 1, 4, 309–328.

Sprout, H. and M. Sprout (1965) *The Ecological Perspective on Human Affairs, with Special Reference to International Politics* (Princeton, NJ: Princeton University Press).
Sprout, H. and M. Sprout (1971) *Toward a Politics of the Planet Earth* (New York: Van Nostrand Reinhold).
Stein, R. and B. Johnson (1979) *Agencies Banking on the Biosphere? Environmental Procedures and Practices of Nine Multilateral Development* (Lexington, MA: D.C. Heath).
Steinberg, P. and S. VanDeveer (eds) (2012) *Comparative Environmental Politics: Theory, Practice and Prospects* (Cambridge, MA: MIT Press).
Stevis, D. (2005) 'The Globalizations of the Environment', *Globalizations* 2, 3, 323–333.
Stevis, D. (2010) 'International Relations and the Study of Global Environmental Politics: Past and Present', in B. Denemark (ed.) The *International Studies Encyclopedia*, volume vii (Hoboken: Wiley-Blackwell), 4476–4507.
Stevis, D. and H. Bruyninckx (2006) 'Looking Through the State at Environmental Flows and Governance', In G. Spaargaren, A. Mol and F. Buttel (eds) *Governing Environmental Flows: Global Challenges to Social Theory* (Cambridge, MA: The MIT Press), 107–136.
Stevis, D. and V. J. Assetto (eds) (2001) *The International Political Economy of the Environment: Critical Perspectives* (Boulder: Lynne Rienner).
Stevis, D., V. Assetto and S. P. Mumme (1989) 'International Environmental Politics: A Theoretical Review of the Literature', In J. P. Lester (ed.) *Environmental Politics and Policy* (Durham: Duke University Press), 289–313.
Stone, C. (1974) *Should Trees have Standing?: Towards Legal Rights for Natural Objects* (Palo Alto, CA: W. Kaufmann).
Swyngedouw, E. (2010) 'Apocalypse Forever? Post-political Populism and the Spectre of Climate Change', *Theory, Culture & Society* 27, 2–3, 213–232.
Taylor, P. and F. H. Buttel (1992) 'How Do We Know We Have Global Environmental Problems? Science and the Globalization of Environmental Discourse', *Geoforum* 23, 3, 405–416.
Thomas, C. (1992) *The Environment in International Relations* (London: Royal Institute of International Affairs).
Thomas, W. L. (ed.) (1956) *Man's Role in Changing the Face of the Earth* (Chicago: University of Chicago Press).
Tickner, A. and O. Waever (eds) *International Relations Around the World* (London: Routledge).
Turner, G. (2008) 'A Comparison of *The Limits to Growth* with 30 Years' Reality', *Global Environmental Change* 18, 397–411.
World Commission on Environment and Development (1987) *Our Common Future* (Oxford: Oxford University Press).
Vogel, D. (1995) *Trading Up* (Cambridge, MA: Harvard University Press).
Vogler, J. (2000) *The Global Commons: Environmental and Technological Governance* (New York: Wiley).
Vogler, J. and M. Imber (eds) (1996) *The Environment and International Relations* (London: Routledge).
Vogt, W. (1948) *Road to Survival* (New York: William Sloane Associates).
Wapner, P. (1996) *Environmental Activism and World Civic Politics* (Albany: SUNY Press).

Wapner, P. (2010) *Living through the End of Nature* (Cambridge, MA: MIT Press).
Ward, B. (1966) *Spaceship Earth* (New York: Columbia University Press).
Weiss, E. B. (1988) 'Our Rights and Obligations to Future Generations for the Environment', *American Journal of International Law* 84, 1, 198–207.
Weiss, E. B. and H. K. Jacobson (eds) (1998) *Engaging Countries: Strengthening Compliance with International Environmental Accords* (Cambridge, MA: MIT Press).
Westing, A. H. (1977) *Weapons of Mass Destruction and the Environment* (London: Taylor and Francis).
Westing, A. H. (1988) *Cultural Norms, War and the Environment* (Oxford: New York).
Willetts, P. (ed.) (1982) *Pressure Groups in the Global System: The Transnational Relations of Issue-Oriented Non-Governmental Organizations* (New York: St. Martin's Press).
Wilson, T., Jr. (1971) *International Environmental Action: A Global Survey* (Cambridge, MA: Dunellen).
Woodhouse, E. J. (1972) 'Re-Visioning the Future of the Third World: An Ecological Perspective on Development', *World Politics* 25, 1, 1–33.
Young, O. R. (1977) *Resource Management at the International Level: The Case of the North Pacific* (London: F. Pinter).
Young, O. R. (1989) *International Cooperation: Building Regimes for Natural Resources and the Environment* (Ithaca: Cornell University Press).
Young, O., F. Berkhout, G. Gallopin, M. Jannsen, E. Ostrom and S. van der Leeuw (2006) 'The Globalization of Socio-Ecological Systems: An Agenda for Scientific Research', *Global Environmental Change* 16, 3, 304–316.
Young, O., L. King and H. Schroder (eds) (2008) *Institutions and Environmental Change: Principal Findings, Applications, and Research Frontiers* (Cambridge, MA: MIT Press).

3
Theoretical Perspectives on International Environmental Politics

Matthew Paterson

Reviews of the theories of international environmental politics (IEP), like those of international relations (IR) more generally, tend to be organized around different perspectives, commonly realism, liberalism/institutionalism/pluralism, structuralism/Marxism, and 'critical theories' (variously Frankfurt school critical theory, poststructuralism, feminism, green thought) (see, for example, Laferrière and Stoett, 1999; Paterson, 2000: Chapters 2–3; Vogler, 2011). Such ways of organizing tend to create the sense of homogenous, internally consistent perspectives, and perhaps more importantly fail to investigate the specifically theoretical aspects of the ideas – that is, they describe the arguments offered by differing perspectives but do not get to the heart of the assumptions underpinning them or ask questions about the internal logic and how one gets to the perspective from these assumptions.

Instead, therefore, this chapter starts from the basic starting points that are often offered as ways to theorize IEP. I investigate the nature of these starting points and the ways that various perspectives arise out of them. These starting points are not all of the same character or nature: some, for example, are ontologies – basic assumptions about what the world is like – while others are normative commitments. They are not all therefore commensurate with each other, making comparison

---

Thanks are due to the participants at the ISA workshop for the first edition of this book in Portland, Oregon, in March 2003, for useful feedback, and especially to the editors of the book for very useful comments on earlier drafts which forced me to think more carefully about some of the arguments here and the overall structure of the chapter.

between them complicated. The perspectives which arise out of these starting points also exceed their limits, and many specific theorists or perspectives could fall into more than one group. Of course, some perspectives we are used to treating as distinct fall into more than one category. My focus is on what each particular theoretical starting point enables specific perspectives to analyze or interpret. Starting with these basic assumptions enables us to show how a perspective has been built on them, what different perspectives have been built on the same assumption, and how persuasive each case might then be taken to be.

It seems to me that there are six principal starting points for enquiry that guide most analyses of IEP. In roughly the order that they have appeared in contemporary debates (see also the historical account offered by Stevis in this book), they are as follows:

- the logic of international anarchy as a driver,
- capital accumulation as the central driver,
- knowledge as central to responses to environmental problems,
- IEP characterized by complexity and plurality,
- structural inequalities as enduring obstacles, and
- the search for sustainability as the overriding question.

These starting points can be understood principally through the way they enable us to address two principal questions of concern within IEP: questions concerning the political origins of environmental degradation and questions about the character of political processes generated by attempts to address that degradation. While all of them address both questions, they do so with differing degrees of emphasis, and I start with what seems to me to be the two that attempt to address both most systematically – the focus on international anarchy and the focus on capitalism. The relationship between these starting points and theoretical perspectives that readers may be more familiar with is presented heuristically in Figure 3.1.

International anarchy

For some, international anarchy is the central starting point for the study of IEP. It provides an overarching logic that explains both the generation of environmental problems and the central political structure within which responses to such problems are generated. The proposition that the central starting point for analyzing IEP is the anarchic structure of international politics is one which unites three traditions which

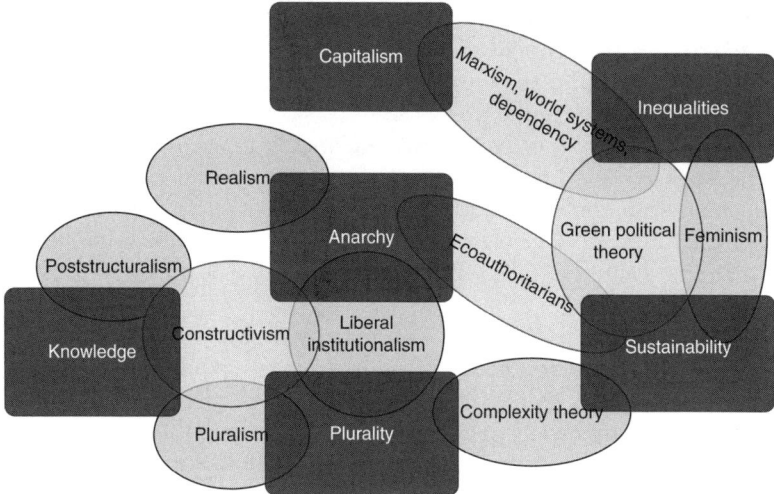

Figure 3.1 Relationship between theoretical starting points and theoretical perspectives in IEP
Note: The six starting points for theoretical enquiry discussed in this chapter are in the rectangles. The ovals indicate the theoretical perspectives that start from these perspectives. Overlaps indicate that perspectives may deploy the logics of more than one starting point, and overlap with each other also. Not all connections are possible to indicate in the figure.

many would suggest are the dominant traditions in the subject: realism, liberal institutionalism, and eco-authoritarianism. Some constructivists in IR are also highly focused on the question of anarchy (notably Wendt, 1999), although in IEP constructivists have tended not to be driven by this concern, but focus rather on ideas, knowledge, and the construction of interests (for example, Bernstein, 2001; Lövbrand, this book; and see section on knowledge below). The 'English School' in IR (for example, Bull, 1977) also starts from the premise of anarchy, and analysts within this tradition have started to develop arguments about environmental responsibility as an emerging norm within international society (for example, Falkner, 2012). Outside IR, and especially in economics, anarchy provides the dominant account for the character of IEP (see, for example, Barrett, 2003; Aldy and Stavins, 2007; Gupta et al., 2007).

Anarchy remains the central guiding assumption underpinning the work of many scholars in IR. It does not refer to continual chaos but simply to the absence of a central authority. Thus world politics is taken to be composed principally of states – political institutions defined in the sense in terms of their 'monopoly of the legitimate use of physical force within a given territory' (Weber, 1968/1919) – more commonly in IR referred to as the condition of sovereignty. States recognize no authority which can legitimately impose its will on them, and also recognize each other as sovereign in this sense.

But by contrast, environmental degradation is typically transnational. 'The earth is one, the world is not', runs a standard slogan informing much environmental analysis. IEP is thus defined as a collective action problem – how do sovereign actors interact when faced with problems which they cannot individually resolve but need to deal with each other to address?

Three principal responses to this anarchy problématique have been articulated in IEP. For ecoauthoritarians such as Ophuls (1977), one of the earliest writers in IEP, this is a fundamental contradiction. International anarchy means that states pursue their own interests, and for Ophuls, this inevitably means that common resources will be overused. The principal theoretical logic is that of the 'tragedy of the commons', articulated famously by Garrett Hardin (1968). This involves an account of collective action problems which emphasizes the uncooperative nature of common (or more precisely open access) property regimes.[1] In an open access resource, the resource itself has overall limits on sustainable use while individual users pursue their own interests. In this situation, each user has an incentive to use the resource more than their 'fair share' to pursue immediate interests in increasing wealth. The result is overuse and degradation of the resource overall. The problem of anarchy is thus not only that it impedes cooperation on environmental problems but also that it helps understand why they exist in the first place. For Ophuls, the consequence is that structural overhaul of global politics becomes necessary, involving the establishment of a world government. If the problem is the lack of central authority and thus the pursuit of individual interests which collide with the collective good of sustainability, then the solution is to shift authority to the global level, so there is no longer any contradiction between the collective interests of the world and the individual interests of the principal global political institution.

Realism, usually referred to as the dominant approach in IR, has many similarities to ecoauthoritarianism. Realists and ecoauthoritarians both agree that resources are overused because of the contradiction between individual state interests and those of the common good, but they spend more effort explaining how this contradiction prevents sufficient cooperation to alleviate the resulting problems in any significant manner. The notion of the 'tragedy of the commons' is very similar structurally to realist interpretations of the Prisoner's Dilemma (PD), which also focuses on the disjuncture between individual and collective interests to explain non-cooperation and suboptimal outcomes.

The difference is that realists tend to eschew the normative conclusion promoted by ecoauthoritarians. They tend to argue that the structural overhaul of world politics is simply impossible to achieve because states will not cede authority to any such global institutions (Bull, 1977: 293–295; Shields and Ott, 1974). Therefore, the world is for realists 'doomed' to unsustainability. On the one hand, on specific environmental problems, sufficient cooperation will never be achieved, as individual state interests prevent such cooperation from emerging.[2] On the other hand, environmental degradation becomes nationalized, resulting in the emergence of environmental security discourse.[3] While this discourse has a range of meanings, one origin is the realist assumption that states are the primary actors and their principal motivation for action is individual state interests, frequently understood in terms of security. In this sense, environmental security refers to the attempt to overcome the anarchy/environment contradiction by rendering the environment in nationalist terms (see Swatuk, this book).

Liberal institutionalism has been the mainstay of much analysis in IEP since the late 1980s. Institutionalists agree with realists that an overarching change in the organizing principles of world politics is impossible, but disagree with both realists and ecoauthoritarians concerning the implications of international anarchy. The starting point is a critique of realist accounts of collective action problems, especially developed theoretically in terms of game theory. The principal theoretical development, most famously by Robert Axelrod (1984), was to show that the standard model of PD was entirely consistent with substantial cooperation between actors, if one assumes both that the game was iterated allowing strategic interaction ('tit for tat', or conditional cooperation strategies) to elicit cooperation from other actors and that communication between actors may generate trust. Both are absent from the classical PD situation articulated by realists, but both are present in most situations of international cooperation, including regarding environmental questions. Liberal institutionalists then strengthen this argument by arguing that states act as absolute gains maximizers (especially Keohane, 1989), and thus the potential 'zone of agreement' is much larger than the single point assumed by realists focusing on relative gains. If 'cooperation under anarchy' (Oye, 1986) is possible, a further consequence is that the role of institutions is significantly greater than realists accept. For realists, international institutions are fundamentally epiphenomena, while for liberals they can play significant roles in forging cooperation between states, acting as entrepreneurial leaders, helping to find points of agreement, reducing transaction costs, facilitating information flows, building trust, and so on. None of these alter the fundamentally state-centric nature of international politics for institutionalists, but they are significant in promoting interstate cooperation. This is the case even when some analysts (for example, Haas et al., 1993; Ostrom, 1990; Young, 1997) appear to make the institutions themselves the focus of the analysis. These institutions are for them fundamentally inter*state* institutions which arise because of the logic of anarchy.

Institutionalists refer to the structured patterns of interstate cooperation which result as international regimes.[4] IEP has been a significant site of regime research, contributing greatly to empirical research and theoretical refinement. Research elaborating different 'regime stages' – formation, development, implementation – has been developed through research in IEP (for example, Haas et al., 1993; Rowlands, 1995; Young, 1989, 1994, this book). There is work that has elaborated general explanations of regimes (a triad of emphases on power, interests, and cognitive factors is common; for example, Rowlands, 1995) as well as more 'meso' level analyses of specific factors favoring regime success (Bernauer, 1995; Hahn and Richards, 1989). The concept of regime effectiveness has been elaborated principally in relation to environmental research, in part because of the specific conceptual problems in evaluating effectiveness of environmental regimes (Hovi et al., 2003; Miles et al., 2002; Mitchell, 2002; Young, 1999, this book; Grundig, 2012). More recent debates within this perspective have started to focus on the question of interlinkages or interplay between different institutional arrangements in world politics, to address a variety of questions, such as whether these institutions work in complementary or contradictory fashion to each other, whether the interplay is well managed (for example, Chambers, 2008; Oberthür, 2009; Oberthür and Gehring, 2006), and in particular whether this emerging landscape of overlapping institutions should be understood as a problem of fragmentation (Biermann et al., 2009) or as a 'regime complex' (Keohane and Victor, 2011).

For each of these perspectives, anarchy means that the actors in world politics are for the most part assumed to behave in a manner to be understood in terms of rational choice (for the major exception, see the work of Oran Young, 1989, 1994, this book). They are actors who pursue their individual interests, who do not (except for tactical or strategic reasons) take the interests of other actors into account in deciding what to do or how they should act. The two perspectives differ in what they think this means in terms of concrete actions; for realists and ecoauthoritarians, this means sufficient cooperation to alleviate environmental problems is impossible to achieve, while for liberal institutionalists, interdependence (in environmental matters as elsewhere in world politics) makes it rational in many instances to cooperate.

Constructivists, and English School writers, however, start with similar assumptions about anarchy, but rather see agents (states and others) as norm-bound actors, engaged in the reflexive and intersubjective production and internationalization of norms. In this they share much with sociological institutionalists (see also Gupta and Mason, this book). IEP is thus, while still in important ways shaped and constrained by the logic of anarchy as understood by the three perspectives above, nevertheless to be understood fundamentally via the evolution of norms which are both the product of state interaction and the shaper of state identities and thus action. A common story among both constructivist (Reus-Smit, 1996; Bernstein, 2001; Eckersley, 2004b) and English School (Falkner, 2012) writers in IEP is that these norms have shifted substantially since the 1970s, in two principal ways. First, there has been a progressive internalization of environmental norms by states, from the Stockholm Conference of 1972 onwards, such that there is now, in Falkner's words, a strong norm of 'global environmental responsibility' (2012: 504). Second, there has been a shift in environmental norms themselves from 'limits to growth' to 'liberal environmentalism', as such norms have been shaped by other prevailing norms in international society, principally to do with free market organizing principles of the global economy (Bernstein, 2001).

Capitalist logics

While some regard anarchy to be the driving logic behind IEP, for others the driver is the character of capitalist society. This similarly provides an explanation both for the origins of environmental degradation and for the character of responses to environmental problems.

For Marxists, the central starting point is the dynamics of capitalist society, which is understood as fundamentally constituted in terms of class relations – the division of society into differing classes according to their access to the means of production. The specific class relation to capitalism is wage labor – that workers contract a portion of their time, or the labor needed for a specific piece of work, to an employer. This generates a way of extracting surplus value from labor which is particularly efficient, and thus creates the enormous technological dynamism and unprecedentedly rapid economic growth of capitalist societies (see, for example, Harvey, 1989).

At the same time, accumulation (or crudely, economic growth, although these are not precisely the same thing) is widely regarded to lie at the origins of many of the environmental problems the world faces. Whether one subscribes strictly to a 'limits to growth' thesis, specific patterns of growth in material production and consumption clearly underlie particular patterns of environmental degradation, from greenhouse gas emissions (GHGs) and climate change to toxic waste generation. Capitalism generated what many have called a 'metabolic rift' – a shift that occurred in the Industrial Revolution from broadly cyclical processes based on animal or renewable (wind, water) sources of power to broadly linear processes based on non-renewable sources of power (notably coal) (Clark and York, 2005). Capital accumulation has fundamentally been based on extracting resources, deploying them, and not having to worry about the wastes that are produced, in these linear processes. Schnaiberg refers to this dynamic as the 'treadmill of production' (Schnaiberg et al., 2002), and researchers within this tradition have emphasized how little has changed in terms of the overall pattern of environmental degradation produced by capital accumulation despite the rhetoric of 'ecological modernization' (ibid; cf Mol, 2001). Taking these two aspects of capitalism – its class character and its growth dynamic – as a starting point generates what are sometimes called eco-Marxist approaches to environmental politics (for example, O'Connor, 1994; Benton, 1996; Bellamy Foster, 2000; Hornborg et al., 2007).

The political point here is that the dynamism of capitalist societies is unstable – it is highly uneven and unequal in the distribution of its benefits and downsides, but class relations also both create political conflict and contribute to the instability of accumulation because of a lack of effective demand for the products of capitalist industry. Thus one of the central functions of states in capitalist societies is to create the political conditions for promoting capital accumulation. As a consequence, promoting growth has become the political imperative for elites throughout the world, even in countries that are not ideologically inclined to capitalism.[5] Conversely, those who organize growth (capital, as a class, businesses, as individual enterprises) gain structural power with respect to policy making, thus structuring environmental policy making in particular directions. For some, environmental degradation and its politics thus appear as a 'second contradiction of capital' (O'Connor, 1991), between the mode of production and the (ecological) conditions of production. Increasingly, capital reaches limits in terms of how environmental degradation itself creates costs for capital and society as a whole, and as environmental movements articulate growth itself as a problem. This then creates contradictions for policymakers, as they simultaneously face the need to intervene to promote growth and at the same time to legitimize themselves to environmentally aware electorates to secure their rule, which frequently involves intervention to limit growth (Hay, 1994).

These features of capitalist societies provide explanations for a range of processes in IEP. They enable explanations of the origins of environmental degradation, with the principal drivers of environmental degradation identified as economic growth and the externalization of environmental costs. In addition to providing an explanation for the social conflicts and inequalities which dominate many environmental policy arenas, it enables an explanation of the way that particular environmental policy projects are structured by these class conflicts, the dominant ideologies through which class dominance is legitimized, and large-scale transformations in the global economy.

This generates a number of research foci (although not all of the authors mentioned below would subscribe to the perspective just outlined). One such focus is the power of and importance of business in IEP, increasingly important for many in the field (for example, Clapp, 1998, this book; Clapp and Helleiner, 2012; Falkner, 2008; Levy and Newell, 2005; Newell, 2000; Stevis, this book). It is important in part because of its lobbying activity in relation to specific environmental issues or regimes. But it is also important because of its role in the recent emergence of extensive private governance initiatives, such as investor-led governance activities, public–private partnerships, and the like (for example, Pinkse and Kolk, 2009). For the most part, this literature suggests that business activity slows down progress in responding to environmental problems, although in certain circumstances business interests may favor action (a classic case being the ozone depletion regime, where the interests of the two main producers of chlorofluorocarbons (CFCs), Dupont and ICI, started lobbying for elimination of CFCs, in part because they realized they could have a greater market share in the replacements – see Oye and Maxwell, 1994).

A second focus is on the inter-relation between IEP and broader patterns of global politics, especially global political economy. Debates over 'global governance' from this perspective involve discussions about how IEP is at the forefront of producing new forms of governance for capitalism and at the same time closely connected in patterns of resistance to such 'globalization', where attempts to move toward sustainable societies intermingle closely with resistance to corporate power and projects (for example, Paterson et al., 2003). The first process can be seen, for example, in the International Organization for Standardization (ISO) and its relationship to the World Trade Organization (WTO) (Clapp, 1998), in WTO trade–environment debates (Eckersley, 2004a; Williams, 2001), in the use of biotechnology (often legitimized in relation to biodiversity, especially in the Convention on Biological Diversity) as a way to embed Western intellectual property regimes globally, in the formation of the World Business Council for Sustainable Development as a key organization of globalizing capital (Sklair, 2000; van der Pijl, 1998), and most recently in the shaping of the responses to climate change around the construction of carbon and other environmental markets – novel ways to turn an environmental problem into a series of new property rights and financial commodities (see, for example, Lohmann, 2006; Swyngedouw, 2007; Newell and Paterson, 2010). The second process can be seen in the plethora of resistance movements – Northern environmentalists, Southern subsistence farmers, fair trade movements, and at times coalitions of mostly Southern governments (as in recent developments within Latin America in particular), many of whose agenda come together in a concern to resist neoliberal forms of economic management, in part because of their socio-ecological impacts, and to build alternative forms of economy and policy (Klein, 2002; Starr, 2000), such as most recently in the climate justice movement (Roberts and Parks, 2009; Routledge, 2011). Such movements simultaneously arise out of the injustices in IEP, discussed in a later section, but also more immediately out of the way that accumulation disrupts the daily lives and subsistence potential of many around the world.

Knowledge processes

The centrality of knowledge is clear in environmental policy debates, where ideas concerning uncertainty, risk, claims about expertise, and the importance of 'sound science' predominate in considerations about how policy decisions can be best made. Others have also made a different, and often contrasting, argument, that underlying environmental degradation is precisely the universalizing, reductionist sort of knowledge often regarded as integral to science. This centrality of knowledge claims to environmental politics has meant that a range of perspectives in IEP have emerged taking knowledge processes as the starting point (see also Lövbrand, this book). I will analyze three principal perspectives here, which I will label (recognizing the oversimplification involved) rationalist, constructivist, and critical.

The rationalist project takes the importance of science to 'good' environmental policy as self-evident and seeks to identify the conditions under which scientific advice is taken seriously and acted on by policymakers. This sort of IR focuses on how scientific knowledge and 'rational management' are essential for successful responses to global environmental problems. Good scientific knowledge is necessary both to be able to identify environmental problems and to provide the tools to respond effectively. Thus a prevalent argument is that international cooperation on environmental problems depends on sufficient availability of scientific information to be able to assess the rationality and effectiveness of various strategies as well as the existence of an epistemic consensus among the relevant scientific experts (Andresen and Ostreng, 1989; Andresen et al., 2000; Mitchell et al., 2006).

Peter Haas (1989, 1990) first articulated what remains a popular way of analyzing IEP through such a lens in his epistemic communities model, which he developed as part of his analysis of the Mediterranean Action Plan. For Haas and others, epistemic communities are the agents pursuing the solutions to collective action problems as in the liberal institutionalist accounts of cooperation. Haas and colleagues define epistemic communities as 'a network of individuals or groups with an authoritative claim to policy-relevant knowledge in their domain of expertise' (Adler, 1992: 101). They are regarded as central to IEP because they are the actors who policymakers turn to under the conditions of uncertainty and the requirements for expert knowledge pervading environmental politics, especially for 'new' issues that established state bureaucracies have little experience of dealing with (see also Betsill, this book).

The relationship between science and policy is conceptualized differently by other works which nevertheless take knowledge to be central to IEP. Using Foucault, Karen Litfin (1994) subjects the claims made by scientists regarding their neutrality to critique and finds that the relations between scientists and politicians are more complex than in the rationalist model. The rationalist perspective tends to assume that improvements in scientific knowledge lead to improvements in environmental policy, whereas Litfin shows that they are just as likely to harden existing policy positions and be used to political advantage.[6] Others take this skepticism further and are highly critical of scientists involved in environmental policymaking, verging on conspiracy theory. Boehmer-Christiansen's (1995a, 1995b; Boehmer-Christiansen and

Kellow, 2002) work on the Intergovernmental Panel on Climate Change (IPCC) and climate change science-policy processes exemplifies such tendencies. Boehmer-Christiansen argues that IPCC scientists have been involved in a set of strategies of exaggerating climate change threats to get it on the political agenda and secure increased funding for high-profile and high-prestige research activities, and then of emphasizing remaining scientific uncertainties for the same purpose. However, one doesn't need to hold to the conspiratorial tone of her arguments to recognize that scientists have interests of their own in addition to their cognitive/professional commitments, and that there is a complex interplay between this and the use of scientists by political elites to legitimize policy outcomes.

The Foucauldian arguments have been extended more recently, especially in relation to climate change, to focus on the productive nature of scientific knowledge in shaping the character of policy responses. This work shows in particular how specific ways that climate change is conceptualized and framed scientifically have important implications for climate change governance. For example, the emergence of the Global Warming Potential, a means of comparing the climate impacts of different GHGs, was nevertheless important to shaping later developments of market-led responses to climate change (Paterson and Stripple, 2012), while differing accounting regimes for measuring GHG emissions open up new avenues for ways of governing climate change (Lövbrand and Stripple, 2011).

Within this sort of approach a significant challenge has come from those who tend to regard the Foucauldian approach as rather totalizing, and wanting to emphasize more on the fragmentary and 'assembled' character of science–political interactions. Scholars within this vein, often known as 'actor network theory', focus on detailed studies of particular ways that scientific claims and technologies are put together with particular effects. Lovell and Liverman (2010), for example, explore the ways that carbon offset projects are assembled via particular technologies of accounting, verifying, and so on, while MacKenzie (2009) explores how precisely it is that different things (forests, coal fired power stations) are 'made the same' in carbon markets via techniques such as the Global Warming Potential. Goeminne (2012; see also Jacques, 2012 and Forsyth, 2012) explores the complexities of the science–policy interaction regarding climate change in ways that go beyond assuming a virtuous climate scientist versus a corrupt denier (or the inverse, as in Boehmer-Christiansen's arguments mentioned above).

A third perspective takes a critical point of view with regard to the role of science in producing environmental degradation and the marginalization of alternative forms of knowledge about the world involved in the hegemony of (Western) science (for example, Broadhead, 2002). Many critics interpret modern scientific rationality and scientific institutions as underlying structural causes of environmental problems. There are two principal aspects to this argument. Firstly, modern science was founded on the dualistic assumption of human separation from and domination over the rest of the natural world, and for many scientists its purpose has been precisely to further this separation and domination. Many writers suggest that this has led to anti-ecological attitudes and practices because the rest of the natural world has been reduced to an object for human instrumental use, whereas conceiving nature as an end in itself would produce less ecologically damaging behavior. It is also because of the way in which science has adopted a reductionist methodology, where phenomena are reduced to their constituent parts, and analyzed as individuals. Science has therefore been less well focused (perhaps at least until recently, see, for example, Worster, 1994) on the interactions between things; yet it is primarily in these interactions that environmental problems emerge (Merchant, 1980; Plumwood, 1993).

The second aspect to this critical perspective is that the emergence of science has transferred legitimacy concerning knowledge about environmental problems to particular elites. Science has become a way in which control over environments has been transferred from individuals or communities to experts, who increasingly live away from the environments which they are charged with managing, and thus have no personal interest in whether the management of those environments is sustainable, or whether it meets the needs of those who do depend on it. But if successful responses to environmental problems rely on those who depend on resources being able to control how they are used, then at the very least the particular organization of modern science (for example, elitist rather than democratic) is problematic from an environmental point of view (Banuri and Apffel-Marglin, 1993; Beck, 1995: Chapter 7; Ecologist, 1993: 67–69, 183–186). One common argument is that this concentration of power among those adopting a Western scientific episteme involves the marginalization of other forms of knowledge claims, such as indigenous knowledge claims, based more on direct experience and a more 'embedded' account of human interactions with ecosystems (Banuri and Apffel-Marglin, 1993; Martello, 2001). There is a corresponding argument that appropriate forms of environmental action are rooted in such embedded forms of knowledge, with of course a counter-argument that this frequently romanticizes indigenous peoples and their knowledge systems.

Plurality and complexity

For some, IEP exemplifies the shift in the structure of international politics away from the 'anarchy' emphasized by realists and others to the emergence of new actors in the international system. Underlying this approach is the fundamental assumption of international interdependence, as articulated in the 1970s by pluralists such as Keohane and Nye (1977) and Rosenau (1990). Interdependence signifies not only the mutual dependence of states with each other but also a multiplicity of inter-societal contacts which break down the exclusively state-centered nature of international politics. 'New' actors come onto the international scene, notably multilateral organizations, transnational corporations (TNCs), and non-governmental organizations (NGOs) (Betsill, this book).

In IEP, most attention from writers in this vein has been on NGOs, often branching out into more conceptual discussions about 'transnational (or global) civil society'. Wapner (1996) and Lipschutz (with Mayer, 1996) are the paradigm cases here.[7] The central theoretical claim is not only that NGOs have become more important in affecting interstate regimes[8] but that they are important both in producing new forms of governance in IEP and in providing new models of politics. Wapner's (1996) threefold image of World Wide Fund for Nature (WWF)/Friends of the Earth (FoE)/Greenpeace as different political models is illustrative here – WWF representing traditional interstate models of governance, Greenpeace being associated with notions of world government, and FoE advocating a form of environmental politics focused on localism with transnational networking between local communities. Another commonly mentioned example here is the Forest Stewardship Council (FSC), representing a model of governance that is fundamentally deterritorialized (Cashore et al., 2004). The Council, established by WWF, attempts to govern the practices of forestry companies through a labeling scheme with which it then works with major timber retailers and the construction industry to create consumer demand for FSC-approved wood.

The other important component in this approach is the claim that governance practices in IEP (as in global politics more generally) are 'bifurcating', 'fragmegrating', 'glocalizing', or undergoing some other such transformation (the phrases are Rosenau's) (Biermann, this book). Rosenau (1993) has made this claim specifically with regard to IEP, and Hempel's *Environmental Governance* (1996) is centered on this idea. Notions of 'multilevel governance' (for example, Vogler, 2003) convey a notion that the patterns of authority in global politics, driven at least in part by IEP processes, are moving upwards to regional and global levels, and at the same time downwards to subnational and local levels. For others, this is more understood as a process of privatization of governance (Clapp, 1998; Levy and Newell, 2005; Paterson, 2013), although this is more commonly connected to the role of transnational business, specifically in seeking to govern for its own interests (see previous discussion on capitalist logics). For others still, however, the notion of 'transnational governance' is more apt or all-encompassing than either the multilevel or the privatization frame, suggesting the myriad of ways that authority is generated across borders (for example, Bulkeley and Betsill, 2003; Andonova et al., 2009; Andonova and Mitchell, 2010; Bulkeley et al., 2012). Governance from this perspective then ceases to follow a 'sovereign' model with final authority residing at a specific point, instead being distributed at various levels, operating across a range of levels, and organized through networks rather than hierarchies.

Work starting from this focus tends also to be more open regarding the nature of agents. While both those focusing on anarchy and on capitalism tend to (although it is not strictly necessary for them to do so) operate with rationalist accounts of agents – pursuing predefined, clearly identifiable interests – most pluralists tend to be more constructivist and interested in the process by which interests (of states or other actors) are shaped. In the environmental field Steven Bernstein (2001) has elaborated the constructivist position best, focusing on shifts in dominant norms regarding the environment (see also, however, Hoffmann, 2005, or Pettenger, 2007). His point of departure is to argue that regime research based in liberal institutionalism is unable to account for the specific content of the regimes themselves, and for broad shifts in the norms underpinning environmental governance. He describes a shift from norms about environmental governance from limits to growth in the 1970s, to sustainable development in the 1980s, to liberal environmentalism in the 1990s, and shows how these underpin the various specific environmental regimes that emerge in the respective periods. While most of Bernstein's analysis is focused on the shifts in state identities as a result, it remains an account where the relations between states, NGOs, businesses, and other actors are all relatively fluid.

More recently, some authors take these observations of the proliferation of actors and sites of governing activity a stage further, to embed them in a notion of IEP as a complex adaptive system. Hoffmann has been at the forefront of this work (notably 2005, 2010; see also Levin et al., 2012). Regarding ecosystems as themselves complex adaptive systems is well established; indeed many of the key concepts in notions of complexity (emergence, path dependence, tipping points, and so on) have been articulated originally in ecological science. But the many components in human social and political systems can themselves be regarded as operating in a similar fashion, with the plurality of actors thus existing in systemic relations to each other. This generates the proposition that the pursuit of environmental governance is less about the specific activities of any given actor, since no actor is able on their own to generate broad shifts in systemic behavior. Rather, the focus is on how combinations of practices across actors may work to undermine entrenched patterns that generate environmental degradation, and generate shifts toward new, more benign dynamics.

Structural inequalities

While agreeing that there are many other actors in IEP than simply states, others are unconvinced that this plurality of actors is usefully thought of in terms of pluralism or complexity. Rather, the relations among these groups are structured and relational. Marxists, dependency theorists, feminists, and most Greens, all, in differing ways, start with the assumption that the world is politically organized in terms of structural inequalities (of class, gender, race, core/periphery, principally) which are at the root of the conflicts which pervade attempts to resolve environmental problems.

The most immediate and obvious of these inequalities in IEP concerns global economic inequalities broadly along 'North–South' lines (see, for example, Miller, 1996; Rice, 2007; Roberts and Parks, 2007; Parks and Roberts, 2010). The starting point here is the assumption that the dynamics of IEP are driven by conflict between states along a fault line between North and South which both structures the possible bargains between states on specific issues and is itself normally regarded as structural in nature. The common, if often implicit, theoretical backgrounds here are dependency theory and world systems theory, which suggests that the world economy has developed historically in such a way that inequalities are integral to its operation and tend to be self-reproducing. This offers an explanation of IEP in terms of the structuring of the global political economy into core and periphery, with resulting geopolitical conflict over IEP as well as the structuring of certain types of environmental degradation (deforestation and biodiversity/biotechnology are the paradigm cases) by such global inequalities. We can see the former of these, for example, in the debates about the toxic waste trade (Clapp, 1994), ozone depletion (Miller, 1996), climate change (Agarwal and Narain, 1991; Paterson 1996: Chapter 3; Roberts and Parks, 2007), biodiversity (Guha and Martinez-Alier, 1997; Shiva, 1993), or generally over 'sustainable development' (Redclift, 1987) which are fundamentally structured by North–South inequalities.

Normatively, such analyses are often closely connected to a concern with justice as a central ethical question for students and practitioners alike in IEP. The systemic inequalities involved in both the production of environmental degradation and its impacts, as well as the way that the global economy constrains the actions of developing countries in particular with regard to environmental problems, have helped produce the dominance of the discourse of distributive justice in arguments about how responses to global environmental problems can be legitimized. This is perhaps most widely discussed in relation to climate change (for example, Athanasiou and Baer, 2002; Grubb, 1995; Okereke and Charlesworth, this book; Shue, 1992; Page, 2006; Vanderheiden, 2008; Okereke, 2008; Parks and Roberts, 2010) but is also more generally discussed in relation to IEP (Hampson and Reppy, 1996; Low and Gleeson, 1998).

In addition to these inequalities that largely follow national boundaries, other inequalities endemic within many societies also structure IEP in important ways. These inequalities, broadly along the lines of gender, race, and class, are taken as starting points by feminists, environmental justice advocates, and Marxists. Feminists locate the origins of environmental degradation in power relations in patriarchal societies, emphasizing the gendered nature of such inequalities. For many, the origins of this are in the dualistic philosophy of Western society (Merchant, 1980; Plumwood, 1993, Detraz, 2012: Chapter 6; see also Detraz, this book), with mind/matter, nature/culture, male/female as principal dualisms, and the transformation of social relations between men and women set in train in early modernity associated with such philosophical shifts. This both structures everyday relations, meaning that men can often insulate themselves from the environmental impacts of their activities, and acts as a conceptual block, producing the reductionist forms of knowledge which prevent attempts to look holistically at ecosystems. Concretely, this starting point has produced analyses of the gendering of specific environmental debates, such as for desertification and deforestation, with gendered divisions of labor meaning that both the impacts of environmental change are significantly gendered and that male power helps to produce such change in the first place (Dankelman and Davidson, 1988; Sontheimer, 1991). It has also produced much work on the gender politics of environmental movements (for example, Bretherton, 1998, 2003).

Environmental justice movements, starting in the United States but spreading elsewhere, have tended to focus on the way that racial inequalities have been used as well as intensified by environmental degradation (see also Okereke and Charlesworth, this book). The most prominent trigger for such movements has been over the location of toxic waste dumps, which have disproportionately been placed in ethnic minority communities (Bullard, 1990; Szasz, 1994). At the same time, mainstream environmental NGOs have widely been regarded to have failed to develop campaigns to deal with this sort of environmental injustice. These inequalities have thus structured both how environmental degradation is organized and legitimized, and how movements to campaign against such degradation have emerged.

These racial inequalities often also intersect with class inequalities. Both in a loose use of the term 'class', regarding the extreme income inequalities which are prevalent in many countries, as well as regarding a more precise usage concerning the relation to the means of production, environmental politics are conditioned for many by class relations. Early developments of such arguments (Enzensberger, 1974; Harvey, 1993) have tended to regard environmentalism with some suspicion, suggesting that it is a middle-class movement, one of whose effects (even if unintentionally) is to 'pull up the ladder' behind them and prevent working-class people from enjoying the benefits of wealth that they themselves enjoy. This is similar in structure to arguments about 'ecocolonialism' in relation to North–South inequalities (Agarwal and Narain, 1991).

More generally, this focus on structural inequalities emphasizes that attempting to respond to a range of environmental problems without thinking carefully about those inequalities and how they structure the possible bargains between different social groups, the burdens they may disproportionately impose on some over others, and so on will doom such responses to failure. One important question we can ask of this line of inquiry in the light of recent developments concerns shifts in the patterns of inequality. Patterns of emissions of major pollutants (GHGs are paradigmatic) are shifting fairly quickly, with major developing countries (the 'BASIC' countries) becoming major emitters while emissions from the North stabilize or in some cases start to decline. For some, patterns of economic inequality are also becoming less and less readily captured by state-centric indices, meaning that while inequality is still important to environmental politics, it interacts in a more complex manner with the state-centric character of diplomatic activity.

Sustainabilty

Finally, there are groups of authors who, rather than coming out of some more general perspective in IR, arise out of environmentalist concerns and aim to build a theory of global politics from this perspective. Green politics is now the name most often given to such an enterprise, but ecoauthoritarians such as Ophuls also started from this point of view, and others (notably Sprout and Sprout, 1965; Pirages, 1978) also engaged global politics in this manner.

This approach should be distinguished from literatures on sustainable development (see Happaerts and Bruyninckx, this book). Sustainable development (and its sister discourse, ecological modernization[9]) tends to work with a 'weak' notion of sustainability. The key distinction between weak and strong notions of sustainability is that the former assumes that substitution between 'natural' and 'human' capital is by and large possible, while a strong notion rejects this and insists that sustainability requires that societies work within limits to the use of a range of natural resources (see, for example, Ekins, 2000). Weak notions of sustainability do not, in my view, generate a distinct perspective on IEP, being containable within institutionalist or constructivist frames as the norms that underpin international environmental governance, as in Bernstein (2001) or Harrison (2000).

It is only with 'strong' notions of sustainability that we get distinctively different perspectives arising, as this generates fundamental challenges to existing political institutions. Strong sustainability insists on the non-substitutability of human and natural capital, the importance of 'critical natural capital' of particular sorts of ecological disruptions that are irreversible and threaten whole ecosystems. As such, it generates a focus on the scale and character of human use of ecological resources and services, and an argument that such use needs to be radically cut back.

Here, the intention is to attempt to think ecologically, starting with a question such as 'what does sustainability require politically?' (It is perhaps a surprise to many that most of the literatures on IEP are fundamentally not concerned with such a question.) We have a range of analyses consistent with this basic question, but they are by no means homogenous in how they address it. Laferrière and Stoett (1999) employ ecological critiques of existing theories of IR. Kütting (2000) approaches the subject through a critique of regime theory, focusing on notions of time and complexity. Princen (2005) focuses on the value of sufficiency, which he suggests is required to underpin a politics of sustainability. Paterson (2000) attempts to develop an approach that starts from what it is about global power structures that engenders environmental degradation and thus where a politics of sustainability might start.

For most who do address this question directly, an attempt to think through the political implications of limits to economic growth, reducing the throughput of non-renewable resources, reducing global inequalities, and the complexities of many ecosystems involves radical challenges to most contemporary political institutions. This generates, for example, empirical analyses and concepts such as the 'ecological footprint'. For some (Eckersley, 1992; Naess, 1989; Plumwood, 1993) it also involves a deep philosophical shift, a rejection of anthropocentric ethics which puts priority on meeting human wants and needs over those of either other organisms or ecosystems, in favor of ecocentric ethics which prioritize the needs of non-humans and emphasize the (potential) compatibility of these with human interests.

While in the 1970s, this combination of views tended to result in ecoauthoritarian arguments as outlined above, it has become more common to argue for substantial decentralization of authority and social organization, with political institutions embedded in a pattern of global relations in a 'post-sovereign' manner (Dalby, 2002: Chapter 7; Ecologist, 1993; Helleiner, 1996; Paterson, 1996a; W. Sachs, 1993; Ostrom, 2010). The general argument is that the dynamics of unsustainability are characterized by a disjuncture not so much between the territorial state and a global environmental crisis, but between an over-centralized polity and the inevitably local character of environmental problems. The argument for decentralization of power is expressed particularly well by Dryzek (1987). He shows that small-scale political institutions have short feedback loops, meaning that the distance and time between problems appearing and responses being developed are much shorter than with larger-scale institutions. This is in addition to the advantages of small-scale institutions in fostering direct democracy (or at least much less heavily mediated representative democracies), meaning that those making decisions regarding sustainability and those affected by them can be understood to share many more institutions than in large representative systems.

There is of course a connection back to institutionalist arguments here, similar to those developed by Keohane above. Ostrom (1990), Berkes (1989), and McCay and Acheson (1987) all make similar arguments about the importance of scale in determining patterns of successful cooperation over resource use. But such analyses tend to be organized around rational choice assumptions. Political analyses of sustainability emphasize not only that the scale of political and social institutions needs to be radically re-organized but that such reorganization also involves changes in the character of political institutions (direct instead of representative democratic forms), of property relations (communal rather than private), and of social norms (sufficiency rather than accumulation oriented) (Ecologist, 1993).

These arguments have been reinforced most recently by the emergence of complexity ideas focused on the interaction of social and ecological systems (see also above on complexity). These arguments have generated a novel set of arguments about how human social institutions respond to ecological change, focusing on notions such as resilience and adaptive co-management (for example, Plummer et al., 2013) or transitions management (Kemp et al., 2007; Meadowcroft, 2009). These accounts of governance are designed not to treat the institutional arrangements as simply the product of rational human design but rather as a product of complex co-evolution with ecological change and social responses to it.

Conclusions

The strengths and weaknesses of these analyses rest to a large extent on the nature of the questions we want to ask of IEP. In many contexts, where we may want to attempt to influence particular patterns of political behavior, the limited, cautious, careful sets of propositions as developed by regime theory can clearly be useful in working out the dynamics of regime building, and thus how and where actors may try to intervene to affect outcomes. But regime theory makes fundamental assumptions about both 'how' intervention might take place (through the further development of interstate institutions) and 'where' it might take place (through the interaction of states), which are open to challenge. Structuralist perspectives would argue that institutionalists miss a crucial point concerning the importance of political coalitions that legitimize environmental measures because of the capacity of those measures to enable capital accumulation. Also, for structuralists (as for pluralists or those emphasizing on notions of complexity), the sites of intervention may in many instances be different, as governance of the environment occurs by corporations directly, or by different sorts of international organizations than those with which we are familiar (the ISO instead of the United Nations Environment Programme, for example). So even in a pragmatic, policy-oriented mode, institutionalism does not have a monopoly on usefulness.

But of course, there is no reason why social scientists should be tied to such a pragmatic project in any case. To interrogate critically the basic assumptions of theoretical positions, such as the principles of sovereignty, anarchy, and interdependence, is an important component of social scientific enquiry. We can and should thus ask basic questions of the adequacy of taking these six assumptions as starting points. Does it really make sense to characterize world politics in terms of interstate anarchy, whichever analysis one then generates about the implications of anarchy? Is scientific knowledge so critical to environmental politics that it makes sense to develop whole theories out of a focus on it? Are the inequalities in power and wealth structural in nature as Marxists and others argue, and if so, do they determine outcomes in IEP so strongly as they suggest? Such critical interrogations intersect of course with the point above, which is that what fundamentally matters is the question we ask of IEP.

The range of perspectives available to scholars working in IEP has broadened significantly in the last two decades, with the development of many novel perspectives and debates going beyond a debate about whether or not the states system can respond adequately to the environmental crisis. Closely related to this development is that the IR dimensions of environmental politics are now much more closely connected to debates in other fields in political science (comparative politics and political theory, in particular) and to debates beyond the discipline of politics. For the former of these, Robyn Eckersley's *The Green State* (2004b) remains a shining example of how the debates in IR and IEP (particularly some of the analyses by structuralists and constructivists as discussed above) are having an important impact outside IR, but also of how ecological political theory can be used to inform the development of research in IEP. Particularly useful to how our attention might be directed in the future is the attempt to integrate the insights of critical perspectives into more 'practically' oriented research. So, for example, research which retained a critical edge on the possibilities of sustainability in a continually growing world but which at the same time focused on how ecological modernization processes is starting to transform global capitalism, and how such processes might be advanced more fully is of considerable importance in the coming years. Such work has already begun (for example, Newell and Paterson, 2010, for my own attempt at this), but the recent financial crisis and global recession have thrown into relief the complicated dynamics involved – reshaping international inequalities, calling into question marketized forms of environmental governance, constraining action on many environmental problems even while it ameliorates others (by directly reducing emissions of GHGs, for example), generating new alliances between different actors, and so on. The need to think thoroughly about the political drivers, both unsustainability and our attempts to avoid it, remains an important intellectual and practical task.

Notes

1. This is not the place necessarily to go into great detail on the distinction here. Hardin referred to a mythical English commons as a property regime where users were under no restrictions on use of property. He thus failed to distinguish between regimes where there were community-based restrictions on use (commons), and those where no property was asserted in a resource (open access). His account of the commons is now usually referred to as an open access resource. On this conceptual debate, see Berkes (1989), Ostrom (1990), and Vogler (2000). Open access resource is thus the more precise conceptual term. For a fuller account along the lines here, see Paterson (2000: Chapter 2).
2. Soroos' analysis of the climate change regime (2001) is a good example of such realist logic, as well as a good example of the limits of the approach (Paterson, 2001).
3. See also Swatuk (this book) on environmental security. Of course not all authors using the 'environmental security' label start from realist premises. Some, for example, focus explicitly on 'security for the biosphere' rather than 'the nation' (Dalby, 2002). But the dominant discourse of environmental security remains realist in orientation, and even critical writers on the topic tend to take as their starting point precisely a critique of the realist frame.
4. I focus on the arguments of rationalist institutionalists in this section. I understand the insights of sociological institutionalists as substantially similar to those of constructivists or English School writers, treated later in this section. For more on this distinction, see also Gupta and Mason (this book).
5. I do not mean to get into a largely fruitless debate about whether socialist/communist states are 'really' capitalist or not. My point is that such states did (and the remaining ones still do) organize their politics around accumulation, at least in part due to military competition with Western states. The principal contemporary alternative ideological orientation that nevertheless supports the point made here is the notion of 'development'. Southern states, which might ideologically position themselves as non- or anti-capitalist,

nevertheless articulate their political programs toward a project that at its core promotes accumulation as understood here.
6. For analyses similar to Litfin's, see Miller and Edwards (2001); Dimitrov (2003); Harrison and Bryner (2003); Jasanoff and Wynne (1998); Parson (2003); or Shackley and Wynne (1996).
7. Lipschutz's (2004) work has more recently become less pluralist and more structuralist in orientation. It is worth noting also that some within this perspective do focus on TNCs (Garcia-Johnson, 2000), but it would be fair to say that most attention is on environmental NGOs.
8. A claim that is already commonplace in IEP and IR more widely. See for example, Princen and Finger (1994), or Clark, Friedman and Hochstetler (1998). It was also discussed even before regime theory became widely adopted in IEP (Kay and Jacobson, 1983). This work is more appropriately in my view seen as a secondary literature to institutionalist debates about regimes, as NGOs in this view play a role in supporting and furthering interstate regimes. The central theoretical point about pluralism is that interstate forms of environmental governance are no longer the only form of governance available.
9. On ecological modernization, see, for example, Dryzek et al. (2003); Hajer (1995); Mol (2001); or Weale (1992).

Annotated bibliography

Hardin, Garrett (1968) 'The Tragedy of the Commons', *Science*, 162, 1243–1248.

The classic statement from which much later theorizing in IR (and elsewhere) has developed. Hardin uses a metaphor from a mythical medieval English village to provide an explanation for environmental degradation that it arises primarily out of the incentives to overuse resources created by open access (he misnamed them commons) property regimes. The conclusions from these arguments have tended to be arguments for the restraint of freedoms by states, the development of global forms of state-like authority (especially by William Ophuls), and more generally for transformations of property rights away from open access toward private or state property regimes.

The Ecologist (1993) *Whose Common Future? Reclaiming the Commons* (London: Earthscan).

This is a wonderfully rich book, designed both as critique of the UNCED process and to provide the framework for understanding on what political principle sustainable societies might be organized. Their central argument is a direct inversion of Hardin's, and argues that commons are in fact this principle. Most of the book contains densely analyzed cases from around the world of commons regimes working in practice, showing how the interrelations of property regime, orientation toward community, the embeddedness of knowledge, and the orientation toward financial wealth interact to produce sustainable forms of economy and society.

Young, Oran R. (1989) *International Cooperation: Building Regimes for Natural Resources and the Environment* (Ithaca NY: Cornell University Press).

This is the standard text (with a series of updates since then) for liberal institutionalist 'regime theory' in IEP. Young provides the overview of the core concepts in regime theory – the explanations for why states enter into regimes, how regimes effect state behavior, how regimes enable cooperation, in particular – and applies them to cases to illustrate his argument. Much IEP literature focused on regimes and governance is a footnote to this book.

Laferrière, Eric and Peter Stoett (1999) *Ecological Thought and International Relations Theory* (London: Routledge).

This is the fullest survey of the relationship between IR theory and ecological thought in print. It provides a useful overview of the complexities of Green thought, and then runs through chapters organized around a series of perspectives in IR theory (realism, pluralism, structuralism, critical theories). The treatment of each theory is subtle and sophisticated, for example eschewing an overly simplistic treatment of realism as an ecological villain for a more subtle narrative showing elements in the realist canon which provide a base for potential dialog with Greens.

Lohmann, L. (2006) 'Carbon Trading: A Critical Conversation on Climate Change, Privatization and Power', *Development Dialogue*, 48, 1–356.

This provides an extremely rich account of IEP as driven by the logic of capital accumulation. It does so via the exploration of the appropriation of climate change by forces that have used it to create a series of novel markets. It exposes the contradictions and injustices produced by this sort of market-oriented environmental governance.

Wapner, Paul (1996) *Environmental Activism and World Civic Politics* (Albany: State University of New York Press).

This is a wonderfully succinct, clearly argued, book which outlines a pluralist case that global environmental governance is undergoing important shifts away from being centered on states. Wapner focuses on the activities of three key international environmental NGOs: Greenpeace, Friends of the Earth, and the World Wide Fund for Nature. He shows that they embody different principles of world political organization, and are developing strategies for acting politically which work toward those principles. This is the book which has spawned a series of debates and research on NGOs, and on multilevel and 'deterritorialized' governance.

Works cited

Adler, E. (1992) 'The Emergence of Cooperation: National Epistemic Communities and the International Evolution of the Idea of Nuclear Arms Control', *International Organization*, 46, 1, 101–146.

Agarwal, A., and S. Narain (1990) *Global Warming in an Unequal World – a Case of Environmental Colonialism* (New Delhi: Centre for Science and Environment).

Aldy, J., and R. Stavins (2007) *Architectures for Agreement: Addressing Global Climate Change in the Post-Kyoto World* (Cambridge: Cambridge University Press).

Andonova, L., M. Betsill, and H. Bulkeley (2009) 'Transnational Climate Change Governance', *Global Environmental Politics*, 9, 2, 52–73.

Andonova, L., and R. Mitchell (2010) 'The Resclaing of Global Environmental Politics', *Annual Review of Energy and Resources*, 35, 255–282.

Andresen, S., T. Skodvin, A. Underdal, and J. Wettestad (2000) *Science and Politics in International Environmental Regimes: Between Integrity and Involvement* (Manchester: Manchester University Press).

Andresen, S., and W. Østreng (eds) (1989) *International Resource Management* (London: Belhaven).

Athanasiou, T., and P. Baer (2002) *Dead Heat: Global Justice and Global Warming* (New York: Seven Stories Press).

Axelrod, R. (1984) *The Evolution of Cooperation* (New York: Basic Books).

Banuri, T., and F. Apffel-Marglin (eds) (1993) *Who will Save the Forests? Political Resistance, Systems of Knowledge, and the Environmental Crisis* (London: Zed Books).

Barrett, S. (2003), *Environment and Statecraft: The Strategy of Environmental Treaty-Making* (Oxford: Oxford University Press).

Beck, U. (1995) *Ecological Politics in an Age of Risk* (Cambridge: Polity).

Benton, T. (ed.) (1996) *The Greening of Marxism* (New York: Guilford Press).

Berkes, F. (ed.) (1989) *Common Property Resources: Ecology and Community-Based Sustainable Development* (London: Belhaven).

Bernauer, T. (1995) 'The Effect of International Environmental Institutions: How We Might Learn More', *International Organization*, 49, 2, 351–377.

Bernstein, S. (2001) *The Compromise of Liberal Environmentalism* (New York: Columbia University Press).

Biermann, F., P. Pattberg, H. Van Asselt, and F. Zelli (2009) 'The Fragmentation of Global Governance Architectures: A Framework for Analysis', *Global Environmental Politics*, 9, 14–40.

Boehmer-Christiansen, S. (1995a) 'Britain and the International Panel on Climate Change: The Impacts of Scientific Advice on Global Warming Part I: Integrated Policy Analysis and the Global Dimension', *Environmental Politics*, 4, 1, 1–18.

Boehmer-Christiansen, S. (1995b) 'Britain and the International Panel on Climate Change: The Impacts of Scientific Advice on Global Warming Part II: The Domestic Story of the British Response to Climate Change', *Environmental Politics*, 4, 2, 175–196.

Boehmer-Christiansen, S., and A. Kellow (2002) *International Environmental Policy: Interests and the Failure of the Kyoto Process* (London: Edward Elgar).

Bretherton, C. (1998) 'Global Environmental Politics: Putting Gender on the Agenda?', *Review of International Studies*, 24, 1, 85–100.

Bretherton, C. (2003) 'Movements, Networks, Hierarchies: A Gender Perspective on Global Environmental Governance', *Global Environmental Politics*, 3, 2, 103–119.

Broadhead, L. (2002) *International Environmental Politics: The Limits of Green Diplomacy* (Boulder, CO: Lynne Rienner Publishers).

Bulkeley, H., L. Andonova, K. Bäckstrand, M. Betsill, et al. (2012) 'Governing Climate Change Transnationally: Assessing the Evidence from a Database of Sixty Initiatives', *Environment and Planning C: Government and Policy*. 30, 4, 591–612.

Bulkeley, H., and M. M. Betsill (2003) *Cities and Climate Change: Urban Sustainability and Global Environmental Governance* (London: Routledge).

Bull, H. (1977) *The Anarchical Society: A Study of Order in World Politics* (London: MacMillan).

Bullard, R. (1990) *Dumping in Dixie* (Boulder, CO: Westview).

Cashore, B., G. Auld, and D. Newsom (2004) *Governing Through Markets: Forest Certification and the Emergence of Non-State Authority* (New Haven: Yale University Press).

Chambers, W. B. (2008) *Interlinkages and the Effectiveness of Multilateral Environmental Agreements* (Toyko: United Nations University Press).

Clapp, J. (1994) 'Africa, NGOs and the International Toxic Waste Trade', *Journal of Environment and Development*, 3, 2, 17–46.

Clapp, J. (1998) 'The Privatization of Global Environmental Governance: ISO 14000 and the Developing World', *Global Governance*, 4, 3, 295–316.

Clapp, J., and E. Helleiner (2012) 'International Political Economy and the Environment: Back to the Basics?', *International Affairs*, 88, 3, 485–501.

Clark, A. M., E. J. Friedman, and K. Hochstetler (1998) 'The Sovereign Limits of Global Civil Society: A Comparison of NGO Participation in UN World Conferences on the Environment, Women and Human Rights', *World Politics*, 51, 1, 1–35.

Dalby, S. (2002) *Environmental Security* (Minneapolis: University of Minnesota Press).

Dankelman, I., and J. Davidson (1988) *Women and Environment in the Third World* (London: Earthscan).

Detraz, N. (2012) *International Security and Gender* (Cambridge: Polity Press).

Dimitrov, R. S. (2003) 'Knowledge, Power, and Interests in Environmental Regime Formation', *International Studies Quarterly*, 47, 123–150.

Dryzek, J. (1987) *Rational Ecology: Environment and Political Economy* (Oxford: Basil Blackwell).

Dryzek, J. S., D. Downes, C. Hunold, D. Schlosberg, with H. Hernes (2003) *Green States and Social Movements* (Oxford: Oxford University Press).

Eckersley, R. (1992) *Environmentalism and Political Theory: Towards an Ecocentric Approach* (London: UCL Press).

Eckersley, R. (2004a) 'The Big Chill: the WTO and Multilateral Environmental Agreements', *Global Environmental Politics*, 4, 2, 24–50.

Eckersley, R. (2004b) *The Green State: Rethinking Democracy and Sovereignty* (Cambridge, MA: MIT Press).

Ekins, P. (2000) *Economic Growth and Environmental Sustainability: The Prospects for Green Growth* (London: Routledge).

Enzensberger, H. M. (1974) 'A Critique of Political Ecology', *New Left Review*, 84, 3–31.

Falkner, R. (2008) *Business Power and Conflict in International Environmental Politics* (Basingstoke: Palgrave).

Falkner, R. (2012) 'Global Environmentalism and the Greening of International Society', *International Affairs*, 88, 3, 503–522.

Forsyth, T. (2012) 'Politicizing Environmental Science Does Not Mean Denying Climate Science Nor Endorsing It Without Question', *Global Environmental Politics*, 12, 2, 18–23.

Foster, J. B. (2000) *Marx's Ecology: Materialism and Nature* (New York: Monthly Review Press).

Garcia-Johnson, R. (2000) *Exporting Environmentalism* (Cambridge, MA: MIT Press).
Goeminne, G. (2012) 'Lost in Translation: Climate Denial and the Return of the Political', *Global Environmental Politics*, 12, 2, 1–8.
Grubb, M. (1995), 'Seeking Fair Weather: Ethics and the International Debate on Climate Change', *International Affairs*, 71, 3, 463496.
Grundig, F. (2012) 'Dealing with the Temporal Domain of Regime Effectiveness: A Further Conceptual Development of the Oslo-Potsdam Solution', *International Environmental Agreements: Politics, Law and Economics*, 12, 2, 111–127.
Guha, R., and J. Martinez-Alier (1997) *Varieties of Environmentalism: Essays North and South* (London: Earthscan).
Gupta, S., D. Tirpak, N. Burger, J. Gupta, N. Höhne, A. I. Boncheva, G. M. Kanoan, C. Kolstad, J. Kruger, A. Michaelowa, S. Murase, J. Pershing, T. Saijo, and A. Sari (2007) 'Policies, Instruments and Co-operative Arrangements', in B. Metz, O. Davidson, P. Bosch, R. Dave, and L. Meyer (eds) *Climate Change 2007: Mitigation. Contribution of Working Group III to the Fourth Assessment Report of the Intergovernmental Panel on Climate Change* (Cambridge, UK and New York: Cambridge University Press), 745–807.
Haas, P. M. (1989) 'Do Regimes Matter? Epistemic Communities and Mediterranean Pollution Control', *International Organization*, 43, 3, 377–403.
Haas, P. M. (1990) *Saving the Mediterranean: The Politics of International Environmental Cooperation* (New York: Columbia University Press).
Haas, P. M., R. O. Keohane, and M. A. Levy (eds) (1993) *Institutions for the Earth: Sources of Effective Environmental Protection* (Cambridge, MA: MIT Press).
Hahn, R., and K. Richards (1989) 'The Internationalization of Environmental Regulation', *Harvard International Law Journal*, 30, 421–446.
Hajer, M. (1995) *The Politics of Environmental Discourse: Ecological Modernisation and the Policy Process* (Oxford: Clarendon).
Hampson, F. O., and J. Reppy (eds) (1996) *Earthly Goods: Environmental Change and Social Justice* (Ithaca, NJ: Cornell University Press).
Hardin, G. (1968) 'The Tragedy of the Commons', *Science*, 162, 1243–1248.
Harrison, N. E. (2000) *Constructing Sustainable Development* (Albany, NY: SUNY Press).
Harrison, N. E., and G. Bryner, (eds) (2003) *Science and Politics in the International Environment* (Lanham, MD: Rowman & Littlefield Publishers, Inc).
Harvey, D. (1990) *The Condition of Postmodernity* (Oxford: Blackwell).
Harvey, D. (1993) 'The Nature of the Environment: The Dialectics of Social and Environmental Change', In L. Panitch, and R. Miliband (eds) *The Socialist Register 1993* (London: Merlin Press), 1–51.
Hay, C. (1994) 'Environmental Security and State Legitimacy', In M. O'Connor (ed.) *Is Capitalism Sustainable? Political Economy and the Politics of Ecology* (New York: Guilford Press), 217–231.
Helleiner, E. (1996) 'International Political Economy and the Greens', *New Political Economy*, 1, 1, 59–78.
Hempel, L. (1996) *Environmental Governance: The Global Challenge* (Washington, DC: Island Press).
Hoffmann, M. J. (2005) *Ozone Depletion and Climate Change: Constructing A Global Response* (State University of New York Press).

Hoffmann, M. J. (2010) *Climate Governance at the Crossroads: Experimenting with a Global Response After Kyoto* (New York: Oxford University Press).
Hornborg, A., J. R. McNeill, and J. Martinez-Alier (eds) (2007) *Rethinking Environmental History: World System History and Global Environmental Change* (Lanham, MD: Altimera Press).
Hovi, J., D. F. Sprinz, and A. Underdal (2003) 'The Oslo-Potsdam Solution to Measuring Regime Effectiveness: Critique, Response, and the Road Ahead', *Global Environmental Politics*, 3, 3, 74–96.
Jacques, P. J. (2012) 'A General Theory of Climate Denial', *Global Environmental Politics*, 12, 2, 9–17.
Jasanoff, S., and B. Wynne (1998) 'Science and Decision-Making', in S. Rayner and E. Malone (eds) *Human Choice and Climate Change: The Societal Framework* (Columbus, OH: Batelle Press).
Kay, D., and H. Jacobson (eds) (1983) *Environmental Protection: the International Dimension* (Totowa, NJ: Allanheld, Osmun).
Kemp, R., D. Loorbach, and J. Rotmans (2007) 'Transition Management as a Model for Managing Processes of Co-evolution Towards Sustainable Development', *International Journal of Sustainable Development & World Ecology*, 14, 78–91.
Keohane, R. O. (1989) *International Institutions and State Power: Essays in International Relations Theory* (Boulder, CO: Westview Press).
Keohane R. O., and D. G. Victor (2011) 'The Regime Complex for Climate Change', *Perspectives on Politics*, 9, 7–23.
Keohane, R. O., and J. S. Nye (1977) *Power and Interdependence: World Politics in Transition* (Boston: Little Brown and Company).
Klein, N. (2002) *Fences and Windows: Dispatches from the Frontlines of the Globalization Debate* (London: Flamingo).
Kütting, G. (2000) *Environment, Society and International Relations: Towards More Effective International Agreements* (London: Routledge).
Laferrière, E., and P. Stoett (1999) *Ecological Thought and International Relations Theory* (London: Routledge).
Levin, K., B. Cashore, S. Bernstein, and G. Auld (2012) 'Overcoming the Tragedy of Super Wicked Problems: Constraining Our Future Selves to Ameliorate Global Climate Change', *Policy Sciences*, 45, 2, 123–152.
Levy, D., and P. Newell (eds) (2005) *Business in International Environmental Governance* (Cambridge, MA: MIT Press).
Lipschutz, R. (2004) *Global Environmental Politics: Power, Perspectives, and Practice* (Washington, DC: Congressional Quarterly Press).
Lipschutz, R. D. with J. Mayer (1996) *Global Civil Society and Global Environmental Governance: The Politics of Nature from Place to Planet* (Albany: SUNY Press).
Litfin, K. (1994) *Ozone Discourses: Science and Politics in Global Environmental Cooperation* (New York: Columbia University Press).
Lohmann, L. (2006) 'Carbon Trading: A Critical Conversation on Climate Change, Privatization and Power', *Development Dialogue*, No. 48, 1–356.
Lövbrand, E. and J. Stripple (2011) 'Making Climate Change Governable: Accounting for Carbon as Sinks, Credits and Personal Budgets', *Critical Policy Studies*, 5, 2, 187–200.
Lovell, H., and D. Liverman (2010) 'Understanding Carbon Offset Technologies', *New Political Economy*, 15, 2, 255–273.

Low, N., and B. Gleeson (1998) *Justice, Society and Nature: An Exploration of Political Ecology* (London: Routledge).
MacKenzie, D. (2009) 'Making Things the Same: Gases, Emission Rights and the Politics of Carbon Markets', *Accounting, Organizations and Society*, 34, 3–4, 440–455.
Martello, M. L. (2001) 'A Paradox of Virtue?: "Other" Knowledges and Environment-Development Politics', *Global Environmental Politics*, 1, 3, 114–141.
McCay, B., and J. Acheson (eds) (1987) *The Question of the Commons: The Culture and Ecology of Communal Resources* (Tuscon: University of Arizona Press).
Meadowcroft, J. (2009) 'What about the Politics? Sustainable Development, Transition Management, and Long Term Energy Transitions', *Policy Sciences*, 42, 323–340.
Merchant, Carolyn (1980) *The Death of Nature: Women, Ecology and the Scientific Revolution* (San Francisco: Harper & Row).
Miles, E. L., A. Underdal, S. Andresen, J. Wettestad, J. B. Skjærseth, and E. M. Carlin (2002) *Environmental Regime Effectiveness: Confronting Theory with Evidence* (Cambridge, MA: The MIT Press).
Miller, C., and P. N. Edwards (eds) (2001) *Changing the Atmosphere: Expert Knowledge and Environmental Governance* (Cambridge, MA: The MIT Press).
Miller, M. (1996) *The Third World in Global Environmental Politics* (Buckingham: Open University Press).
Mitchell, R. B. (2002) 'A Quantitative Approach to Evaluating International Environmental Regimes', *Global Environmental Politics*, 2, 4, 58–83.
Mitchell, R. B., W. C. Clark, D. W. Cash, and N. Dickson (eds) (2006) *Global Environmental Assessments: Information, Institutions, and Influence* (Cambridge: MIT Press).
Mol, A. (2001) *Globalization and Environmental Reform: The Ecological Modernisation of the Global Economy* (Cambridge, MA: MIT Press).
Naess, A. (1989) *Ecology, Community and Lifestyle* (Cambridge: Cambridge University Press).
Newell, P. (2000) 'Environmental NGOs, TNCs, and the Question of Governance', in Dimitris Stevis and Valerie D. Assetto (eds) *The International Political Economy of the Environment* (Boulder, CO: Lynne Rienner), 85–107.
Newell, P. and M. Paterson (2010) *Climate Capitalism: Global Warming and the Transformation of the Global Economy* (Cambridge, Cambridge University Press).
Oberthür, S. (2009) 'Interplay Management: Enhancing Environmental Policy Integration among International Institutions', *International Environmental Agreements: Politics, Law and Economics*, 9, 27, 371–391.
Oberthür, S., and T. Gehring (eds) (2006) *Institutional Interaction in Global Environmental Governance* (Cambridge, MA: MIT Press).
O'Connor, J. (1991) 'On the Two Contradictions of Capitalism', *Capitalism, Nature, Socialism*, 2, 3, 107–109.
O'Connor, M. (ed.) (1994) *Is Capitalism Sustainable? Political Economy and the Politics of Ecology* (New York: Guilford Press).
Okereke, C. (2008) *Global Justice and Neoliberal Environmental Governance* (London: Routledge).
Ophuls, W. (1977) *Ecology and the Politics of Scarcity* (San Francisco: WH Freeman & Co).

Ostrom, E. (1990) *Governing the Commons: The Evolution of Institutions for Collective Action* (Cambridge: Cambridge University Press).
Ostrom, E. (2010) 'Polycentric Systems for Coping with Collective Action and Global Environmental Change', *Global Environmental Change*, 20, 4, 550–557.
Oye, K. (ed.) (1986) *Cooperation under Anarchy* (Princeton: Princeton University Press).
Oye, K., and J. Maxwell (1994) 'Self-interest and Environmental Management', *Journal of Theoretical Politics*, 64, 4, 593–624.
Page, E. (2006) *Climate Change, Justice, and Future Generations* (Cheltenham: Edward Elgar).
Parks, B. C., and J. T. Roberts (2010) 'Climate Change, Social Theory and Justice', *Theory, Culture & Society*, 27, 2–3, 134–166.
Paterson, M. (1996) *Global Warming and Global Politics* (London: Routledge).
Paterson, M. (1996a) 'Green Politics', in S. Burchill (ed.) *Theories of International Relations* (London: Macmillan), 277–307.
Paterson, M. (2000) *Understanding Global Environmental Politics: Domination, Accumulation, Resistance* (London: Macmillan).
Paterson, M. (2001) 'Climate Policy as Accumulation Strategy: The Failure of COP6 and Emerging Trends in Climate Politics', *Global Environmental Politics*, 1, 2, 10–17.
Paterson, M. (2013) 'Climate Re-Public: Practicing Public Space in Conditions of Extreme Complexity', in J. Best, and A. Gheciu (eds) *The Return of the Public, but Not As We Knew It: Changing Practices of Global Governance* (Cambridge: Cambridge University Press), 149–172.
Paterson, M., D. Humphreys, and L. Pettiford (2003) 'Conceptualizing Global Environmental Governance: From Interstate Regimes to Counter-Hegemonic Struggles', *Global Environmental Politics*, 3, 2, 1–10.
Paterson, M., and J. Stripple (2012) 'Virtuous Carbon', *Environmental Politics*, 21, 4, 563–582.
Pettenger, M. (ed.) (2007) *The Social Construction of Climate Change: Power, Knowledge, Norms, Discourses* (Farnham: Ashgate).
Pinkse J., and A. Kolk (2009) *International Business and Global Climate Change* (London: Routledge).
Pirages, D. (1978) *Global Ecopolitics: The New Context for International Relations* (North Scituate, MA: Duxbury Press).
Plummer, R., D. Armitage, and R. C. de Lot (2013) 'Adaptive Comanagement and Its Relationship to Environmental Governance', *Ecology and Society*, 18, 1, article 21.
Plumwood, V. (1993) *Feminism and the Mastery of Nature* (London: Routledge).
Princen, T. (2005) *The Logic of Sufficiency* (Cambridge MA: MIT Press).
Princen, T., and M. Finger (eds) (1994) *Environmental NGOs in World Politics* (London: Routledge).
Redclift, M. (1987) *Sustainable Development: Exploring the Contradictions* (London: Routledge).
Reus-Smit, C. (1996) 'The Normative Structure of International Society', in F. O. Hampson and J. Reppy (eds) *Earthly Goods: Environmental Change and Social Justice* (Ithaca, NY: Cornell University Press), 96–121.
Rice, J. (2007) 'Ecological Unequal Exchange: Consumption, Equity, and Unsustainable Structural Relationships within the Global Economy', *International Journal of Comparative Sociology*, 48, 1, 43–72.

Roberts, J. T., and B. Parks (2007) *A Climate Of Injustice: Global Inequality, North-South Politics, and Climate Policy* (Cambridge, MA: MIT Press).
Roberts, J. T., and B. C. Parks (2009) 'Ecologically Unequal Exchange, Ecological Debt, and Climate Justice: The History and Implications of Three Related Ideas for a New Social Movement', *International Journal of Comparative Sociology*, 50, 385–409.
Rosenau, J. N. (1990) *Turbulence in World Politics: A Theory of Change and Continuity* (Princeton, NJ: Princeton University Press).
Rosenau, J. N. (1993) 'Environmental Challenges in a Turbulent World', in K. Conca, and R. Lipschutz (eds) *The State and Social Power in Global Environmental Politics* (New York: Columbia University Press), 71–93.
Routledge, P. (2011) 'Translocal Climate Justice Solidarities', in J. Dryzek, R. Norgaard and D. Schlosberg (eds) *The Oxford Handbook of Climate Change and Society* (Oxford: Oxford University Press), 384–398.
Rowlands, I. H. (1995) *The Politics of Global Atmospheric Change* (Manchester: Manchester University Press).
Sachs, W. (ed.) (1993) *Global Ecology* (London: Zed).
Shackley, S., and B. Wynne (1996) 'Representing Uncertainty in Global Climate – Change Science and Policy – Boundary – ordering Devices and Authority', *Science Technology & Human Values*, 21, 3, 275302.
Shields, L., and M. Ott (1974) 'The Environmental Crisis: International and Supranational Approaches', *International Relations*, 4, 6, 584–607.
Shiva, V. (1993) *Monocultures of the Mind: Perspectives on Biodiversity and Biotechnology* (London: Zed).
Shue, H. (1992) 'The Unavoidability of Justice', in A. Hurrell and B. Kingsbury (eds) *The International Politics of the Environment* (Oxford: Clarendon), 373–397.
Sklair, L. (2000) *The Transnational Capitalist Class* (Oxford: Blackwell).
Sontheimer, S. (ed.) (1991) *Women and the Environment: A Reader* (London: Earthscan).
Soroos, M. (2001) 'Global Climate Change and the Futility of the Kyoto Process', Global *Environmental Politics*, 1, 2, 1–9.
Sprout, H., and M. Sprout (1965) *The Ecological Perspective on Human Affairs* (Princeton: Princeton University Press).
Starr, A. (2000) *Naming the Enemy: Anti-Corporate Movements Confront Globalization* (London: Zed Books).
Swyngedouw, E. (2007) 'Dispossessing H_2O: The Contested Terrain of Water Privatization', in N. Heynen, J. McCarthy, S. Prudham, and P. Robbins (eds) *Neoliberal Environments: False Promises and Unnatural Consequences* (London: Routledge), 51–62.
Szasz, A. (1994) *Ecopopulism: Toxic Waste and the Movement for Environmental Justice* (Minneapolis: Minnesota University Press).
The Ecologist (1993) *Whose Common Future? Reclaiming the Commons* (London: Earthscan).
van der Pijl, K. (1998) *Transnational Classes and International Relations* (London: Routledge).
Vogler, J. (2000) *The Global Commons: Environmental and Technological Governance* (London: John Wiley).

Vogler, J. (2003) 'Taking Institutions Seriously: How Regime Analysis can be Relevant to Multilevel Environmental Governance', *Global Environmental Politics*, 3, 2, 25–39.
Vogler, J. (2011) 'International Relations Theory and the Environment', in G. Kütting (ed.) *Global Environmental Politics: Concepts, Theories and Case Studies* (London: Routledge), 11–26.
Wapner, P. (1996) *Environmental Activism and World Civic Politics* (Albany: State University of New York Press).
Weale, A. (1992) *The New Politics of Pollution* (Manchester: Manchester University Press).
Weber, M. (1968/1919) *Politics as a Vocation* (Philadelphia: Fortress Press).
Wendt, A. (1999) *Social Theory of International Politics* (Cambridge: Cambridge University Press).
Williams, M. (2001) 'Trade and Environment in the World Trading System: A Decade of Stalemate?', *Global Environmental Politics*, 1, 4, 1–9.
Worster, D. (1994) *Nature's Economy: A History of Ecological Ideas,* 2nd edition (Cambridge: Cambridge University Press).
Young, O. R. (1989) *International Cooperation: Building Regimes for Natural Resources and the Environment* (Ithaca, NY: Cornell University Press).
Young, O. R. (1994) *International Governance: Protecting the Environment in a Stateless Society* (Ithaca, NY: Cornell University Press).
Young, O. R. (ed.) (1999) *The Effectiveness of International Environmental Regimes: Causal Connections and Behavioral Mechanisms* (Cambridge: MIT Press).
Young, Oran R. (1997) 'Global Governance: Towards a Theory of Decentralized World Order', in O. R. Young (ed.) *Global Governance: Drawing Insights from the Environmental Experience* (Ithaca, NY: Cornell University Press), 272–299.
Vanderheiden, S. (2008) *Atmospheric Justice: A Political Theory of Climate Change* (New York: Oxford University Press).

4
Methods in International Environmental Politics

Kathryn Hochstetler and Melinda Laituri

Researchers in international environmental politics (IEP) have devoted little attention to their field's methods. With a few exceptions, they have simply carried out their research without exploring which methods are best for the field as a whole. This is a laudable approach to an area of research whose data can range from the cultural discourses in global negotiations to measurements of CO_2 in the atmosphere. On the one hand, the absence of a hegemonic methodological discourse in the field fits its diversity well. On the other hand, the lack of extended reflection about the methodologies appropriate to the field may prevent IEP researchers from thinking more creatively about their research approaches. Greater attention to research design and methodology would help them avoid unnecessary weaknesses in their studies. To that end, this chapter outlines a number of different approaches and specifies how they are used and for which kinds of analytical projects, focusing on issues of research design. It also identifies characteristic pitfalls and critiques of the approaches.

The IEP field as a whole needs a full methodological toolbox. While individual researchers may specialize in particular methods, different kinds of research questions demand different kinds of methodological approaches. Traditional qualitative methods and newer discursive analyses are especially useful for studying the *processes* of IEP, including various kinds of formal and informal negotiations. Interviews, observation, and documentary research provide the data for this set of methods. These methods are also especially useful for identifying the *purposes* and worldviews of different actors and policies. Rational choice approaches focus on similar questions, while using quite different analytical models and assumptions. For identifying empirical *patterns* or tendencies across larger populations, various kinds of quantitative methods are more appropriate. Thus surveys help map public attitudes and statistical methods trace causal relationships over larger numbers of cases. Geospatial and statistical data are especially important for viewing the impact of human choices on the physical environment itself.

Much of the existing literature on methods in IEP focuses on the question of whether qualitative or quantitative techniques are more appropriate for its research questions. The predominant mode of analysis of IEP so far has been largely qualitative, with most researchers selecting a small number of case studies for close study. A review of *Global Environmental Politics*, the journal associated with the Environmental Studies Section of the International Studies Association (ISA), finds that almost 90 percent (88 of 99) of the empirical articles published between 2008 and 2012 used qualitative methods, with the rest split between quantitative and formal approaches.[1] Yet only a few researchers have laid out strategies for successful qualitative study of IEP (for exceptions, see Homer-Dixon, 1996; Mitchell and Bernauer, 1998, 2004). Perhaps because of this dominance of qualitative methods in IEP, many of the more explicit discussions of methods in IEP are arguing for the use of other kinds of methods (for example, Kilgour and Wolinsky-Nahmias, 2004; Lempert et al., 2009; Mitchell, 2002; Sprinz, 2004).

The quantitative/qualitative divide is an important one that will be addressed here, but it is cross-cut by a more fundamental division that maps better onto the different theoretical projects of IEP scholars. This is the epistemological divide between approaches that are aiming for positivist causal explanations of the phenomena they study and those (here called 'critical theorists') that reject at least part of the possibility or desirability of achieving such explanations. Because this methodological review sees two dimensions differentiating current IEP research, its organization departs somewhat from other summaries. An initial set of sections introduces different methodologies as they relate to the positivist project of causal explanation. The first of these introduces the standards and procedures that are relevant for all positivist approaches and shows how they appear in their qualitative form. It is then followed by sections focused on the other primarily positivist methodologies – quantitative/statistical, rational choice, and geospatial technologies – which are divided by the kinds of data they require and how they manipulate that data. The second major part of the chapter examines the alternative standards for research design, evidence, and argument of critical approaches. Again, the standards are introduced in their qualitative form, which is how they are usually used in IEP. A final subsection takes up the critical scholars who use non-qualitative methods and discusses how they differ from their counterparts.

The international relations (IR) field as a whole is similarly divided between positivist and critical approaches. While there are elective affinities between IR theories and the various methodological approaches, there is no one-to-one correspondence among them. Constructivists, for example, often use qualitative methods, as these textual and discursive methods are compatible with the constructivists' focus on meanings. Nonetheless, constructivists are divided between positivist and postpositivist approaches and collectively use most of the methodologies discussed (Klotz and Lynch, 2007). Realists are disproportionately present among users of quantitative and formal methodologies, but they use a full range of the positivist methodologies, as do liberal institutionalists. Critical theorists of various kinds are often drawn to qualitative methodologies, but they also sometimes use other methods (Morrow, 1994).

How are methodological issues different when studying international *environmental* politics as compared to international relations more generally?[2] The clearest point of distinction is that environmental issues organically link to the natural world and its associated physical and natural sciences in ways that, for example, human rights and security issues do not. In the course of an IEP project, the scholar may well need to grapple with the complexities of climate change models, debates about how to measure forest biodiversity, or the inner workings of a two-stroke engine. IEP scholars often find themselves in dialog or institutions with natural scientists, who bring their disciplines' approaches with them. The nature of that dialog depends quite a bit on the methodological approach taken by the IEP scholar. The tendency-finding approaches of statistical and geospatial data analysis offer the IEP scholar the most straightforward opportunities for dialog with natural and physical scientists, as their disciplines tend to value such methods. The other approaches, however, intersect in more complicated ways with non-social scientists. A minority of natural scientists uses positivist qualitative methods, but they may be especially useful for disentangling the complex and interactive relationships that are uncovered in studies that look seriously at both natural and social processes. The arguments below might help provide the basis for a conversation with natural scientists about the value of qualitative methods in a scientific research agenda. Critical scholars have the most tendentious relations with natural scientists, since one of their concerns is to undercut the claimed special expertise of science and its practitioners (see Lövbrand, this book).

Positivist approaches to methodology

Positivist standards for research and positivist qualitative methods

A common empirical approach to IEP has been through what are called positivist – or causal or rationalist – qualitative methods. This approach shares a general orientation toward research standards and aims with a number of non-qualitative methodologies, while collecting and analyzing its data in qualitative (non-numerical) ways (Mahoney, 2010; Mitchell and Bernauer, 1998, 2004). In particular, it embraces the project of developing causal explanations for understanding general patterns of IEP. To this end, positivist qualitative scholars often adopt the language of statistical analysis and define independent, dependent, and control variables, developing hypotheses about the relationships among these. One of the key dilemmas for this approach is that an individual qualitative study does not provide the systematic assessment of proposed causal relationships for which positivism aims, since qualitative methods produce a great deal of information about a relatively small number of cases. Qualitative scholars have argued that their methods play an important role in generating hypotheses and in tracing causal processes for the positivist explanatory project. These points are developed in this section.

Much of the analytical work of positivist qualitative methods is done in the research design phase, where a small number of cases are selected on the basis of proposed causal variables (George, 1979). After the cases are chosen, the researcher systematically collects information about the variables and the relationships among them, using documents and other archival records, interviews, and observation (Fenno, 1990; Hill, 1993; Taylor and Bogdan, 1998). During the data collection phase, the researcher codes the case-specific information as examples of more general categories. In the final analytical stage, the researcher relates the observed and collected data to the hypothesized relationship. An initial step asks whether the independent and dependent variables do take on expected values. In assessing this fit, the depth of case knowledge available in qualitative analysis can provide interesting qualifiers to the frequently more-superficial categories used in quantitative analysis. For example, quantitative researchers have been unable to find systematic causal relationships between democracy and environmental protection despite numerous theories suggesting that the relationship exists; qualitative comparison of the same countries in authoritarian and democratic periods finds signs of a relationship (Hochstetler, 2012).

While assessing correlational fit is the most important analytical step in quantitative methods; it is less important for a qualitative project since there are not enough cases to establish causal variation directly. Qualitative researchers have commonly argued that their most important analytical contribution is process tracing, an operation that statistical analysis cannot itself do (Corell and Betsill, 2008; George, 1979; Steinberg, 2007). Process tracing requires the analyst to break '... down an overarching causal relationship into a set of smaller causal links in a larger causal chain' (Mitchell and Bernauer, 1996: 22). The analysis is often presented as a chronological narrative or schematic map of the causal relationship. Moving to this level can help identify scope conditions for the causal relationship and locate intermediate processes and variables (Homer-Dixon, 1996: 144; Schwartz et al., 2000: 85). Process tracing can also help to argue against alternative hypotheses, as the information-dense format allows considering these alternatives and finding them less consistent (Schwartz et al., 2000: 86).

There are two basic uses of qualitative studies, reflecting different views on the primary purposes of qualitative research. The first use sees qualitative methods as most important early in research programs, for generating hypotheses that can later be evaluated more systematically using many more cases. This pre-statistical conception of qualitative research is well illustrated by research on the effectiveness of international environmental regimes (see Young, this book). After qualitative case studies of individual regimes accumulated over a decade, researchers then began to use that qualitative data to create systematic databases on regimes that had been used for quantitative and systematic comparative analysis (for example, Breitmeier et al., 2006; Young, 1999). Similarly, Ostrom drew on more than 5000 qualitative case studies of communities that managed common pool resources to develop her rational choice models of their effects (Ostrom, 1990: xv). Other approaches send multiple researchers to do simultaneous coordinated qualitative studies of large events like negotiation sessions (Collaborative Event Ethnography) or multi-site/scale research of phenomena like a commodity sector (O'Neill et al., 2013). Mitchell and Bernauer offer helpful suggestions for how to design case studies for use in this manner (Mitchell and Bernauer, 2004).

The other approach to qualitative methods sees them as the likely final methodological stage of many IEP research programs. One area of IEP that largely concurs with this point of view is the extensive body of work on non-state actors in IEP (see Betsill, this book; Betsill and Corell, 2001). Other writers agree that qualitative methods are likely to have enduring value for certain kinds of IEP research problems. Mitchell and Bernauer's list (1998: 6–7) includes the following instances: '(a) important but difficult-to-quantify variables (such as power, interests, or leadership); (b) theoretically important, empirically rare, or previously ignored cases; (c) innovative (but, by their nature, rare) international environmental policy strategies; and (d) causal rather than merely correlational relationships'. Homer-Dixon argues that complex ecological-political systems involve interactive, non-linear, and reciprocal causal relations that may always require qualitative methods, because the case variation cannot be systematically controlled (Homer-Dixon, 1996: 132–134).

Critics of positivist qualitative methods argue that such imprecision and inability to systematically control variation are attributes of the approach rather than in the empirical material (for example, Gleditsch, 1998: 391–392). One project, which attempted to systematize qualitative information to use it for quantitative studies of environmental regime effectiveness, complains that too many qualitative studies fail to select their cases on theoretical grounds, involve idiosyncratic choices of variables and operationalizations, and are consequently not generalizable (Breitmeier et al., 2006). Even proponents of qualitative methods admit that, in practice, qualitative scholars often fail to pay sufficient attention to various selection or practitioner biases or are inadequately precise and critical in their analyses (McKeown, 1999: 178). Such flaws can make their work difficult to generalize, or can raise questions about how representative their sample of cases is.

Beyond these errors of execution, positivist qualitative methods are inherently limited in several ways with respect to the aims such scholars themselves claim. As indicated already, qualitative case studies in and of themselves cannot provide the large-number statistical correlations that are the 'gold standard' of positivist causal analysis. Thus they are limited in the crucial aim of generalization even while they might provide other causal pieces that statistical analyses cannot. The tracing of causal mechanisms also cannot formally weight causal variables, although it can show their interaction (Schwartz et al., 2000: 87–88) and rank them (Steinberg, 2007).

Quantitative methodologies

Quantitative causal analysis shares its basic purposes with positivist qualitative approaches, but uses larger numbers of cases. This allows the researcher to speak with greater confidence about patterns across the phenomenon of interest, and about the probability of a causal relationship. A second difference between quantitative and qualitative positivist research is that quantitative researchers develop statistical models of the hypothesized relationship among their variables. This is often a regression equation that must include the major possible explanations of change in the dependent variable, with its variables defined in 'relatively generic and generalizable' terms (Mitchell, 2002: 74). Thus while a qualitative researcher often defines variables such as regime effectiveness in terms that are quite specific to the case at hand, the quantitative researcher will need to define terms in a way that can be measured comparably across a broad set of cases.

A primary empirical task of the quantitative researcher is to create (or locate existing) databases, coded with detailed data protocols, which collect data on the variables of interest in the cases of interest. This is partially a conceptual issue, as different definitions of a concept may exist in the literature, and the decision to operationalize and collect data on one as opposed to another can produce quite different quantitative results (for example, Hochstetler, 2012, on differing definitions of democracy). There is also a secondary issue more unique to quantitative research, for which the requirement is that all conceptual variables be operationalized in ways that can be represented with data with numerical values, a process that can raise issues of construct validity – 'measuring what we think we are measuring' (King et al., 1994: 25). Some data such as trade statistics, gross national product, or levels of pollutants are widely available in numerical form, but many are not (Mitchell, 2002: 70–71).

Quantitative researchers use a variety of statistical procedures, which are beyond the scope of this chapter. They should match the underlying assumptions of the causal model; thus a simple linear regression can be usefully performed only when there are good reasons to think that its assumptions of linearity and independence of the variables are met. Other procedures such as time-series analysis or multivariate regression are needed for more complicated relationships. What all these forms of analysis share is the use of correlational analysis that tests for general hypothesized relationships between dependent and independent variables using a large number of cases. Statistical correlations are then the foundation for general causal claims about the relationships among the variables in the analysis.

Quantitative analysis is still comparatively rare in IEP. Perhaps not surprisingly, these studies first appeared in some of the subject areas of IR where quantitative data is most readily available, especially in the intersection with economic issues. Thus there are studies that link national income, trade, or other economic data to environmental outcomes (for example, Cheon and Urpelainen, 2012; Copeland and Taylor, 2003; Perkins and Neumeyer, 2009). Conflict scholars have taken the lead in tracking the relationship between natural resources or climate change and conflict (Colgan, 2010; Gleditsch, 1998; Hendrix and Salehyan, 2012). Using quantitative methods in IEP often requires significant work to construct original datasets on relevant variables (for example, Bernauer et al., 2013; Hicks et al., 2008).

Both advocates and critics of a quantitative approach to IEP agree that systematic data collection is a crucial requirement for this approach. Gleditsch even calls for 'a Correlates of War project for the environment' (Gleditsch, 1998: 396). Some responses can be found in new databases on Transboundary Freshwater Disputes and International Forestry Resources and Institutions (O'Neill et al., 2013). Ironically, certain kinds of data are now available in 'big data' quantities that are equally challenging to traditional quantitative methods, a development we discuss in more detail below. Without good data behind them, sophisticated statistical techniques are simply misleading. For proponents of quantitative methods, this is a temporary stage in quantitative analysis, which is being overcome. For critics, the quality of quantitative data is a more enduring problem. Non-quantitative scholars often argue that crucial variables are simply very difficult to operationalize in numerical terms. Mitchell and Bernauer voice this critique as the charge that quantitative scholars only study what can be quantified, use data that are too simple to be valid measures of complex constructs, and produce 'precise but unreliable or irrelevant results with sophisticated statistical techniques but data of poor quality' (Mitchell and Bernauer, 1998: 5; see also O'Neill et al., 2013). More fundamentally, the probabilistic nature of the conclusions based on quantitative research may not help to explain outcomes in single cases of particular interest. Thus quantitative and qualitative approaches illustrate a tradeoff between the ability to reach general conclusions and the ability to explain one case well (Mitchell, 2002: 59).

Rational choice approaches

The label 'rational choice approaches' is used here to refer to a family of related literatures that assume that political phenomena can be explained with reference to the microfoundational choices of an individual 'whose behavior springs from individual self-interest and conscious choice. He or she is credited with extensive and clear knowledge of the environment, a well-organized and stable system of preferences, and computational skills that allow the actor to calculate the best choice (given individual preferences) of the alternatives available' (Monroe, 1991: 4). Choice situations may either be parametric, where an individual faces given external constraints from the structure of the situation, or strategic, where the decisions of two or more individuals are interdependent (Elster, 1986: 7–8). In strategic situations, game theory is the methodological tool used to find the equilibrium choices of the set of interdependent actors. For formal models, significant mathematical modeling skills may be needed. Rational choice scholars argue that one of the advantages of their approach is this requirement to clearly and precisely present the relevant actors/players, their alternative choices and preferences, and the various outcomes (Kilgour and Wolinsky-Nahmias, 2004).

Some of the most-cited works in IEP are in the rational choice tradition, broadly defined. Garrett Hardin's story of the tragedy of the commons is a non-formal analysis of a strategic choice situation (Hardin, 1968). His 'tragedy' of collective failure based on individual best choices has become a motivating metaphor for an entire sub-area of rational choice analysis that addresses such social dilemmas and free-riding problems. Elinor Ostrom's Nobel-winning research builds on that work, challenging it by showing that incentives and institutions can co-evolve in a way that allows sustainable management of common pool resources (Ostrom, 1990). Work on regime effectiveness extended and evaluated her insights in an international context (for example, Young, 1999). Ostrom and Young are within the choice tradition that pays particular attention to the ways that social institutions shape behavior. Political science and IR scholars, especially in the United States, have gone on to develop the mathematical formalization of rational choice models. With a few exceptions (for example, Urpelainen, 2013) these have been rarely used in IEP. Rational choice mostly enters IEP metaphorically, such as discussions of climate politics as a commons problem, with shirkers worried about the credible commitments of others (for example, Keohane and Victor, 2011).

Kilgour and Wolinsky-Nahmias (2004) argue for more use of game theory in the study of IEP and offer concrete strategies for doing so. They suggest that the field of international environmental politics presents many rational choice problems, including the 'management of common resources, environmental negotiation, enforcement of environmental agreements, and the balance of domestic and international incentives' (Kilgour and Wolinsky-Nahmias, 2004: 318). These entail classic dilemmas of bargaining and strategic action. One reason for the scarcity of empirical rational choice studies in IEP may be the difficulty in gathering the information required to develop and test rational choice models. Empirical rational choice analysis requires data about the components of its explanations, such as institutional arrangements, preference structures, and so on. Because other approaches are not as interested in decisional microfoundations of IEP, rational choice scholars again often need to develop their own datasets. An enduring critique specific to the rational choice approach is that there are profound tensions between the many simplifications necessary to produce models that are mathematically tractable and what non-rational choice scholars consider adequate representations of 'empirically encountered situations' (McKeown, 1999: 177–178). This is a debate among scholars who are all committed to positivism, as well as a critique levied by non-positivists. Finally, non-rational choice scholars often criticize the excessive focus on theoretically driven model development, including models which will not be empirically testable in any foreseeable future.

Emerging methods: Computational social science and geospatial information technologies

To this point, we have focused on approaches to research that are commonly used by IEP scholars. We now introduce a series of additional approaches and methods that are broadly positivist and have areas of potential applicability to IEP but have so far been used more often by scholars in other disciplines (often to study the global environment, but not politics). The 21st century has experienced a spatial turn that encompasses the physical and social sciences as well as the humanities where space has become an integrating theme. Coupled with the fast-paced emergence of new internet technologies and data sources, methods and approaches to understanding social systems and policy decisions are rapidly changing. Thematic emphases on space, place, flow, connection, and complexity using mixed methods emphasize interdisciplinarity, synthesis, and holistic understanding (Sui and Delyser, 2012; Carpenter et al., 2009; Tashakkori and Teddlie, 2010). Cross-indexing between themes and populations using geographic coordinates to integrate data and identify geo-relationships is increasingly common. A shift has occurred in the traditional uses of geographic information technologies (GITs) from inventories, assessments, and analyses of physical and socioeconomic landscapes to a focus on qualitative geographic information science (GIS) that includes participatory GIS, feminist GIS, and new methods of interactive visualization (Sui and Delyser, 2012; Aiken and Kwan, 2010; Knigge and Cope, 2009).

New centers have emerged, such as the Center for Spatially Integrated Social Science, providing spatial resources for the social sciences and demonstrating computational social science applications.[3]

These shifts in the research landscape reflect changes in technology as well as changes in approaches and methodologies. The adoption of a systems approach that addresses the relationships between different parts of a system and the cyclical responses focuses on a holistic approach to problem solving. Linking this approach to the availability of large datasets and integrative techniques to combine both qualitative and quantitative data has resulted in the possibility of an expansive research landscape (Lazar et al., 2009; Conte et al., 2012; Young et al., 2006). This research landscape is enriched by interdisciplinary discussion and communication where computer scientists, engineers, social scientists, and mathematicians work together (Goodchild et al., 2000). For example, the National Science Foundation established a National Socio-Environmental Synthesis Center to facilitate the interaction of natural and social scientists with policymakers to prioritize and co-develop research questions.[4]

The uses of GITs have expanded with the variety of digital earth applications available via the Internet. In 1998, US Vice President Al Gore described a virtual globe where freely available scientific and cultural information would be used to understand the world and human activities. The fruition of this vision is exemplified in Digital Earth geobrowsers such as Google Earth, NASA's World Wind, ESRI's ArcGIS Explorer, and Bing Maps 3D, where private/public partnerships have enabled a massive distribution of georeferenced data as well as the ability for citizen scientists and community groups to upload and share local information. GITs underpin these applications through the seamless use of remotely sensed data (RS, that is, satellite imagery captured at an increasingly finer resolution) and global positioning systems (GPS, that is, identifyingly explicit locations using coordinates such as latitude and longitude) organized within geographic information systems (GIS) using increasingly complex and distributed relational databases. One does not need to be a GIS expert to use these technologies where point-and-click capability, internet connectivity, and available data provide the basis for using Digital Earth products.

The integration of GIT with internet technologies and electronic communication has established coupled systems that mesh multiple data streams and innovative methods for data representation, visualization, and sharing (Goodchild and Janelle, 2010). Complementing this integration is the development of social media in the form of blogs, wikis, mash-ups, and social networks utilizing multiple sources and types of data. The informal development of new social processes driven by technology has outpaced the traditional methods and approaches used in the social sciences and humanities. However the emerging interdisciplinary field of computational social science addresses this gap through the development of multiple methods that include automated information extraction, social network analysis, social GIS, and complexity modeling (Cioffi-Revilla, 2010).

Automated data extraction: Automated data extraction refers to a suite of capabilities: content analysis via machine coding of large volumes of text; text mining where discovery of useful knowledge is derived from unstructured or semi-structured text; information extraction via multiple methods of machine coding and human coding of text; and the utilization of 'big data' (Mooney and Nahm, 2005; McCallum and Jensen, 2003; Nature Publishing Group, 2008). Such extraction techniques are not limited to text but also include audio, images, and video (Cioffi-Revilla, 2010). Big data are those datasets that become so large that they are difficult to process, use, and analyze using traditional data management and processing tools, requiring its own data analysis tools.

A suite of 'dataware' products have been created to access online clearinghouses of legal, health, economic, political, and demographic information (Torrens, 2010; for example, Nvivo software), as well as to organize data through collaborative categorization tools (Conte et al., 2012; for example, Web 2.0). These data can be intersected using geography where multiple types of data are linked, organized, and indexed across disciplines and applications using location as the unifying link (Torrens, 2010).

Social networks: Integrating mathematics, statistics, and social science allows for the integration of social networks and spatial information that can be analyzed and visualized using social network analysis and GIS. This emergent area has numerous applications that include alliance networks, terrorist networks, human migration patterns, regional innovation systems, and knowledge spillovers (Cioffi-Revilla, 2012; Ter Wal and Boschma, 2009). Social networks, electronic communication, and Internet access have played a critical role in the organization of the Arab Spring democratic mobilization movements of 2009. These outcomes have been mapped both temporally and spatially to track these diffusion patterns, as well as to examine the electronic footprint of mobile phone calls and crowdsourcing (Conte et al., 2012; Curtin, 2007). Now common in political science and IR, social network analysis is just beginning to be used in IEP (for example, Paterson et al., 2014).

Complexity modeling: Complexity addresses the multiple components of a system, emphasizing the holistic functioning that includes feedback loops, cascading effects, emergent properties, and non-linear relationships. Resilience and adaptation across boundaries and scales are hallmarks of complex systems. Additionally, complexity studies model outcomes and patterns from an interdisciplinary perspective using such tools as agent-based modeling, system dynamics, and simulation modeling (Goodchild and Janelle, 2010). Complexity modeling can include analysis of wealth and poverty in developing societies, political instability, foreign aid distribution, international conflicts, climate change, and natural disasters (Cioffi-Revilla, 2010; Hoffmann, 2011; Laituri and Kodrich, 2008). Agent-based modeling is a signature tool of complexity studies where agents are assigned roles, rules, and behaviors to mimic and model their impact and influence on a given system over time and space. Organizational structures, land use change, and warfare have been examined using agent-based modeling across geographic landscapes (Alessa et al., 2006; Kuznar et al., 2005).

The integrating qualities of GIT facilitate interdisciplinary approaches with the attendant blurring of disciplinary boundaries. In the current configuration of new methods, GIT is situated within the computational social sciences as a central tool and approach. It is particularly salient to IEP and can enhance interdisciplinary research through collaboration and development of projects that reach across disciplinary boundaries and integrate diverse datasets. A study conducted by the Climate Change and Development Division of the Embassy of Switzerland created the Micro Insurance Academy (MIA) to enhance the resilience of vulnerable communities to climate change by developing pro-poor microinsurance solutions. The result was a map of those areas that are more exposed to adverse events, have lesser adaptive capacity, and are more vulnerable to climate risks (Sharma and Jangle, 2011).[5]

Environmental security is another critical area for research and examination. The spatial representation of governance structures provides insights into North–South issues. For example, the concept of water as a human right has been variously adopted around the world and reveals relationships between the state and populations with respect to provisioning of basic services that ensure national security (Laituri and Sternlieb, 2012). Environmental security can be examined through the lens of disasters that link global and local processes. In the case of the Japanese tsunami and accompanying nuclear disaster, linkages were examined across spatial scales that included local, oceanic, and atmospheric connections.[6]

The drawbacks to these approaches are numerous. Data and access to data remain key issues. Data may be sporadically collected and result in inconsistent spatial coverage within a country as well as between countries. There is no standardization between countries for data collection, creation, and maintenance, although there are efforts to address this.[7] Changing political boundaries result in obsolete datasets. Different nations have different policies with regard to public access to data that may be considered sensitive to national security interests. The digital divide is a real phenomenon and is manifested in several ways. There is inconsistent digital data in many developing countries. This is further compounded by a lack of access to hardware and software for both developing countries and marginalized groups such as indigenous peoples. However, the ubiquity of cell phones, smart phones, and alternative methods for accessing digital data and digital communication is occurring ever more quickly (Duggan and Rainie, 2012; Fox and Duggan, 2012). New global datasets and methods for downscaling are providing innovative ways to analyze issues related to climate justice, disease, and refugee movement (Schimmer, 2008; Ceccato et al., 2005; Dennis et al., 2005). The increased access to satellite imagery at finer resolutions has contributed to social analysis that includes the examination of nightlights to estimate population (Small et al., 2011).

Computational social science coupled with GIT creates a powerful approach to addressing issues related to IEP. However, there remain significant impediments to widespread implementation. First, there is a need for new perspectives that cross disciplinary and digital divides to yield new understanding of human behaviors. The capacity to understand dynamic data collected across multiple scales of location and population demands new theories, approaches, and methods that are newly emerging. Venues must be established to support training of new scholars. Second, the shift from the physical and biological sciences to the social sciences of GIT calls for an understanding and examination of issues related to privacy and anonymity of research subjects. Third, adoption of computer-intensive applications entails a significant investment in infrastructure for distributed monitoring, permissions, and encryption in fields that traditionally have fewer funding avenues than engineering and physical science departments. Fourth, big data collected from such sources as Twitter and Facebook may not be truly reflective of the population at large (Giles, 2012; Conte, 2012; Lazar et al., 2009). These drawbacks should not prevent researchers from using GIT in conjunction with IEP. If anything, they represent an exciting new area for research and collaboration across boundaries of various kinds, including disciplinary. GIT provides a medium for analysis that is dynamic, complicated, and difficult.

Critical approaches to methodology

Critical qualitative methods

Critical – or reflexive – qualitative methods share one crucial area of intersection with positivist ones: they also find their data in the form of documents and other archival records, interviews, and observation. Differences in how positivist and critical scholars design research programs and analyze their data during and after collecting it distinguish the two approaches, however. Most importantly, most critical scholars question the value and possibility of the positivist enterprise of cumulatively building a body of replicable, unbiased, causal explanations of generalized phenomena that constitute an objective reality independent of language or theory. As a group, they work instead to show the power relations inherent in both academic and policy constructions of phenomena, identifying dominant constructions and undermining them. Critical theorists are especially interested in how theoretical and policy problems are defined as problems, and also in how solutions reach the status of solutions (Stevis and Assetto, 2001). Many critical theorists aim to go beyond critique to seek political solutions that challenge power (Wapner, 2008).

A number of different approaches challenge positivism, including interpretive, poststructural, structuralist, and feminist approaches. This section emphasizes their common threads, with somewhat more attention to the variants that stress the role of ideas and identity, as virtually all scholars who follow those variants use qualitative methods. The next section separates out the approaches that are more structuralist and materialist as they are most likely to use non-qualitative methods. These divisions are preliminary and possibly controversial, and deserve much more attention from scholars working in these areas. There are comparatively few explicit discussions of methodology in the critical traditions (with the exception of interpretivism – which is not necessarily critical – and feminism) and few written by IEP scholars (Methmann et al., 2013). Some critical scholars have argued that the lack of methodological discussions too easily grants positivists the right to define the standards of good research (Milliken, 2001).

Critical research projects often begin by identifying a particular conventional understanding for further analysis. The careful strictures on representativeness and variation control of positivist analysis are not relevant here, as any specific context will embody the power dynamics of meaning construction. Rather than seeing cases as representative examples of generalized patterns in an objective reality, the critical theorist sees specific contexts as providing glimpses of underlying systemic relations. Thus 'case selection' relies on locating particularly telling contexts that illuminate these processes and describing those, even though many critical theorists are uneasy about the word 'empirical' and its implicit claim that there are data independent of theory:

> Description, while not inherently critical, becomes so if it makes us look again, in a fresh way, at that which we assume about the world because it has become overly familiar.... In this way, new spaces are opened for thinking about the past and the present and, therefore, how we construct the world.
> (Fierke, 2001: 122)

Thus while a positivist might consider alternate readings of a particular event with the intent of figuring out which is correct, critical theorists will often simultaneously engage multiple readings, with the intent of showing the purposes and interests that each reading serves. Some poststructuralists argue that there can never be a 'correct' reading, while critical and feminist theorists often make politically grounded choices about the most useful interpretations of a situation (Fanow and Cook, 1991).

Critical theorists usually use qualitative techniques for recovering alternate descriptions of the world. One common strategy is to use qualitative interviews to recreate the voices of those left out in dominant accounts of elite decisions – to interview women, indigenous peoples, or peasants to hear their experiences of large dams, free trade agreements, agricultural policies, and so on (for example, Conca, 2006; McCormick, 2009). Such interviews can provide vivid narratives that stress considerations beyond national and international power dynamics or global market forces. Direct observation of meetings and negotiations, another classic qualitative method, can also deepen understanding of the different positions and identities expressed by participants as they interact. Participatory Action Research even makes research 'subjects' active participants in the research itself (O'Neill et al., 2013). The very immediacy of these kinds of methods can raise dilemmas for critical scholars, however. The researcher may face both political and ethical dilemmas about how to analyze and publish the research (Punch, 1986).

Qualitative documentary research is very common among critical scholars. Documentary research carries less of the political risks of the more direct methods, but the comparative distance of documentary research raises a different issue: most documents are created and stored by dominant actors and so tend to reflect dominant constructions. As a result, critical theorists disagree with the classic use of archives that sees them as a 'repository of "facts" ' (Spivak, 1985: 248). Archives and documents are still important for the critical theorist, however, who uses them while remaining alert to these issues. One analytical strategy is to focus more on unprocessed rather than processed or final documents, as the former are more likely to contain the traces of alternate points of view that undercut official positions that interest the critical scholar. Another strategy is to use the official documents and their presentations of reality as analytical subjects in their own right, turning attention to how discourses create systems of meaning, produce actors and other phenomena related to the discourse, and are played out in practice (Milliken, 2001: 138–139). Increasingly, documents of the less powerful are readily available. For example, international negotiations have been challenged in recent years by the ability of NGOs to disseminate alternate accounts immediately through electronic means, and the Internet generally is a powerful tool for those who have access to it – as well as a rich source of data for researchers.

Any issue in IEP could be studied from a critical perspective. Topics related to international political economy and globalization are prominent for critical theorists, with their companion topic of resistance to neoliberal globalization (Newell, 2012; Stevis and Assetto, 2001). Critical theorists have researched the relationship between the environment and numerous kinds of social inequity, especially through studies of environmental justice (see Okereke and Charlesworth, this book) and a gender perspective on the international environment (see Detraz, this book). Another line of focus is on the discourses and identities created in environmental politics and the way cultural dynamics lead to particular environmental outcomes (Dryzek, 1997; Hajer, 1995; Litfin, 1994).

Critical IEP scholars need to go on developing more explicit methodological discussions that would give greater guidance to new scholars and offer clearer standards for evaluating their research. Jettisoning the positivist aims of the mainstream means that critical theorists are often working outside what is recognized as good research, especially as indicated by adjectives like systematic or scientific. Thus common criticisms include critical scholars' failure to produce precise, neutral, and generalizeable findings. As this chapter has shown, positivists and critical theorists have genuinely different approaches to research and it is unlikely that they will ever evaluate research in the same ways. More explicit methodological discussion would at least allow positivists to try to evaluate critical research by its own standards and would encourage critical theorists to reflect on what kinds of research would move their theoretical agendas forward.

Structuralist and non-qualitative critical approaches

A final approach to IEP research joins some of critical theory's critiques of positivism while accepting more of positivism's research aims and non-qualitative tools. As it is a hybrid approach with little explicit methodological discussion, the boundaries of this kind of approach are not very clear. Most scholars in this category are a variant of critical theorists, who agree with the argument outlined above that positivists' refusal to take power into account makes their empirical conclusions biased. In their view, positivists implicitly accept the existing international order and do not consider alternatives to it, but seek to manage it. On the other hand, this category of critical scholars is more inclined to both structuralist and materialist views of the world, which limit their attention to language and interpretation compared to other critical theorists (Buttel, 2002: 39). Instead, they posit the existence of underlying material processes that produce empirical outcomes that can be described and measured. This means they are likely to accept the project of cumulative building of causal explanations, so long as the research does consider those underlying relations of power. It also means that they are more willing to use non-qualitative methods, since they believe that there should be visible and systematic material manifestations of the power dynamics they identify.

Critical theorists will often look to the structural dynamics of capitalism as the foundation of their models. In one example of this kind of approach, Marian Miller traces the process and effects of the 'enclosure of knowledge' to a fundamental logic of capitalism: 'Integral to the logic of capitalism is the need for repetitive expansion, involving a continuous reinvestment of profits to create more profits' (Miller, 2001: 113). From this starting point, she develops a model of capitalist expansion based on the enclosure of land and knowledge, which she then further develops empirically with studies of these linked processes in Kenya and India. World Systems theory focuses on the structural forces produced by 'the historical legacy of a country's "incorporation" into the global economy' (Roberts and Grimes, 2002: 167), as well as other materialist and historical concerns. Researchers have linked this causal argument to environmental outcomes using a variety of quantitative and qualitative methods in addition to modeling likely outcomes of structural processes (for example, Goldfrank et al., 1999). Roberts (1996), for example, concluded on the basis of aggregate cross-national data analysis that World System position was a better predictor than GNP/capita of which countries would sign international environmental treaties. More recently, he has used World Systems arguments about US hegemonic decline to explain its laggard stance in climate negotiations (Roberts, 2011). Similarly, Bergesen and Parisi (1999) used quantitative data to conclude that World System position was a useful predictor of rates of toxic emissions.

This work lacks the extended methodological discussions and body of actual research that would allow full evaluation of the approach. Efforts to join critical epistemologies and ontologies with methods more often associated with positivism might fall prey to the problems of both, display the virtues of both, or fall somewhere in between.

Conclusion

IEP scholars have a wide variety of methodological tools they can use to study their diverse topics of interest. While these tools are divided between positivist and critical epistemologies and between qualitative and quantitative methods, all of them make distinct contributions to the field as a whole. All also raise some serious potential problems that researchers should try to avoid, when possible. Since these drawbacks often appear as reversed tradeoffs when one method is compared to another, the field – and individual research projects – benefits from the use of multiple methods to approach any given topic. Beyond its general call for pluralism, this chapter has identified several directions for further methodological development that seem especially likely to be fruitful.

The largest gap calls on critical IEP scholars to engage in more explicit methodological discussions. Such approaches are quite common in the study of IEP, but the methodological discussions have not kept pace. The overarching purpose of these discussions should be to provide guidance for scholars looking to begin critical projects. Among the crucial questions that should be addressed are the purposes and use of empirical materials in such approaches, including directives for the kinds and nature of evidence to be presented. In addition to allowing scholars who use these approaches to proceed with clearer self-understanding, such discussions would allow positivists to evaluate critical research on its own terms and to see the internal consistency of critical theory (or variants of critical theory) as an approach to research.

One of the more promising developments in recent methodological discussions has been the effort to consider the relationships of different positivist methodologies to each other. This has been done in two ways. The first way is exemplified in the volume edited by Sprinz and Wolinsky-Nahmias (2004), which consists of explicit methodological discussion of the varying contributions of qualitative, quantitative, and rational choice methodologies to causal explanation and the possible pitfalls of each. This chapter has similar aims. The second approach has taken place within particular substantive areas of research and consists of collective discussion of their progress to date in methodological terms. Qualitative case studies may be aggregated for quantitative analysis, or works may consider the contrasting conclusions of qualitative and quantitative work (for example, Hochstetler, 2012; Young, 1999). These efforts can find support in an emerging literature that suggests strategies for combining qualitative and quantitative methods in single research projects or in an area of research. Mixed method or 'nested' analysis considers the potential contributions of different methodological approaches to an empirical problem from the outset, and uses distinct methods together in a deliberate way (Lieberman, 2005; Small, 2011). This goes beyond simple methodological pluralism to a more coordinated and strategic use of different methodological tools (Sil and Katzenstein, 2010).

Finally, this chapter identifies several potentially relevant methodologies that have not been used very often in IEP research to date, notably computational social science, geospatial information technologies, and rational choice methodologies. Methodological choices should not be made on the basis of novelty, of course, and these approaches do have the drawback of requiring both methodological training and additional data in forms that may not yet exist. Yet as the heavy preponderance of a single approach in the field's leading journal, *Global Environmental Politics*, shows, IEP scholars are much more likely to be too conservative than too experimental in their methodological choices. More self-conscious attention to those choices would be salutary for the field.

Notes

1. http://www.mitpressjournals.org/loi/glep.
2. For a more detailed discussion, see O'Neill et al. (2013).
3. www.csiss.org.

4. http://www.sesync.org/
5. http://www.geospatialworld.net/Regions/ArticleView.aspx?aid=30395)
6. http://marinedebris.noaa.gov/tsunamidebris/
7. For example, the Federal Geographic Data Committee, www.fgdc.gov/.

Annotated bibliography

Sprinz, D. and Y. N. Wolinsky-Nahmias (eds) (2004) *Models, Numbers, and Cases: Methods for Studying International Relations* (Ann Arbor: University of Michigan Press).

This volume provides a comparative analysis of positivist methodologies. There are chapters on qualitative, quantitative, and rational choice methodologies as general approaches to the study of IR. Additional chapters discuss how these methodologies appear in the study of particular IR topics, including IEP. The parallel presentation of different methodologies will help scholars to identify the most useful ones for their research agendas.

Mitchell, R. and T. Bernauer (1998) 'Empirical Research on International Environmental Policy: Designing Qualitative Case Studies', *Journal of Environmental and Development*, 7, 1, 4–31.

This article is an excellent introduction to positivist qualitative methodologies. In addition to describing the aims of the qualitative approach, the article includes discussions of process tracing and generalization. Perhaps most usefully, it lays out a series of concrete research design tasks for undertaking research on international environmental policy through qualitative case studies.

O'Neill, K., E. Weinthal, K. R. Marion Suiseeya, S. Bernstein, A. Cohn, M. W. Stone and B. Cashore (2013) 'Methods and Global Environmental Governance', *Annual Review of Environment and Resources*, 38, 11.1–11.31.

This review of the current literature on the methods used for the study of global environmental governance is a useful complement to this chapter. It focuses on four challenges: 'complexity and uncertainty, vertical linkages across multiple scales, horizontal linkages across issues areas, and (often rapidly) evolving problem sets and institutional initiatives' (11.5).

Cioffi-Revilla, C. (2010) 'Computational Social Science', *WILEY Interdisciplinary Reveiws: Computational Statistics*, 2, 3, May/June, 259–271.

This article provides an overview of the emerging disciplinary field of computational social science that includes automated information extraction systems, social network analysis, social geographic information systems, complexity modeling, and social simulation models.

Sui, D. and D. DeLyser (2012) 'Crossing the Qualitative-quantitative Chasm I: Hybrid Geographies, the Spatial Turn, and Volunteered Geographic Information (VGI)', *Progress in Human Geography*, 36, 1, 111–124.

This article reviews methods and approaches that combine methodologies to create hybrid approaches to analyzing today's challenging problems. This article examines methods that bridge qualitative/quantitative approaches that address such problems synergistically and holistically.

Works cited

Aiken, S. and M. P. Kwan (2010) 'GIS as Qualitative Research: Knowledge, Participatory Research, and the Politics of Affect', in D. Delyser, S. Herbert, S. Aiken, M. Crang and L. McDowell (eds) *Handbook of Qualitative Geography* (London: SAGE), 287–304.

Alessa, L., M. Laituri and M. Barton (2006) 'An "All Hands" Call to the Social Science Community: Establishing a Community Framework for Complexity Modeling Using Agent Based Models and Cyberinfrastructure', *Journal of Artificial Societies and Social Simulation*, 9, 4.

Bergesen, A. J. and L. Parisi (1999) 'Ecosociology and Toxic Emissions', in W. L. Goldfrank, D. Goodman and A. Szasz (eds) *Ecology and the World-System* (Westport, CT and London: Greenwood Press), 43–58.

Bernauer, T., T. Böhmelt and V. Koubi (2013) 'Is There a Democracy-Civil Society Paradox in Global Environmental Governance?' *Global Environmental Politics*, 13, 1, 88–107.

Betsill, M. and E. Corell (2001) 'NGO Influence in International Environmental Negotiations: A Framework for Analysis', *Global Environmental Politics*, 1, 4, 65–86.

Breitmeier, H., O. R. Young and M. Zürn (2006) *Analyzing International Environmental Regimes: From Case Study to Database* (Cambridge: The MIT Press).

Buttel, F. H. (2002) 'Environmental Sociology and the Classical Sociological Tradition: Some Observations on Current Controversies', in R. E. Dunlap, F. H. Buttel, P. Dickens and A. Gijswijt (eds) *Sociological Theory and the Environment: Classical Foundations, Contemporary Insights* (Lanham, MD: Rowman and Littlefield).

Carpenter, S. R., E. V. Armbrust, P. W. Arzberger, F. S. Chapin III, J. J. Elser and E. J. Hackett (2009) 'Accelerate Synthesis in Ecology and Environmental Sciences', *Bioscience*, 59, 699–701.

Ceccato, P., S. Connor, I. Jeanne and M. Thomsom (2005) 'Application of Geographical Information Systems and Remote Sensing Technologies for Assessing and Monitoring Malaria Risk', *Parassitologia*, 47, 81–96.

Cheon, A. and J. Urpelainen (2012) 'Oil Prices and Energy Technology Innovation: An Empirical Analysis', *Global Environmental Change*, 22, 2, 407–417.

Cioffi-Revilla, C. (2012) 'Computational Social Science', *WILEY Interdisciplinary Reviews: Computational Statistics*, 2, 3, May/June, 259–271.

Colgan, J. D. (2010) 'Oil and Revolutionary Governments: Fuel for International Conflict', *International Organization*, 64, 661–694.

Conca, K. (2006) *Governing Water: Contentious Transnational Politics and Global Institution Building* (Cambridge, MA: The MIT Press).

Conte, R., N. Gilbert, G. Bonnelli, C. Cioffi-Revilla, G. Deffuant, J. Kertesz, V. Loreto, S. Moat, J. P. Nadal, A. Sanchez, A. Nowak, A. Flache, M. San Miguel and D. Helbing (2012) 'Manifesto of Computational Social Science', *European Physical Journal Special Topics*, 214, 325–346.

Copeland, B. R. and M. S. Taylor (2003) *Trade and the Environment: Theory and Evidence* (Princeton: Princeton University Press).

Corell, E. and M. M. Betsill (2008) 'Analytical Framework: Assessing the Influence of NGO Diplomats', in M. M. Bestill and E. Corell (eds) *NGO Diplomacy: The Influence of Nongovernmental Organizations in International Environmental Politics* (Cambridge, MA: The MIT Press), 19–42.

Curtin, K. (2007) 'Network Analysis in Geographic Information Science: Review, Assessment, and Projections', *Cartography and Geographic Information Science*, 34, 2, 103–111.

Dennis, R., J. Mayer, G. Applegate, U. Chokkalingam, C. Colfer, I. Kuriniawan, H. Lachowski, P. Maus, R. Permana, Y. Ruchiat, F. Stolle, Suyanto and T. Tomich (2005) 'Fire, People, and Pixels: Linking Social Science and Remote Sensing to Understand the Underlying Causes and Impacts of Fire in Indonesia', *Human Ecology*, 33, 4, August, 465–504.

Dryzek, J. S. (1997) *The Politics of the Earth: Environmental Discourses* (Oxford: Oxford University Press).

Duggan, M. and L. Rainie (2012) 'Cell Phone Activities 2012', *Pew Research Center's Internet & American Life Project*, http://pewinternet.org/Reports/2012/Cell-Activities.aspx

Elster, J. (1986) 'Introduction', in J Elster (ed.) *Rational Choice* (New York: NYU Press), 1–33.

Fanow, M. M. and J. A. Cook (eds) (1991) *Beyond Methodology: Feminist Scholarship as Lived Research* (Bloomington: Indiana University Press).

Fenno, R. E. Jr (1990) *Watching Politicians: Essays on Participant Observation* (Berkeley: University of California, Institute of Governmental Studies).

Fierke, K. (2001) 'Critical Methodology and Constructivism', in K. Fierke and K. E. Jørgensen (eds) *Constructing International Relations: The Next Generation* (Armonk NY and London: M.E. Sharpe), 115–135.

Fox, S. and M. Duggan (2012) '"Mobile health 2012" Pew Research Center's Internet and American Life Project', http://pewinternet.org/Reports/2012/Mobile-Health.aspx

George, A. L. (1979) 'Case Studies and Theory Development: The Method of Structured, Focused Comparison', in P. G. Lauren (ed.) *Diplomacy: New Approaches in History, Theory, and Policy* (New York: The Free Press), 43–68.

Giles, J. (2012) 'Computational Social Science: Making the Links', *Nature*, 488, 23 August, 448–450.

Gleditsch, N. P. (1998) 'Armed Conflict and the Environment: A Critique of the Literature', *Journal of Peace Research*, 35, 3, 381–400.

Goodchild, M., L. Anselin, R. Appelbaum and B. H. Harthorn (2000) 'Toward Spatially Integrated Social Science', *International Regional Science Review*, 23, 2, 139–159.

Goodchild, M. and D. Janelle (2010) 'Toward Critical Spatial Thinking in the Social Sciences and Humanities', *GeoJournal*, 75, February 1, 3–13.

Hajer, M. A. (1995) *The Politics of Environmental Discourse: Ecological Modernization and the Political Process* (Oxford: Oxford University Press).

Hardin, G. (1968) 'The Tragedy of the Commons', *Science*, 162, 1243–1248.

Hendrix, C. S. and I. Salehyan (2012) 'Climate Change, Rainfall, and Social Conflict in Africa', *Journal of Peace Research*, 49, 39–50.

Hicks, R. L., B. C. Parks, J. T. Roberts and M. J. Tierney (2008) *Greening Aid: Understanding the Environmental Impact of Development Assistance* (Oxford: Oxford University Press).

Hill, M. R. (1993) *Archival Strategies and Techniques* (Newbury Park, CA and London: Sage Publications).

Hochstetler, K. (2012) 'Democracy and the Environment in Latin America and Eastern Europe', in P. F. Steinberg and S. D. VanDeveer (eds) *Comparative Environmental Politics: Theory, Practice, and Prospects* (Cambridge, MA: The MIT Press), 199–229.

Hoffmann, M. J. (2011) *Climate Governance at the Crossroads: Experimenting with a Global Response after Kyoto* (Oxford: Oxford University Press).

Homer-Dixon, T. (1996) 'Strategies for Studying Causation in Complex Ecological-Political Systems', *Journal of Environment and Development*, 5, 2, 132–148.

Keohane, R. O. and D. G. Victor (2011) 'The Regime Complex for Climate Change', *Perspectives on Politics*, 9, 1, 7–23.

Kilgour, D. M. and Y. Wolinsky-Nahmias (2004) 'Game Theory and International Environmental Policy', in D. F. Sprinz and Y. Wolinsky-Nahmias (eds) *Models, Numbers, and Cases: Methods for Studying International Relations* (Ann Arbor: University of Michigan Press), 317–343.

Klotz, A. and C. Lynch (2007) *Strategies of Research in Constructivist International Relations* (Armonk: M.E. Sharpe).

Knigge, L. and M. Cope (2009) 'Grounded Visualization and Scale: A Recursive Examination of Community Spaces', in M. Cope and S. Elwood (eds) *Qualitative GIS: A Mixed Methods Approach* (London: SAGE), 95–114.

Kuznar, L. A. and R. Sedlmeyer (2005) 'Collective Violence in Darfur: An Agent-based Model of Pastoral Nomad/sedentary Peasant Interaction', *Mathematical Anthropology and Cultural Theory: An International Journal*, 1.

Laituri, M. and F. Sternlieb (2012) 'Mapping Institutional Landscapes: Global Efforts to improve Access to Water', *International Journal of Sustainable Development and Planning*, 7, 3, 273–287.

Laituri, M. and K. Kodrich (2008) 'On Line Disaster Response Community: People as Sensors of High Magnitude Disaster Using Internet GIS', *Sensors*, 8, 3037–3055.

Lempert, R., J. Scheffran and D. F. Sprinz (2009) 'Methods for Long-Term Environmental Policy Challenges', *Global Environmental Politics*, 9, 3, 106–133.

Lieberman, E. (2005) 'Nested Analysis as a Mixed-Method Strategy for Comparative Research', *American Political Science Review*, 99, 3, 435–452.

Litfin, K. (1994) *Ozone Discourses: Science and Politics in Global Environmental Cooperation* (New York: Columbia University Press).

Mahoney, J. (2010) 'After KKV: The New Methodology of Qualitative Research', *World Politics*, 62, 1, 120–147.

McCallum, A. and D. Jensen (2003) 'A Note on the Unification of Information Extraction and Data Mining Using Conditional-probability, Relational Models', http://works.bepress.com/david_jensen/3.

McCormick, S. (2009) *Mobilizing Science: Movements, Participation, and the Remaking of Knowledge* (Philadelphia: Temple University Press).

McKeown, T. J. (1999) 'Case Studies and the Statistical Worldview: Review of King, Keohane, and Verba's *Designing Social Inquiry: Scientific Inference in Qualitative Research'*, *International Organization*, 53, 1, 161–190.

Methmann, C., D. Rothe and B. Stephan (2013) *Interpretive Approaches to Global Climate Governance: (De)Constructing the Greenhouse* (New York: Routledge).

Miller, M. (2001) 'Tragedy for the Commons: The Enclosure and Commodification of Knowledge', in D. Stevis and V. Assetto (eds) *The International Political Economy of the Environment: Critical Perspectives* (Boulder: Lynne Rienner), 111–134.

Milliken, J. (2001) 'Discourse Study: Bringing Rigor to Critical Theory', in K. Fierke and K. E. Jørgensen (eds) *Constructing International Relations: The Next Generation* (Armonk, NY and London: M.E. Sharpe), 136–159.

Mitchell, R. (2002) 'A Quantitative Approach to Evaluating International Environmental Regimes', *Global Environmental Politics*, 2, 4, 58–83.

Mitchell, R. and T. Bernauer (1998) 'Empirical Research on International Environmental Policy: Designing Qualitative Case Studies', *Journal of Environmental and Development*, 7, 1, 4–31.

Mitchell, R. and T. Bernauer (2004) 'Beyond Story-Telling: Designing Case Study Research in International Environmental Policy', in D. F. Sprinz and Y. Wolinsky-Nahmias (eds) *Models, Numbers, and Cases: Methods for Studying International Relations* (Ann Arbor: University of Michigan Press), 81–106.

Monroe, K. R. (1991) 'The Theory of Rational Action: Its Origins and Usefulness for Political Science', in K. R. Monroe (ed.) *The Economic Approach to Politics: A Critical Reassessment of the Theory of Rational Action* (New York: Harper Collins), 1–31.

Mooney, R. and U. Y. Nahm (2005) 'Text Mining with Information Extraction', in W. Daelemans, T. du Plessis, C. Snyman and L. Teck (eds) *Multilingualism and Electonic Language Management:* Proceedings of the 4th International MIDP Colloquium, September 2003 (Bloemfontein, South Africa: Van Schaik), 141–160.

Morrow, R. A. with D. D. Brown (1994) *Critical Theory and Methodology* (Thousand Oaks, CA: Sage Publications).

Nature Publishing Group (2008) 'Community Cleverness Required', *Nature*, 455, 7209, 1.

Newell, P. (2012) *Globalization and the Environment: Capitalism, Ecology and Power* (Cambridge: Polity Press).

O'Neill, K., E. Weinthal, K. R. M. Suiseeya, S. Bernstein, A. Cohn, M. W. Stone and B. Cashore (2013) 'Methods and Global Environmental Governance', *Annual Review of Environment and Resources*, 38, 11.1–11.31.

Ostrom, E. (1990) *Governing the Commons* (Cambridge, UK: Cambridge University Press).

Paterson, M., M. Hoffmann, M. M. Betsill and S. Bernstein (2014) 'The Micro Foundations of Policy Diffusion towards Complex Global Governance: An Analysis of the Transnational Carbon Emission Trading Network', *Comparative Political Studies*, 47, 420–449.

Perkins, R. and E. Neumeyer (2009) 'Transnational Linkages and the Spillover of Environmental Protection into Developing Countries', *Global Environmental Change*, 19, 3, 375–383.

Punch, M. (1986) *The Politics and Ethics of Fieldwork* (Newbury Park and London: Sage).

Roberts, J. T. (1996) 'Predicting Participation in Environmental Treaties: A World-System Analysis', *Sociological Inquiry*, 66, 1, 38–57.

Roberts, J. T. (2011) 'Multipolarity and the New World (Dis)Order: US Hegemonic Decline and the Fragmentation of the Global Climate Regime', *Global Environmental Change*, 21, 3, 776–784.

Roberts, J. T. and P. E. Grimes (2002) 'World-System Theory and the Environment: Toward a New Synthesis', in R. E. Dunlap et al. (eds) *Sociological Theory and the Environment: Classical Foundations, Contemporary Insights* (Lanham, MD: Rowman and Littlefield).

Schimmer, R., R. Geerken and B. Kiernan 'Tracking the Genocide in Darfur: Populuation Displacement as Recorded by Remote Sensing', *Genocide Studies Working Paper No. 36* Yale University Genocide Studies Program, http://www.moncamer.ch/DRDC2/images/DRDC/StudiesAndResearch/Tracking_the_Genocide_in_Darfur.pdf

Schwartz, D. M., T. Deligiannis and T. F. Homer-Dixon (2000) 'The Environment and Violent Conflict: A Response to Gleditsch's Critique and Some Suggestions for Future Research', *Environmental Change and Security Project Report*, 6, 77–94.

Sharma, B. and N. Jangle (2011) 'Assessing Vulnerability to Climate Change Using GIS', *Geospatial Media and Communications*, http://www.geospatialworld.net/Regions/ArticleView.aspx?aid=30395

Sil, R. and P. J. Katzenstein (2010) 'Analytic Eclecticism in the Study of World Politics: Reconfiguring Problems and Mechanisms Across Research Traditions', *Perspectives on Politics*, 8, 2, 411–431.

Small, C., C. Elvidge, D. Balk and M. Montgomery (2011) 'Spatial Scaling of Stable Night Lights', *Remote Sensing of Environment*, 113, 269–280.

Small, M. L. (2011) 'How to Conduct a Mixed Methods Study: Recent Trends in a Rapidly Growing Literature', *Annual Review of Sociology* 37, 55–84.

Spivak, G. C. (1985) 'The Rani of Sirmur: An Essay in Reading the Archives', *History and Theory*, 24, 347–372.

Sprinz, D. F. (2004) 'Environment Meets Statistics: Quantitative Analysis of International Environmental Policy', in D. Sprinz and Y. N. Wolinsky-Nahmias (eds) *Models, Numbers, and Cases: Methods for Studying International Relations* (Ann Arbor: University of Michigan Press), 177–192.

Steinberg, P. (2007) 'Causal Assessment in Small-N Policy Studies', *Policy Studies Journal*, 35, 2, 181–204.

Stevis, D. and V. J. Assetto (eds) (2001) *The International Political Economy of the Environment: Critical Perspectives* (Boulder: Lynne Rienner).

Sui, D. and D. DeLyser (2012) 'Crossing the Qualitative-Quantitative Chasm I: Hybrid Geographies, the Spatial Turn, and Volunteered Geographic Information (VGI)', *Progress in Human Geography*, 36, 1, 111–124.

Tashakkori, A. and C. Teddlie (2010) 'The Past and Future of Mixed Methods Research: From Triangulation to Mixed Model Design', in A. Tashakkori and C. Teddlie (eds) *Handbook of Mixed Methods in Social and Behavioral Research*, second edition (Thousand Oaks, CA: SAGE), 671–701.

Taylor, S. J. and R. Bogdan (1998) *Introduction to Qualitative Research Methods: A Guidebook and Resource* 3rd ed. (New York: Wiley).

Ter Wal, A. L. J. and R. A. Boschma (2009) 'Applying Social Network Analysis in Economic Geography: Framing Some Key Analytic Issues', *Annals of Regional Science*, 43, 3, 739–756.

Torrens, P. (2010) 'Geography and Computational Social Sciences', *GeoJournal*, 75, 133–148.

Urpelainen, J. (2013) 'Promoting International Environmental Cooperation through Unilateral Action: When Can Trade Sanctions Help?' *Global Environmental Politics*, 13, 2, 26–45.

Wapner, P. (2008) 'The Importance of Critical Environmental Studies in the New Environmentalism', *Global Environmental Politics*, 8, 1, 6–13.

Young, O. R. (ed.) (1999) *The Effectiveness of International Environmental Regimes: Causal Connections and Behavioral Mechanisms* (Cambridge, MA: The MIT Press).

Young, O. R., E. F. Lambin, F. Alcock, H. Haberl, S. I. Karlsson, W. J. McConnell, T. Myint, C. Pahl-Wostl, C. Polsky, P. S. Ramakrishnan, H. Schroeder, M. Scouvart and P. H. Verburg (2006) 'A Portfolio Approach to Analyzing Complex Human-Environment Interactions: Institutions and Land Change', *Ecology and Society*, 11, 2, 31–45.

Part II
Major Research Areas

5
International Political Economy and the Environment

Jennifer Clapp

In this era of economic globalization, there has been remarkable growth in the volume and value of global trade, investment, and finance. These international economic relationships have important implications for the natural environment, as all have been identified as having some linkage to environmental quality. The extent to which these international economic relationships contribute to environmental problems or to solutions for environmental problems is the subject of extensive debate (see, for example, Boyce, 2008; Gallagher, 2009; Clapp and Dauvergne, 2011; Newell, 2012; Clapp and Helleiner, 2012). Some see the relationship as largely positive, with environmental benefits being attached to the economic growth that global economic transactions seek to facilitate. For these thinkers, environmental policies should be able to address any negative outcomes that may arise in ways that do not impede global economic activity. Others, however, see mainly negative environmental implications arising from global economic relationships and the economic growth that is associated with it. For them, it is important that environmental policies do restrict global economic transactions. A third view which seeks to bridge the divide is also gaining prominence, arguing that while there are some potential negative aspects of global economic relations for the environment, a balanced management of the global economy can bring both economic and environmental benefits.

This chapter outlines the relationship between the international political economy (IPE) and the global environment, as well as the debates that surround that relationship. The IPE–environment interface is multifaceted and complex. It encompasses the linkages between global economic interactions and the emergence of environmental problems on the one hand, as well as the global economic institutions and actors that implement policies to address those problems through international governance mechanisms on the other (Clapp and Helleiner, 2012). In some cases, some of the same actors (for example, transnational corporations or TNCs) are involved in both economic activities that may contribute to environmental harm and in shaping the governance of global economic institutions that seek to regulate those activities. This can lead to conflicts of interest and also highlights the importance of power relationships in the study of the interface between IPE and the environment.

Tracing the roots of debate over IPE and the environment

It was in the early 1990s that debates over the global political economy and the environment erupted in full force. This was in large part a product of concerns over trade policy and its potential impact on the environment at that time, especially the negotiation of the North American Free Trade Agreement (NAFTA) and the General Agreement on Tariffs and Trade (GATT) Tuna-Dolphin challenge (Williams, 2001; Esty, 1994). There was concern that trade and investment interests would override environmental considerations and these cases brought the question to the fore. The result was a large volume of literature on the topic, from a variety of viewpoints.

The debate over the relationship between the IPE and the environment overlaps to some extent but is not entirely encompassed within traditional debates in the field of International Relations (IR) (see Stevis, this book). In other words, the roots of various positions in both policy and theory on the links between trade, investment, and finance, on the one hand, and environment, on the other, originate in a number of fields apart from IR. These include, most importantly, ecology and economics. Paterson and Stevis (both in this book) recognize the need to look beyond the three 'traditional' camps in IR when considering global environmental politics. When examining debates about IPE and the environment, we could also add the neoclassical economic view, which has been very active on the issue. I argue that there are three main 'camps' in the debate over IPE and the environment. I explain these below using Paterson's categories with the addition of neoclassical economics.

Neoclassical economists are a dominant voice in both policy and academic debates, and tend to see the expansion of global trade, investment, and finance as on the whole positive for the quality of the natural environment. This position derives from a fundamental assumption taken by neoclassical economic thinking, that of comparative advantage, which dates back to the writings of David Ricardo and assumes that international trade benefits all partners in material terms. In today's global economy, neoclassical economists also argue that transnational investment and financing are positive for the material gain of nations, as they are seen mainly as activities to facilitate trade and enhance economic growth. Neoclassical economists have accommodated environmental concerns within this view by arguing that material gain that arises from global economic interactions can be used to finance environmental improvements. To back up this argument, these thinkers rely on evidence from studies which show that an inverted U relationship exists in OECD countries between income growth and a cleaner environment (also known as the Environmental Kuznets Curve, or EKC). In other words, as incomes rise, environmental problems may initially get worse, but then improve as incomes rise further. The rationale is that a wealthier population will demand a cleaner environment, that governments will respond by enacting stricter environmental laws, and that firms react by introducing 'greener' products (Grossman and Krueger, 1995; Dinda, 2004; Van Alstine and Neumayer, 2008). From this perspective, the global political economy and the environment are mutually supportive, such that growth can go on indefinitely, and environmental improvements will result. For these neoclassical economic thinkers, the liberalization of trade, investment, and finance lead to increased global economic integration, and in turn generate more wealth with which to protect the environment.

Other views have argued vigorously that the global political economy and environment are not mutually supportive. Ecological economists and many radical thinkers have been very vocal critics of the neoclassical economic position. This critical camp includes thinkers drawing on the ecoauthoritarian, sustainability analysis, structural conflict, and accumulation schools outlined by Paterson (this book). The thinkers from each of these schools have a deep skepticism about the impact of economic growth on the environment, as well as skepticism of the actors promoting it, and thus do not buy into the EKC argument presented by neoclassical economists (Arrow et al., 1995; Stern, 2004). They argue instead that the liberalization of trade, investment, and finance, all aimed to increase economic activity, will have disastrous results for the natural environment. Economic growth is based on nature, for both resources as inputs into industry and sinks for the wastes that result, and nature is not limitless. Economic growth, they argue, has in fact already surpassed its sustainable limit. Ecological economists use the laws of thermodynamics to demonstrate that these limits are real and are being surpassed (Daly, 1996; Daly and Farley, 2010). Some have argued that we have already experienced a 'peak globalization' and that a relocalization will become a necessity (Curtis, 2008). Others in this camp also focus on the ways in which global economic relationships create or perpetuate inequalities that have negative environmental and social implications (Muradian and Martinez Alier, 2001; Newell, 2012; Unmüßig et al., 2012). These thinkers have called for 'de-growth' in the industrialized world as a first step toward rectifying these problems (Kallis et al., 2012).

These two camps have tended to dominate the debate over the global political economy and the environment over the past decade, but a third view has gained prominence in this period. This third view originates in large part from what Paterson calls the liberal institutionalist school within traditional IR and also includes ecological modernization thinkers. Thinkers in these schools of thought argue that in some cases common ground can be found between neoclassical economists and their critics when it comes to the environment (Neumayer, 2001; Gallagher, 2009). Institutionalists, who focus on structured cooperation between states, argue that it is possible to establish strong rules to govern the global economy in ways that ensure that the environment does not suffer while also reaping benefits from economic globalization. In making this argument, institutionalists largely agree with neoclassical economists about the possibility that the growth of the global economy can have positive impacts on the environment, and they also agree with the critics that in some instances they are not mutually supportive. Thus, their policy advice is to create global rules to avoid the cases where serious harm occurs, and encourage those where positive outcomes are likely. Likewise, ecological modernization thinkers are optimistic that governance arrangements can be made in a way that encourages a decoupling of environmental harm from economic growth (Mol, 2001). Both of these middle-ground approaches fed into Green Economy thinking, pushed by the United Nations Environment Program (UNEP) for Rio + 20 in 2012 (UNEP, 2011).

Each of these three camps in the broader debate over the global political economy and the environment has also engaged in the more specific debates over trade, investment, and finance and their relationship to the environment.[1] The theoretical and policy debates in each of these areas of the global political economy are discussed below.

Trade and the environment

International trade is a very important part of the global political economy. World trade has expanded greatly in recent years, growing from 25 percent of global GDP in 1960 to 57 percent in 2006. During the 1950–2008 period, there was a 32-fold increase in global exports of goods. There has also been a massive increase in the value of world trade, from US$58 million in 1948 to over US$18 trillion in 2011 (WTO data, see www.wto.org). Much of the literature on trade and the environment tends to engage either with the debates over the environmental merits and demerits of international trade or with debates over the way in which environmental issues should or should not be incorporated into international trade governance and vice versa.

The impact of trade on the environment

There has been a long-standing and heated debate over the impact of trade on the environment (for overviews, see Esty, 1994; Neumayer, 2001; Gallagher, 2008). Neoclassical thinkers have put forward several arguments to back their view that trade's impact on the environment is positive overall. A principal argument made by those taking this view hinges on the relationship between trade and economic growth, as noted above. Neoclassical thinkers further stress that trade has positive environmental implications related to efficiency. Advocates of free trade argue that more efficient allocation of scarce resources results from a policy based on the specialization of production and trade compared to a policy based on national self-sufficiency. Further, trade barriers themselves create distortions that result in inefficiencies (Bhagwati, 1993). Trade restrictions such as tariffs, quotas, and export subsidies, this view argues, can lead to the domestic underpricing of resources, which encourages their overuse. Again, this works against the goals of efficiency and conservation. From this perspective, then, the key is better environmental policy, not restriction of trade.

Taking issue with these arguments are ecological economists and other critical thinkers, who argue that the assumptions about trade on which the liberal economic view is based are fundamentally flawed. From their perspective, economic growth that should result from trade is itself a large part of the environmental problem. This is because it results in more 'throughput' in the economy, ultimately resulting in greater environmental degradation. In other words, more physical materials from the earth are used in production, and additional waste is created (Daly and Farley, 2010). Any efficiency gains that arise from trade, then, are outstripped by this growth, ultimately destroying the planet (W. Sachs, 1999; Jackson, 2009).

A number of ecological economists point to the changing 'social metabolism' of an economy, meaning the material and energy flows that cycle through the economy to support its activities, as evidence of the damaging impact of throughput. These economists argue that a kind of 'metabolic transition' is underway in many countries, as their economies shift from biomass energy sources to fossil fuels. This transition, which dramatically increases both fossil energy use and waste, typically occurs as economies industrialize, a process prompted by global trade and investment (Krausmann et al., 2008; Haberl et al., 2011). India, China, and Brazil, for example, have seen rising consumption of fossil energy and mineral resources linked to their rising industrial activity in the global economy (Harris, 2011; Singh et al., 2012). Industrialized countries have historically been able to consume high levels of materials and energy by acquiring them through international trade, thereby casting ecological shadows on other parts of the world (Dauvergne, 2008). Similarly, the changing social metabolism of rising Southern countries also has global environmental repercussions. Indeed, as South–South trade and investment increases, there has been concern about the environmental implications of rising powers' global search for land, energy, and minerals (Hochstetler, 2013).

Critics also identify numerous other problems with trade. Specialization for trade purposes may not only distribute pollution unevenly but also redistribute natural resources globally, both of which can result in unsustainable and ecologically unequal exchange (Roberts et al., 2007; Roberts and Parks, 2009). Pockets of the developing world, for example, produce a greater amount of toxic products and export their natural resources more than would be the case without trade (W. Sachs, 1999; Clapp, 2001). Critics also point to the development of a 'race to the bottom'. Rather than environmental standards rising with trade liberalization, some countries fear raising standards, and in fact may lower them, in a bid to attract investment and improve their trade competitiveness (Daly, 1993; Porter, 1999). Additionally, growth in trade inevitably means more pollution from transportation of goods around the world (Curtis, 2008; Daly and Farley, 2010; Newell, 2012: 66). Policies that regulate trade on environmental grounds are a perfectly legitimate policy response for these thinkers.

While the debate between free traders and anti-free trade critics with respect to the environment has been polarized for years, the literature seeking to forge a middle ground is becoming increasingly prominent.

This literature points out the commonality between the two extremes to map out scenarios when trade can be beneficial for the environment. It also proposes global governance mechanisms to achieve this goal (for example, Weinstein and Charnovitz, 2001; Esty, 2001; Neumayer, 2001; Charnovitz, 2007). These thinkers argue, for example, that there are certain obvious cases where free trade should not be encouraged, such as the trade in toxic substances and dangerous chemicals. These problems, they argue, can be managed effectively through trade measures incorporated into environmental agreements and environmental measures incorporated into trade agreements. Yet, at the same time they do see a role for trade liberalization when clear efficiency gains will result, for example, the removal of subsidies, such that resources are managed more effectively.

Global governance for trade and environment

The theoretical debate over trade and environment as discussed above is extremely important because it underpins debates on global governance over trade and environment issues. The mechanisms of global governance that deal with trade and environment issues include international agreements on trade as well as international environmental agreements, as discussed below.

International trade agreements have historically paid little attention to the interface between trade and environment. The GATT, first negotiated in 1947, made no explicit mention of 'environment'. In 1995 the WTO replaced the GATT Secretariat, and the new organization now oversees a series of trade agreements, of which GATT is just one. Although the WTO preamble does stress the need to promote 'sustainable development', there is no specific language in the WTO agreements allowing exemptions to trade rules in the name of environmental protection. GATT rules do, however, stress that countries cannot discriminate against products based on country of origin or how they are produced. What this means is that countries cannot apply trade restrictions on goods based on process and production methods (PPMs) if the products are otherwise identical. This rule keeps countries from applying trade restrictions on goods that are produced in ways that are known to be environmentally damaging (Esty, 1994: 49–51).

Only Article XX of the GATT has the potential to allow for trade restrictions on environmental grounds (Charnovitz, 2008). Article XX sets out circumstances where states are eligible for exceptions to the GATT rules and include measures undertaken to protect human, animal, or plant life or health, or to ensure the conservation of natural resources. But these exceptions are qualified. Such measures must be shown to be 'necessary' before they can be exempted from GATT rules and typically must be accompanied by equally strict domestic measures. Moreover, Article XX says little about global environmental problems. It is thus very difficult for states to be exempted from GATT rules for measures taken to address environmental issues outside of their borders or that affect the global commons (see Esty, 1994; Neumayer, 2001). Given all of these qualifications to Article XX, it should not be surprising that its use has been very limited in practice. Several states have tried to justify trade restrictions for environmental purposes by claiming exemption under Article XX (for example, the Tuna-Dolphin disputes in 1991 and 1994). Attempts to invoke Article XX have typically been struck down by the GATT dispute panels on the grounds that multilateral solutions should have been available (Vogel, 2000). Some say that the WTO is not inherently anti-environment, however, and that recent rulings show that it is making efforts to incorporate environmental issues more fully (DeSombre and Barkin, 2002; Charnovitz, 2007).

The other main governance area in which the interface between trade and environment is prominent is in multilateral environmental agreements (MEAs). About 20 of the 250 or so MEAs specifically incorporate trade provisions, indicating that parties to those agreements felt that the best way to address those environmental issues was with some sort of restriction on trade (Clapp and Dauvergne 2011: 152). These include, for example, the Basel Convention (hazardous wastes), the Cartagena Protocol (biosafety), the Kyoto Protocol (climate change), the Montreal Protocol (ozone layer), the Rotterdam Convention (pesticides), and the Stockholm Convention (persistent organic pollutants). The rules that restrict trade vary according to the agreement and the issue at hand. Some restrict trade in dangerous items, while others restrict trade between parties and non-parties of items they seek to control as a mechanism to encourage countries to sign on to those agreements. Many of these MEAs also include other control measures that may potentially affect trade, such as provisions for technology transfer and prior informed consent (see Stilwell and Tarasofsky, 2001).

Although these environmental agreements have incorporated trade provisions, it is not clear which takes legal precedence: trade law or environmental law. In 1994 the GATT/WTO constituted a trade and environment committee to discuss, among other issues, the relationship of global trade rules to MEAs and in 2001 trade ministers agreed to clarify the relationship between WTO rules and MEAs at the start of the Doha Round of trade talks. Although the need to clear up this issue was reinforced at the World Summit on Sustainable Development in 2002, there is still a lack of clarity. This was made evident in a dispute at the WTO over the European Union's (EU) de facto moratorium on genetically modified organisms (1998–2003) which was challenged by the United States, Canada, and Argentina as contravening various provisions of the GATT. Although the EU claimed it was merely exercising precaution, as it is allowed to do as a signatory to the Biosafety Protocol, the United States, as a key challenger, chose to take up the matter at the WTO instead. The WTO ruling came down on the side of the challengers on many counts, and in the end the EU ended its moratorium (Lieberman and Gray, 2008). It is somewhat ironic that early concerns centered on whether international environmental agreements might contravene WTO rules, rather than the other way around. Some have argued for a World Environment Organization (WEO) to counter the power of the WTO (Biermann, 2000, this book; Barkin, 2008), but as yet this idea has not been taken up at the international level.

Regional trade agreements and organizations are also important mechanisms of governance that touch on trade and environment issues. Unlike the GATT and WTO, the NAFTA, negotiated in the early 1990s, does explicitly attempt to incorporate environmental concerns not just in its preamble but also directly into the main text of the agreement. It does this by mentioning specific international environmental treaties that should take precedence over trade rules incorporated into the agreement, provided they are carried out in the least trade-distorting manner. The treaties mentioned include the Convention on the International Trade in Endangered Species, the Basel Convention, and the Montreal Protocol, as well as four bilateral treaties (Soloway, 2002). The NAFTA also established an environmental side agreement, the North American Agreement on Environmental Cooperation (NAAEC), which aims to ensure that states comply with and enforce their own national environmental laws, and it establishes a mechanism for settling environmental disputes. A Commission on Environmental Cooperation (CEC) was also established to oversee the NAAEC. This body allows citizen input into reporting of international environmental violations (Barkin, 2008). Despite the efforts to head off conflicts between trade and the environment within NAFTA, the results remain controversial, with critics arguing that the outcomes are at best ambiguous (Wise, 2008).

The EU is another important regional organization grappling with both trade and environmental issues. The EU is different from the NAFTA in that it promotes both political and economic integration, including the harmonization of environmental laws (Selin and VanDeveer, 2008). Since the 1980s the EU has developed and adopted EU-wide policies for environmental protection, including those regarding waste, air pollution, water, nature protection, and climate change. These laws were developed in a parallel process to the EU's economic integration and are thus not widely seen to be explicitly tied to trade performance (Stevis and Mumme, 2000: 31). Some EU states, however, primarily those with weaker environmental standards to begin with, such as Greece, Portugal, Spain, and Italy, were more hesitant than others to adopt EU-wide environmental policies because of their concern over the economic implications of more stringent environmental regulations. It is for this reason that there is flexibility built into the EU's environmental regulations to allow for some degree of difference in requirements according to member states' economic and environmental situations (Selin and VanDeveer, 2008: 197). Despite these differentials, however, the EU is widely seen to be a successful case of upward harmonization of environmental laws within the context of enhanced economic integration.

Transnational investment and the environment

There has been extensive growth in both the number of TNCs and the amount of foreign direct investment (FDI) in the past few decades. In 2007 there were over 79,000 TNC parent firms globally, up from just 7,000 in 1970 and some 850,000 foreign affiliate firms – that is corporations affiliated with a TNC – operating around the world. FDI flows have also grown in this period. In 1970 the level of FDI inflows stood at US$9.2 billion, and by 2011 this figure was US$1.5 trillion (Clapp and Dauvergne, 2011: 163–165). With such a significant weight in the global economy, it is important to examine the impact of TNCs and FDI on the environment. The literature on transnational investment and the environment has thus far focused mainly on debates over whether firms relocate to take advantage of differential environmental standards, and also on debates over whether firms' voluntary 'greening' has been effective.

Environmental standards and transnational investment
Since the 1970s there has been heated debate over whether international investment, particularly FDI undertaken by TNCs, is negative or positive for the environment. Those from the critical camp argue that TNCs invest most heavily in jurisdictions where environmental regulations

are weakest (Karliner, 1997; Korten, 1995). This kind of behavior leads to several phenomena. First, it can trigger 'industrial flight' from countries that raise environmental standards. Second, it can lead to 'pollution havens' when some countries, primarily developing ones, set out to lower their environmental standards in a deliberate attempt to attract foreign investment. Not unrelated to the first two, we may see 'double standards' appear where different branches of the same TNC have different sets of environmental standards depending on where they are operating (Clapp and Dauvergne, 2011). For these thinkers, this only contributes to the race to the bottom, whereby states lower standards not just to improve trade competitiveness but also to attract FDI.

Most neoclassical economists do not see the environmental impact of TNCs as being a significant enough issue to worry about. They argue that different environmental standards in different countries are normal and part of a country's own capacity to absorb pollution. Proof of industrial flight and pollution havens for them is elusive, as statistical studies based on aggregate data on pollution costs and investment patterns have failed to conclusively show that they exist (Copeland, 2008). TNCs, these thinkers argue, in fact help to raise environmental performance in developing countries by transferring cleaner, state-of-the-art technologies, compared with local firms that use outdated and more polluting methods. This transfer of cleaner technology, they argue, results in a 'pollution halo' effect (Zarsky 1999; Zeng and Eastin, 2007).

Scholars on both sides of the debate have stressed that we need to open up this debate and move beyond the narrow focus on pollution havens and halos to consider the broader issues linking foreign investment and environmental impact, including in sectors outside of manufacturing, such as forestry and mining (Wheeler, 2002; Hall, 2002). Site practices of TNCs, regardless of whether they located their operations to take advantage of regulatory differentials, is also a concern in a number of sectors, including the oil extraction and electronics industries (Clapp and Dauvergne, 2011: 174–179). There has also been some suggestion from the middle-ground thinkers that policy actions, even in the absence of widely agreed evidence that pollution havens exist, would be beneficial, and could be acceptable from both sides of this debate (Neumayer, 2001). A focus on a different concept, the 'regulatory chill', is helpful here. A regulatory chill refers to situations where states fail to raise environmental standards for fear of losing investment and weakening trade competitiveness. Whether such outcomes would result from raising standards is not the point. What matters is that states act in certain ways based on the belief that they may lose investment if they impose stricter standards (Porter, 1999; Zarsky, 2006). If environmental regulations stay static, then we will see a continuation of poor-quality regulations and an entrenchment of regulatory differences between countries. Aiming policy initiatives to ward against such a regulatory chill would be an easy start toward improving the situation.

TNCs and global environmental governance

There is much discussion on the role TNCs play in the formation of global environmental governance (see also Betsill, this book; Biermann, this book). TNCs have operated through a number of channels in their bid to take part in the negotiation of global environmental governance (Clapp and Meckling, 2013; Levy and Newell, 2005). Lobbying the state at the domestic level before state delegations head to international environmental negotiations has traditionally been a key strategy for industry. In this way they are able to exert significant influence over governments' positions from behind the scenes through lobbying activities at the national level. Business players are also increasingly lobbying at the international level as well, via industry associations and industry representatives who gain observer status at such meetings. The presence of these actors at global negotiations is now a regular feature, as we have seen in the case of the waste trade, climate change, ozone depletion, biosafety, and POPs.

Corporate lobby groups have also been active in other global forums for sustainable development, such as the Rio Earth Summit in 1992, the World Summit on Sustainable Development (WSSD) in 2002, and the Rio + 20 UN Conference on Sustainable Development in 2012. At the 1992 Rio Earth Summit, industry groups formed the Business Council on Sustainable Development as a coordinated voice for industry. Similarly, at WSSD, and Rio + 20, industry groups established the Business Action for Sustainable Development (BASD) to form a single industry voice (Clapp, 2005; see BASD website http://basd2012.org). At these major international meetings on sustainable development, the main message from industry was that voluntary initiatives, rather than outside regulation of TNCs, would be the most effective and efficient way to promote sustainable development.

A more diffuse but no less important way that TNCs have influenced global environmental governance is via their 'structural power', stressed by the critical camp (Fuchs, 2005). They exert this power by carving out a role for themselves in terms of defining 'sustainable development' (Sklair, 2001). In defining the mainstream version of sustainable development in a way that enables them to maintain goals of economic globalization, including continued openness to global investment and faith in industry efforts to save the environment, they can escape much regulation. The structural power of capital – nationally and globally – has been seen by some to be key in terms of explaining the influence that industry does indeed seem to have over government policy more broadly (Levy and Newell, 2005). Some argue that the current era of increased global economic competition has meant that many states have pursued domestic policy outcomes that would be acceptable to corporations in order to keep or attract investment in their country. This trend is reinforced by the emergence of a system of investor protection within a growing number of bilateral investment agreements that has enabled corporations to make legal claims against states that alter environmental regulations in ways that impose new costs on those firms (Van Harten, 2013). In such circumstances, governments are likely to be reluctant to tighten environmental regulations.

Transnational corporate actors also influence global environmental governance through other international forums. Private firms have been involved in recent years in the establishment of private forms of governance such as voluntary codes of environmental conduct at both the national and international levels (Auld et al., 2008). This includes participation in establishing industry-based environmental codes of conduct, such as the ISO 14000 and Responsible Care. The ISO 14000 environmental management standards is an interesting case in this regard, as it is the most widely recognized set of global standards of this sort (Clapp, 1998a). But debates continue as to whether these standards will really make a difference to firms' environmental behavior, whether they will help improve environmental practice among firms in developing countries, and whether the drafting process was democratic and legitimate (Prakash and Potoski, 2006; Clapp, 2008). For example, because the ISO 14000 standards only ask firms to abide by the environmental laws in the country in which they operate, they do little to raise standards. Moreover, the standards do not include performance criteria, but only changes to management practices. While they may not be the strongest in terms of leading to environmental improvements, the ISO 14000 environmental management standards have been recognized as legitimate standards by the WTO.

It is often assumed that business players generally oppose strong global environmental rules because they impose costs on firms, but deciphering the business position on a particular environmental issue is not always so straightforward, and sometimes business actors are in conflict with one another over regulatory approaches to environmental issues

(Falkner, 2008; Newell, 2012). In some cases corporate actors push for weak global environmental rules, but in some cases they are content to go along with strong rules pushed for by NGOs and states. Economic considerations are often key to understanding the positions taken by corporate actors, though uncovering these motivations is often complex (Falkner, 2008). In the cases of ozone-depleting substances and persistent organic pollutants, industry actors have largely been united in favor of strict rules calling for a ban on the production and trade of these harmful substances, largely because these industries can gain economically from the sale of substitutes. But in the case of waste recycling and biosafety, the entrenched industries' chances at gaining from substitutes are slim, so they have a much stronger stake in opposing strong rules that they see as harming the very core of their industry (Clapp, 2003). In the case of climate change, industry response has been varied among different firms, indicating that individual firms have different economic and political interests in this issue (Rowlands, 2000; Falkner, 2008).

Global finance and the environment

There are numerous linkages between global finance and the environment. Although they may be less visible and therefore harder to trace than those of trade and investment, the effects are no less important. Global financing comprises many types of global transactions, including international development finance as well as private financial transactions. Development finance includes both bilateral (government to government) loans and grants, and financing from multilateral agencies. While development assistance as a whole has seen considerable 'greening' since the 1980s, some argue that bilateral aid has fared somewhat better in this regard than multilateral aid (Roberts et al., 2009). Recent years have also seen growing concern about the environmental implications of private international finance, including financial market transactions and private bank lending and investment. Financial markets have also become entwined in global environmental governance through the rise of carbon markets as well as financial investment products associated with offsets for biodiversity and conservation.

Multilateral finance

The largest multilateral lending agency is the World Bank, which in 2011 lent US$43 billion to developing countries to support both projects and policy initiatives (World Bank website: www.worldbank.org). Since the 1980s, environmental groups have critiqued World Bank lending to developing countries over concern that the projects being funded by the bank were not taking environmental considerations adequately into account, which resulted in a number of environmentally unsustainable projects. The World Bank's procedures for project design and evaluation had few if any references to the natural environment (Reed, 1997). Although all development assistance at the time had similar disregard for the environment, the World Bank became a focus of environmental groups because of its visibility and power to influence other sources of lending.

The World Bank's focus on lending for large-scale infrastructure projects – such as dams, power projects, and roads – as well as migration schemes and industrial agricultural projects was seen by critics to be causing a great deal of environmental harm. Although the World Bank had some environmental policies in place by the mid-1980s, often these policies were not followed (Wade, 1997: 634–637). Environmental groups followed the World Bank's environmental record closely, assisted in many cases by local groups in developing countries. Cases such as the Polonoreste road project in Brazil, which was linked by environmental groups to massive deforestation, as well as the Narmada Dams Scheme in India, which led to resettlement of thousands of people, were high-profile targets of these campaigns (Rich, 1994). The Environmental Defense Fund (now Environmental Defense) played a key role in the United States in terms of exposing the World Bank's record in an attempt to force it to adopt more environmentally friendly lending policies.

As a result of these campaigns, the World Bank began to make a specific effort to 'green' its project-lending policies by the late 1980s (Wade, 1997: 680). It undertook a major restructuring at this time and created a new Environment Department and increased its environment staff over the course of the early 1990s (Reed, 1997). Although the Bank began to 'green' itself at that time, critics noted that the newfound enthusiasm for the environment was tenuous (Fox and Brown, 1998). NGOs continued to call for more substantial reforms within the Bank, both in its environmental lending and consultation with stakeholders (Goldman, 2005). Environmental groups, for example, took on the World Bank by pointing out that the institution continued to provide loans in support of the construction of coal-fired power plants when at the same time the Bank recommended a reduction in reliance on fossil fuels and warned that the world's poor suffer disproportionately from climate change (Rich, 2009). NGOs also pointed out that the climate focus of the Bank's *World Development Report* for 2010 made no mention of the billions it had invested in fossil fuel projects in a host of developing countries, including India, South Africa, and Botswana. Over the 2005–2008 period, World Bank lending for fossil fuel-based projects nearly tripled to over US$4 billion, which directly contradicted a 2004 recommendation by the World Bank-commissioned Extractive Industries Review (EIR) that the Bank phase out lending for coal and oil by 2008 (Halifax Initiative, 2008).

The Global Environment Facility (GEF) is a multilateral funding channel for developing countries that is specifically geared toward projects with environmental benefits. Administered by the World Bank along with the United Nations Development Program (UNDP) and the UNEP, the GEF provides grant funding for developing countries that covers the 'incremental costs' of meeting their global environmental obligations under international environmental agreements. These 'incremental costs' are in effect the 'extra' costs incurred by developing countries to undertake projects that have global benefits (Streck, 2001: 73). The GEF initially only granted funds for international waters, climate change, ozone depletion, and biodiversity projects that have clear global benefits. It has since added efforts to reduce persistent organic pollutants and land degradation to its list. Environmental NGOs and developing countries were very critical of the GEF in its early years (Fairman, 1996). The World Bank dominated the GEF's operations, and there were complaints of minimal consultation on project design and implementation with affected local communities. The GEF was restructured to take these concerns into account (Streck, 2001), but critics remained skeptical (Horta et al., 2002).

Since 2006, a number of new funds, many with overlapping mandates, have emerged to provide multilateral support for climate change and have overshadowed the GEF. According to some critics, the effectiveness of this additional funding is likely to be compromised by the confusion from such a crowded funding landscape (Porter et al., 2008). In particular, there appears to be a shift away from the GEF and back toward the World Bank as the primary organization for provision of developing country financing to address climate change. Critics are concerned that new World Bank funds will operate without proper consultation and input from potential stakeholders, including recipient countries, civil society groups, and indigenous peoples. These new climate-focused funding mechanisms were also set up outside of the United Nations Framework Convention on Climate Change (UNFCCC) process, and their existence is perceived to undercut the multilateral process. The new climate funds also provide loans rather than grants, meaning that the poorest developing countries may not have adequate access to them since they require repayment. Critics have also pointed out inconsistencies in the focus of the funds. The Clean Technology Fund, for example, underplays investment in renewable energy technology by allowing investment in more efficient fossil fuel-based energy projects (Halifax Initiative, 2008; Porter et al., 2008).

Private international finance

Private international finance has enormous implications for the natural environment but has received far less attention than multilateral development financing in the global environmental politics literature (Helleiner, 2011). Part of the reason for the relative inattention to private international finance is the fact that it is much harder to track the financial transactions to the ultimate impact on the environment, and partly for this reason, there are fewer international governance mechanisms for holding financial actors accountable. The volume of private finance, however, is huge. Some US$3 trillion changes hands daily, dwarfing the value of global trade, investment, and multilateral finance. Many of the daily financial transactions are speculative investments in currencies, but global private finance also includes investment in stocks, bonds, and financial investment products known as derivatives. Large-scale investors, such as sovereign wealth funds, pension funds, hedge funds, and insurance companies, invest regularly in sectors with environmental implications, including fossil fuels, agricultural commodities, minerals, and real estate. Financial derivatives exist for all of these kinds of investments, including products such as commodity index funds that track prices of a bundle various commodities, from oil to minerals to livestock to food (Clapp and Helleiner, 2012).

As we witnessed in the 2007–2008 financial crisis, global financial markets are highly unstable and prone to bubbles and crashes as investors flock to certain investments and then retreat when conditions change. This instability in financial markets only reinforces short-term thinking among financial investors who are out to earn quick returns on their investments before markets turn. Critics of global financial markets have pointed out that the short-term mentality means that in effect investors have more incentive to cut down a forest and invest the money from its sale into financial investment products with high immediate returns than they do to invest in a system that ensures that the forest will be harvested in a sustainable manner over a number of years. Because this short-term mentality prevails, sustainable investment practices tend to be discriminated against by financial markets

(Schmidheiney and Zorraquin, 1996; Helleiner, 2011; Thistlethwaite, 2011; Bracking, 2012).

Growing concern about the short-termism of financial markets and its potential environmental effects has led to calls for international governance frameworks to ensure that firms following sustainable practices are rewarded within the financial system rather than driven out of business for taking on environmental measures (Thistlethwaite, 2011). As early as 1992, the UN Environment Program founded its Finance Initiative to engage banks and other financial institutions directly in discussions promoting environmentally sound financial investment (UNEP FI website: http://www.unepfi.org/). The UNEP Finance Initiative works with over 200 financial institutions that have signed on to several voluntary statements that promote positive financial rewards for sustainable investments. Because these measures are voluntary, they serve mainly to raise awareness among financial institutions, and it is difficult to say with certainty what their overall influence has been.

A number of other initiatives aimed at sustainable banking and financial investment have also emerged over the past decade. The Equator Principles, launched in 2003, is a voluntary code of conduct that governs project financing from private sector banks. Banks that adhere to these principles pledge to withhold funds for projects in developing countries that may cause environmental or social harm (Wright, 2012). As of 2013, some 78 financial institutions have signed onto these principles (Equator Principles website: http://www.equator-principles.com/). Critics of the Equator Principles charge that the principles are too weak and lack transparency and are limited only to project finance, when in fact the world of global finance is much broader (Missbach, 2004). The Principles for Responsible Investment (PRI) was established in 2006 as an initiative of the UNEP FI and the UN Global Compact. The PRI aims to influence large-scale investors by asking them to voluntarily incorporate environmental, social and governance issues into their investment decision-making. In 2012 the PRI had over 1000 signatories across 50 countries with assets of over $30 trillion (PRI website: http://www.unpri.org/). Similarly, the Carbon Disclosure Project (CDP) aims to enable financial investors to discriminate between firms based on their efforts to address climate change. The CDP seeks to produce a standardized metric for carbon emission reporting by firms and provides those reports to investors in the hope that it will shape their investment decisions toward low-emitting firms. Some 5000 firms currently provide data to CDP (CDP website: https://www.cdproject.net/). Critics charge that the CDP is weak, however, in that it is voluntary and lacks an enforcement mechanism for disclosure (Kolk et al., 2008; Clapp and Thistlethwaite, 2012).

Alongside efforts to green finance through governance initiatives aimed at financial institutions, the rise of carbon trading as a key governance mechanism for climate change through the Clean Development Mechanism, the European Emissions Trading Scheme, and voluntary carbon markets (including carbon offset markets) has put a great deal of faith in the power of private finance to create and trade financial derivatives based on carbon reduction (Newell and Paterson, 2010). With the failure to put in place a post-Kyoto agreement, however, and in the wake of the financial crisis, these markets have flagged in recent years, raising serious questions about the viability of a kind of 'climate capitalism' to save the day (Lohmann, 2010; Newell and Paterson, 2010). In addition to the establishment of carbon markets, there are now emergent markets for biodiversity and other conservation-based offsets and derivatives. These developments have given rise to concern over the broader implications of the 'financialization' of nature, which might only value certain kinds of nature and its services with possibly perverse outcomes (Sullivan, 2013).

As these various initiatives to 'green' global finance have emerged in the international governance landscape, a broader debate has emerged over whether this kind of tinkering with financial markets in the broader capitalist world economy will be enough to make a difference for the environment (Newell and Paterson, 2010). Advocates see such initiatives as a positive way to reorient the flow of money toward sustainable activities within the current global economy, while critics are deeply skeptical of such voluntary measures given the complexity and lack of transparency in financial markets.

Case study: International transfer of hazardous wastes from rich to poor countries

The debates regarding the global political economy and the environment are large and multifaceted. Trade, investment, and financial issues overlap in many ways, and in looking at a specific environmental problem, it is important to keep this in mind. I have chosen here to focus on the problem of the international transfer of hazardous waste from rich to poor countries. This case study is useful in that it is closely linked to various global economic relationships in different ways (Clapp, 2001). It highlights questions of trade, investment, and to a lesser extent financing issues.

Trade is the most explicit link to the transfer of hazardous wastes from rich to poor countries. It became apparent in the 1980s that toxic wastes from industrial countries were being shipped to a number of developing countries. Africa, Asia, and Latin America were all affected by this practice. So the problem was identified as a trade issue. But the problem has ties to global finance, as it was the indebtedness of these countries in the 1980s that led them to be targets for hazardous waste, as accepting it brought desperately needed sources of foreign exchange. But while it was initially seen by some in these countries as a means to earn their way out of debt, it was soon recognized that the price was simply not worth it. Moreover, public outcry in both rich and poor countries over this practice exploded in the 1980s.

In the face of public criticism once this practice came to light, the international community negotiated the *Basel Convention on the Transboundary Movement of Hazardous Wastes and their Disposal* in 1989. This MEA seeks to control the international trade in hazardous waste by only allowing the trade to occur with prior notification and consent. But it did not ban the trade in hazardous wastes outright. Regional- and national-level laws also emerged, many of which were more stringent than the provisions laid out in the Basel Convention. The rules of the Basel Convention enabled many developing countries to refuse waste imports, and by the mid- to late 1990s the blatant dumping of the 1980s had dwindled to a trickle. But a new problem appeared on the scene with respect to the waste trade. This was the export of wastes destined for recycling operations in the developing world. Although wastes exported for recycling are covered by the rules of the Basel Convention, it was difficult to enforce the prior notification and consent procedure when the wastes were labeled as 'products' to be recycled rather than as wastes to be dumped. But the recycling of hazardous wastes in poor countries is more often than not carried out in environmentally unsound conditions, which many consider to be as bad if not worse than simple disposal of the wastes (Puckett, 1994; Clapp, 2011).

Two kinds of recycling of toxic wastes emerged in particular by the late 1990s. One was the export of decommissioned ships containing toxic materials to developing countries (Clapp, 2011). Most of these end-of-life ships were exported from rich industrialized countries to countries such as India and Bangladesh, where they were broken down and their reusable portions recycled. The problem is that these older ships often contain highly toxic materials such as polychlorinated biphenyls (PCBs) and asbestos. Because the ships were technically in use when exported and often not designated as waste material until they reached their final destination, they frequently escaped control under the rules of the Basel Convention (Kanthak, 1999). The export of electronic waste (e-waste), including discarded computers, mobile phones, and other electronic equipment that contain toxic materials, has also been on the rise since the late 1990s. Rapid technological change and innovation has fueled the growth in e-waste trade, and developing countries in Asia and Africa have seen imports of e-waste rise despite bans in those countries on the import of toxic waste (Iles, 2004; Selin and VanDeveer, 2006; Clapp, 2011). E-waste often is able to skirt the rules of the Basel Convention because it is frequently exported under the pretense of reuse, when in practice they are mined for parts or simply sent to landfills. The conditions under which these e-wastes are recycled and disposed in these locations are typically very dangerous and environmentally unsound (BAN, 2002, 2005).

The surfacing of the recycling problem in the 1990s prompted environmental NGOs to call for changes to the rules of the Basel Convention to ban the trade in wastes, for both disposal and recycling, between rich countries (explicitly the Organization for Economic Cooperation and Development (OECD) countries) and poor countries. Developing countries were largely in agreement that the Basel Convention should incorporate more explicit measures to halt this practice. However, industry lobby groups, along with several key OECD countries, were opposed. They did not want to see stricter trade rules in an MEA like the Basel Convention and even argued that it contravened GATT/WTO rules to ban the waste trade (Krueger, 1999; O'Neill, 2000). Despite their protests, the conference of parties to the Basel Convention adopted an amendment in 1995 which bans the export of toxic wastes from OECD to non-OECD countries for both disposal and recycling. Since the amendment was adopted, environmental groups have lobbied hard for states to ratify it, as three quarters of the parties must ratify it before it comes into force. Industry groups have focused their energies on encouraging countries not to ratify the agreement. As of early 2013, there are 74 ratifications. The ban has not come into force, however, due to confusion over whether the required number of ratifications should be three quarters of the number of countries present when the ban was adopted in 1995, or three quarters of the parties at the present time. Under the first rule, the ban would have been adopted with the 62nd ratification. But because of the confusion, the Amendment is still not in force (see BAN 2006; Basel Convention webpage http://www.basel.int).

The transfer of hazards between rich and poor countries also has important links to investment. While much of the literature on pollution havens and industry flight argues that firms do not relocate to developing countries for environmental reasons, most of these studies do make an exception to this argument when it comes to the most highly hazardous industries (Leonard, 1988). In other words, it appears that the dirtiest and most toxic industries do tend to move to jurisdictions with weak environmental regulations. Those from a more radical perspective have been making this argument for years. They have argued that hazards were moving both as a result of higher regulation being imposed in rich countries (industry flight) as well as due to weaker rules in poor countries (pollution havens) (Castleman, 1985). Moreover, double standards in these industries were blatant. The accident at the Union Carbide pesticide plant in Bhopal, India, in 1984 is a prime example (Morehouse, 1994; Rajan, 2001). The problem has been present since at least the 1970s, but the concern over it has heightened in recent years. Since the Basel Convention amendment which bans the export of waste to developing countries, many are worried that the problem of hazardous industry transfer could only intensify. If firms that produce toxic wastes are unable to send their wastes to developing countries, they may be even more encouraged to move their entire production process to these countries (Clapp, 1998b). Evidence of this type of hazard transfer has been seen in a number of case studies, most prominently the *maquiladora* firms in Mexico (Frey, 2003).

Hazardous waste transfers from rich to poor countries via exports and transnational investment demonstrate clearly the linkages between the global political economy and the environment. The globalization of the production process and the footloose nature of transnational investment is a key factor behind movement of hazards around the world. Developing countries have been particularly vulnerable to this type of trade and investment due to weaker environmental regulations and/or lack of enforcement of such regulations. The liberalization of trade and investment policies in indebted countries has played a role in opening up these countries to new trade and investment of this sort.

Conclusion

This chapter has outlined the theoretical debates and policy dilemmas that have dominated the field of IPE and the environment. While there have been strongly opposing views expressed by neoclassical economists and critical thinkers over the past decade regarding the environmental benefits and downsides of international trade, investment, and financing, liberal institutionalists and ecological modernization thinkers have advocated a middle ground. This middle view has been expressed especially regarding environmental policy issues linked to the global economy. Advocated by this view is enhanced state (and in some cases private sector) management of international trade, investment and finance where clear environmental harm results from these activities. Active management, this view argues, can effectively decouple global economic activity from environmental impact. Such an approach, these advocates argue, is necessary in the face of a deadlock between the opposing sides in the theoretical debate, which they see as hampering progress in the policy realm. This middle-ground approach was very much present at the Rio +20 Conference in 2012, in the form of the 'Green Economy.' Whether this approach will make significant improvements to the state of the world's environment remains to be seen. Assessing the influence of this theoretical development is an important area for future research on the links between the IPE and the environment. Careful study of the effectiveness of the various approaches in practice will not only provide an empirical base to judge their merits but also inform the theoretical debates and help them to move forward.

In addition to the rich theoretical avenues for further research, there are several policy areas that merit further investigation by scholars of the global economy–environment interface. The first is the enormous flows of private international finance and their environmental implications. In addition to the need for empirical work on emerging governance initiatives that seek to curtail the potentially negative environmental implications of financial investment, there is also a need to study the phenomenon more broadly. In particular, work on how economic value and risks are constructed within financial markets would help to gain a better understanding of the environmental implications of these markets (Clapp and Helleiner, 2012: 493). The recent rise of new nature-based financial derivatives has to date been relatively under-examined in the literature and deserves greater attention (Sullivan, 2013; Bracking, 2012).

Secondly, there is a need for further work not just on corporate social responsibility efforts but also on corporate accountability. The past decade has seen a growth in research on multi-stakeholder governance initiatives that include industry participants, but whether corporations can be held accountable within these mechanisms remains unclear. With the promotion of a 'green economy' that includes corporate participation in promoting positive synergies between the economy and environment, including on a global scale, it will be important for researchers to watch and investigate the role of specific TNCs in these initiatives as well as the development of legal mechanisms to hold firms liable in instances of environmental harm.

Note

1. In Clapp and Dauvergne (2011) these various camps of thinkers are defined as market liberals, institutionalists, bioenvironmentalists, and social greens. Here the neoclassical economic thinkers equate to the market liberals; the liberal institutionalists equate to the institutionalists; and the critical camp equates to the social greens and bioenvironmentalists.

Annotated bibliography

Clapp, J. and P. Dauvergne (2011) *Paths to a Green World: The Political Economy of the Global Environment*, 2nd edition (Cambridge, MA: MIT Press).

This book systematically examines the linkages between the global economy and the environment through the lenses of four worldviews: market liberals, institutionalists, bioenvironmentalists, and social greens. The book explains the contrasting viewpoints and policy dilemmas that dominate the field, covering globalization, growth, trade, investment, and finance.

Gallagher, K. (2008) *Handbook on Trade and the Environment* (Cheltenham: Edward Elgar).

This handbook provides a comprehensive collection of state-of-the art-articles on various themes related to the trade and environment debates. Topics covered include pollution havens, WTO rules on the environment, corporate codes of environmental conduct, and regional trade arrangements.

Neumayer, E. (2001) *Greening Trade and Investment: Environmental Protection Without Protectionism* (London: Earthscan).

This book provides a comprehensive and detailed analysis of trade and investment issues as they relate to the environment. It covers the environmental implications of WTO trade rules as well as the links between the NAFTA investment provisions and the environment. The book also suggests areas for reform of international trade and investment regimes which would take the environment more seriously into account.

Newell, P. (2012) *Globalization and the Environment* (Cambridge: Polity).

This book provides an examination of the political dimensions of the relationship between economic globalization and the natural environment. The book highlights power relationships in the trade, investment, and financial sectors of the global economy and how these influence environmental outcomes.

Works cited

Arrow, K., B. Bolin, R. Costanza, P. Dasgupta, C. Folke, C. S. Holling, B. Jansson, S. Levin, K. Mäler, C. Perrings and D. Pimentel (1995) 'Economic Growth, Carrying Capacity, and the Environment', *Science*, 268, 520.

Auld, G., S. Bernstein and B. Cashore (2008) 'The New Corporate Social Responsibility', *Annual Review of Environment and Resources*, 33, 413–435.

Barkin, S. J. (2008) 'Trade and Environment Institutions', in K. Gallagher (ed.) *Handbook on Trade and the Environment* (Cheltenham: Edward Elgar), 318–326.

Basel Action Network (2002) 'Exporting Harm: The High-Tech Trashing of Asia', available online at: http://www.ban.org/E-waste/technotrashfinalcomp.pdf

Basel Action Network (2005) 'The Digital Dump: Exporting Re-Use and Abuse to Africa', available online at: http://www.ban.org/Library/TheDigitalDump.pdf

Basel Action Network (2006) 'A Call for an Interpretation of Article 17 by the Parties for Rapid Entry into Force of the Basel Ban Amendment', available online at: http://www.ban.org/deposit-box/

Bhagwati, J. (1993) 'The Case for Free Trade', *Scientific American*, 269, 5, 42–49.

Biermann, F. (2000) 'The Case for a World Environment Organization', *Environment*, 42, 9, 22–31.

Boyce, J. K. (2008) 'Globalization and the Environment: Convergence or Divergence?', in K. Gallagher (ed.) *Handbook on Trade and the Environment* (Cheltenham: Edward Elgar), 97–115.

Bracking, S. (2012) 'How Do Investors Value Environmental Harm/Care? Private Equity Funds, Development Finance Institutions and the Partial Financialization of Nature-based Industries', *Development and Change*, 43, 1, 271–293.

Charnovitz, S. (2007) 'The WTO's Environmental Progress', *Journal of International Economic Law*, 10, 3, 685–706.

Charnovitz, S. (2008) 'An Introduction to the Trade and Environment Debate', in K. Gallagher (ed.) *Handbook on Trade and the Environment* (Cheltenham: Edward Elgar), 237–245.

Clapp, J. (1998a) 'The Privatization of Global Environmental Governance: ISO 14000 and the Developing World', *Global Governance*, 4, 3, 295–316.

Clapp, J. (1998b) 'Foreign Direct Investment in Hazardous Industries in Developing Countries: Rethinking the Debate', *Environmental Politics*, 7, 4, 92–113.

Clapp, J. (2001) *Toxic Exports: The Transfer of Hazardous Wastes from Rich to Poor Countries* (Ithaca, NY: Cornell University Press).

Clapp, J. (2003) 'Transnational Corporate Interests and Global Environmental Governance: Negotiating Rules for Agricultural Biotechnology and Chemicals', *Environmental Politics*, 12, 4, 1–23.

Clapp, J. (2005) 'Global Environmental Governance for Corporate Responsibility and Accountability', *Global Environmental Politics*, 5, 3, 23–34.

Clapp, J. (2008) 'Global Mechanisms for Greening TNCs: Inching Toward Corporate Accountability?', in K. Gallagher (ed.) *Handbook on Trade and the Environment* (Cheltenham: Edward Elgar), 159–170.

Clapp, J. (2011) 'Toxic Exports: Despite Global Treaty, Hazardous Waste Trade Continues', in N. Gilman, J. Goldhammer and S. Weber (eds) *Deviant*

Globalization: Black Market Economy in the 21st Century (London: Continuum, 2011), 166–179.
Clapp J. and E. Helleiner (2012) 'International Political Economy and the Environment: Back to the Basics?', *International Affairs*, 88, 3, 485–501.
Clapp, J. and J. Meckling (2013) 'Business as a Global Actor', in R. Falkner (ed.) *Handbook of Global Climate and Environment Policy* (Chichester: Wiley-Blackwell), 286–303.
Clapp, J. and J. Thistlethwaite (2012) 'Private Voluntary Programs in Environmental Governance: Climate Change and the Financial Sector', in K. Ronit (ed.) *Business and Climate Policy: Potentials and Pitfalls of Voluntary Programs* (UN University Press), 43–76.
Clapp, J. and P. Dauvergne (2011) *Paths to a Green World: The Political Economy of the Global Environment*, 2nd Edition (Cambridge, MA: MIT Press).
Copeland, B. (2008) 'The Pollution Haven Hypothesis', in K. Gallagher (ed.) *Handbook on Trade and the Environment* (Cheltenham: Edward Elgar), 60–70.
Curtis, F. (2008) 'Peak Globalization: Climate Change, Oil Depletion and Global Trade', *Ecological Economics*, 69, 427–434.
Daly, H. (1993) 'The Perils of Free Trade', *Scientific American*, 269, 5, 50–57.
Daly, H. (1996) *Beyond Growth: The Economics of Sustainable Development* (Boston: Beacon).
Daly, H. and J. Farley (2010) *Ecological Economics* (Washington, D.C.: Island Press).
Dauvergne, P. (2008) *The Shadows of Consumptiom* (Cambridge, MA: MIT Press).
DeSombre, E. R., and J. S. Barkin (2002) 'Turtles and Trade: The WTO's Acceptance of Environmental Trade Restrictions', *Global Environmental Politics*, 2, 1, 12–18.
Dinda, S. (2004) 'Environmental Kuznets Curve Hypothesis: A Survey', *Ecological Economics*, 49, 4, 431–455.
Esty, D. (1994) *Greening the GATT: Trade, Environment and the Future* (Washington D.C.: Institute for International Economics).
Esty, D. (2001) 'Bridging the Trade-Environment Divide', *Journal of Economic Perspectives*, 15, 3, 113–130.
Fairman, D. (1996) 'The Global Environment Facility: Haunted by the Shadow of the Future', in R. Keohane and M. Levy (eds) *Institutions for Environmental Aid* (Cambridge, MA: MIT Press), 55–87.
Falkner, R. (2008) *Business Power and Conflict in International Environmental Politics* (Basingstoke: Palgrave Macmillan).
Fox, J. and L. D. Brown (eds.) (1998) *The Struggle for Accountability: The World Bank, NGOs and Grassroots Movements* (Cambridge, MA: MIT).
Frey, R. S. (2003) 'The Transfer of Core-Based Hazardous Production Processes to the Export Processing Zones of the Periphery: The Maquiladora Centers of Northern Mexico', *Journal of World-Systems Research*, 9, 2, 317–354.
Fuchs, D. (2005) 'Commanding Heights? The Strength and Fragility of Business Power in Global Politics.' *Millennium: Journal of International Studies*, 33, 3, 771–801.
Gallagher, K. (2008) *Handbook on Trade and the Environment* (Cheltenham: Edward Elgar).
Gallagher, K. (2009) 'Economic Globalization and the Environment', *Annual Review of Environment and Resources*, 34, 279–304.

Goldman, M. (2005) *Imperial Nature. The World Bank and Struggles for Social Justice in the Age of Globalization* (New Haven: Yale University Press).

Grossman, G. and A. Krueger (1995) 'Economic Growth and the Environment', *Quarterly Journal of Economics*, 110, 2, 353–377.

Halifax Initiative (2008) 'The World Bank, Climate Change and Energy. Issue Brief', October, available online at: http://www.halifaxinitiative.org/index.php/Factsheets/1116.

Hall, D. (2002) 'Environmental Change, Protest and Havens of Environmental Degradation: Evidence from Asia', *Global Environmental Politics*, 2, 2, 20–28.

Harris, P. G. (2011) 'Peace, Security and Global Climate Change: The Vital Role of China', *Global Change, Peace & Security*, 23, 2, 141–145.

Helleiner, E. (2011) 'Greening Global Financial Markets?', *Global Environmental Politics*, 11, 2, 51–53.

Hochstetler, K. (2013) 'South-South Trade and the Environment: A Brazilian Case Study', *Global Environmental Politics*, 13, 1, 30–48.

Horta, K., R. Round and Z. Young (2002) 'The Global Environmental Facility: The First Ten Years – Growing Pains or Inherent Flaws?', Report for Environmental Defense and The Halifax Initiative.

Iles, A. (2004) 'Mapping Environmental Justice in Technology Flows: Computer Waste Impacts in Asia', *Global Environmental Politics*, 4, 4, 76–107.

Jackson, T. (2009) *Prosperity without Growth* (London: Earthscan).

Kallis, G., C. Kerschner and J. Martinez-Alier (2012) 'The Economics of Degrowth', *Ecological Economics*, 84, 172–180.

Kanthak, J. (1999) *Ships for Scrap: Steel and Toxic Wastes for Asia* (Hamburg: Greenpeace).

Karliner, J. (1997) *The Corporate Planet, Ecology and Politics in the Age of Globalization* (San Francisco: Sierra Club).

Kolk, A., D. Levy and J. Pinkse (2008) 'Corporate Responses in an Emerging Climate Institutionalization and Commensuration of Carbon Disclosure', *European Accounting Review*, 17, 4, 719–745.

Korten, D. C. (1995) *When Corporations Rule the World* (West Hartford and San Francisco: Kumarian Press and Berrett-Koehler Publishers).

Kreuger, J. (1999) *International Trade and the Basel Convention* (London: Earthscan).

Krut, R. and H. Gleckman (1998) *ISO 14001: A Missed Opportunity for Global Sustainable Industrial Development* (London: Earthscan).

Leonard, H. J. (1988) *Pollution and the Struggle for the World Product* (Cambridge: Cambridge University Press).

Levy, D. and P. Newell (eds) (2005) *The Business of Global Environmental Governance* (Cambridge, MA: MIT Press).

Lieberman, S. and T. Gray (2008) 'The World Trade Organization's Report on the EU's Moratorium on Biotech Products: The Wisdom of the US challenge to the EU in the WTO', *Global Environmental Politics*, 8, 1, 33–52.

Lohmann, L. (2010) 'Uncertainty Markets and Carbon Markets: Variations on Polanyian Themes', *New Political Economy*, 15, 2, 225–254.

Missbach, A. (2004) 'The Equator Principles: Drawing the Line for Socially Responsible Banks? An Interim Review from an NGO Perspective', *Development*, 47, 3, 78–84.

Mol, A. (2001) *Globalization and Environmental Reform: The Ecological Modernisation of the Global Economy* (Cambridge, MA: MIT Press).

Morehouse, W. (1994) 'Unfinished Business: Bhopal Ten Years After', *The Ecologist*, 24, 5, 164–168.
Muradian, R. and J. Martinez-Alier (2001) 'Trade and the Environment: From a "Southern" Perspective', *Ecological Economics*, 36, 281–297.
Neumayer, E. (2001) *Greening Trade and Investment: Environmental Protection Without Protectionism* (London: Earthscan).
Newell, P. (2012) *Globalization and the Environment: Capitalism, Ecology and Power* (Cambridge: Polity).
Newell, P. and M. Paterson (2010) *Climate Capitalism* (Oxford: Oxford University Press).
O'Neill, K. (2000) *Waste Trading among Rich Nations: Building A New Theory of Environmental Regulation* (Cambridge, MA: MIT).
Porter, G. (1999) 'Trade Competition and Pollution Standards: "Race to the Bottom" or "Stuck at the Bottom"?', *Journal of Environment and Development*, 8, 2, 133–151.
Porter, G., N. Bird, N. Kaur and L. Peskett (2008) *New Finance for Climate Change and the Environment* (Washington, D.C.: WWF and Heinrich Böll Foundation), available online at: http://assets.panda.org/downloads/ifa_report.pdf.
Prakash, A. and M. Potoski (2006), *The Voluntary Environmentalists: Green Clubs, ISO 14001, and Voluntary Environmental Regulations* (Cambridge: Cambridge University Press).
Puckett, J. (1994) 'Disposing of the Waste Trade: Closing the Recycling Loophole', *The Ecologist*, 24, 2, 53–58.
Rajan, S. R. (2001) 'Toward a Metaphysic of Environmental Violence: The Case of the Bhopal Gas Disaster', in N. L. Peluso and M. Watts (eds) *Violent Environments* (Ithaca, NY: Cornell University Press), 380–398.
Reed, D. (1997) 'The Environmental Legacy of the Bretton Woods: The World Bank', in O. Young (ed.) *Global Governance: Drawing Insights from the Environmental Experience* (Cambridge, MA: MIT Press), 227–245.
Rich, B. (1994) *Mortgaging the Earth: The World Bank, Environmental Impoverishment, and the Crisis of Development* (London: Earthscan).
Rich, B. (2009) *Foreclosing the Future: Coal, Climate and Public International Finance* (Washington, D.C.: Environmental Defence), available online at: http://www.edf.org/documents/9593_coal-plants-report.pdf.
Roberts, J. T. and B. Parks (2009) 'Ecologically Unequal Exchange, Ecological Debt and Climate Justice', *International Journal of Comparative Sociology*, 50, 3–4, 385–409.
Roberts, J. T., B. C. Parks, M. J. Tierney and R. L. Hicks (2009) 'Has Foreign Aid Been Greened?' *Environment: Science and Policy for Sustainable Development*, 51, 1, 8–21.
Rowlands, I. (2000) 'Beauty and the Beast? BP's and Exxon's Positions on Global Climate Change', *Environment and Planning C: Government and Policy*, 18, 3, 339–354.
Sachs, W. (1999) *Planet Dialectics: Explorations in Environment and Development* (London: Zed).
Schmidheiny, S. and F. Zorraquin (1996) *Financing Change: The Financial Community, Eco-Efficiency and Sustainable Development* (Cambridge, MA: MIT Press).
Selin, H. and S. VanDeveer (2006) 'Raising Global Standards: Hazardous Substances and E-Waste Management in the European Union', *Environment*, 48, 10, 7–18.

Selin, H. and S. VanDeveer (2008) 'The Politics of Trade and Environment in the European Union', in K. Gallagher (ed.) *Handbook on Trade and the Environment* (Cheltenham: Edward Elgar), 194–203.

Singh, S. J., F. Krausmann, S. Gingrich, H. Haberl, K. Erb, P. Lanz, J. Martinez-Alier and L. Temper (2012) 'India's Biophysical Economy, 1961–2008. Sustainability in a National and Global Context', *Ecological Economics*, 76, 60–69.

Sklair, L. (2001) *The Transnational Capitalist Class* (London: Blackwell).

Soloway, J. (2002) 'The North American Free Trade Agreement: Alternative Models of Managing Trade and the Environment', in R. Steinberg (ed.) *The Greening of Trade Law: International Trade Organizations and Environmental Issues* (Lanham, MD: Rowman and Littlefield), 155–188.

Stern, D. (2004) 'The Rise and Fall of the Environmental Kuznets Curve', *World Development*, 32, 8, 1419–1439.

Stevis, D. and S. Mumme (2000) 'Rules and Politics in International Integration: Environmental Regulation in NAFTA and the EU', *Environmental Politics*, 9, 4, 20–42.

Stilwell, M. and R. Tarasofsky (2001) *Towards Coherent Environmental and Economic Governance: Legal and Practical Approaches to MEA-WTO Linkages* (Gland: WWF), available at: http://www.ciel.org/Publications/Coherent_EnvirEco_Governance.pdf.

Streck, C. (2001) 'The Global Environment Facility – a Role Model for International Governance?', *Global Environmental Politics*, 1, 2, 71–94.

Sullivan, S. (2013) 'Banking Nature? The Spectacular Financialisation of Environmental Conservation', *Antipode*, 45, 1, 198–217.

Thistlethwaite, J. (2011) 'Counting the Environment: The Environmental Implications of International Accounting Standards', *Global Environmental Politics*, 11, 5, 75–97.

Thistlethwaite, J. (2012) 'The ClimateWise Principles: Self-regulating Climate Change Risks in the Insurance Sector', *Business & Society* 51, 1, 121–147.

UNEP (2011) *Towards a Green Economy: Pathways to Sustainable Development and Poverty Eradication* (Geneva: UNEP).

Unmüßig, B., W. Sachs and T. Fatheuer (2012), *Critique of the Green Economy: Toward Social and Environmental Equity* (Berlin: Heinrich Böll Foundation).

Van Alstine, J. and E. Neumayer (2008) 'The Environmental Kuznets Curve', in K. Gallagher (ed.) *Handbook on Trade and the Environment* (Cheltenham: Edward Elgar), 49–59.

Van Harten, G. (2005) 'Private Authority and Transnational Governance: The Contours of the International System of Investor Protection', *Review of International Political Economy*, 12, 4, 600–623.

Vogel, D. (2000) 'International Trade and Environmental Regulation', in N. J. Vig and M. E. Kraft (eds) *Environmental Policy: New Directions for the Twenty-First Century*, 4th edition (Washington DC: CQ Press).

Wade, R. (1997) 'Greening the Bank: The Struggle over the Environment, 1970–1995', in J. Lewis and R. Web (eds) *The World Bank: Its First Half Century*, Vol. 2 (Washington DC: Brookings Institute), 611–734.

Weinstein, M. and S. Charnovitz (2001) 'The Greening of the WTO', *Foreign Affairs*, 80, 6, 147–156.

Wheeler, D. (2002) 'Beyond Pollution Havens', *Global Environmental Politics*, 2, 2, 1–10.

Williams, M. (2001) 'In Search of Global Standards: The Political Economy of Trade and the Environment', in D. Stevis and V. Assetto (eds) *The International Political Economy of the Environment: Critical Perspectives* (Boulder: Lynne Rienner), 39–61.

Wise, T. (2008) 'The Environmental Costs of Mexico-USA Maize Trade under NAFTA', in K. Gallagher (ed.) *Handbook on Trade and the Environment* (Cheltenham: Edward Elgar), 126–135.

Wright, C. (2012) 'Global Banks, the Environment and Human Rights: The Impact of the Equator Principles on Lending Policies and Practices', *Global Environmental Politics,* 12, 1, 56–77.

Zarsky, L. (1999) 'Havens, Halos and Spaghetti: Untangling the Evidence about Foreign Direct Investment and the Environment', in OECD (ed.) *Foreign Direct Investment and the Environment* (Paris: OECD).

Zarsky, L. (2006) 'From Regulatory Chill to Deepfreeze?' *International Environmental Agreements: Politics, Law and Economics,* 6, 4, 395–399.

Zeng, K. and J. Eastin (2007) 'International Economic Integration and Environmental Protection: The Case of China', *International Studies Quarterly,* 51, 4, 971–995.

6
Gender and International Environmental Politics
Nicole Detraz

Environmental issues are regarded by many as urgent, complex, and transboundary, but are they gendered? Environmental scholars, institutions, and policymakers are increasingly recognizing that they are. There are a variety of important connections between gender and international environmental politics (IEP). For example, people often experience environmental problems differently because of socially constructed ideas about the appropriate roles and responsibilities of men and women. Men may be expected to migrate in the face of environmental degradation in order to provide for their families (Boehm, 2008; Terry, 2009), or women might be at greater risk to die in natural disasters if it is considered inappropriate for them to leave their houses alone (Hunter and David, 2011). While it is important that we avoid essentializing people's experiences, it is also vital that we recognize consistent, gendered patterns in the area of environmental politics.

This edited book addresses a number of central concepts and topics within IEP, including sustainable development, environmental justice, and environmental security. This chapter addresses each of these topics, along with debates about population growth and movement, using gender lenses. These topics have been chosen because of their centrality in IEP debates, along with their importance for policymaking. The chapter considers how looking at the environment through gender lenses pushes us to change our questions and sphere of analysis in ways that positively impact policymaking processes. Each of these topics is central within IEP debates and has important gender components that are often ignored in both scholarship and policymaking. The case of the politics of biodiversity in South American Amazonian states provides an interesting space to explore real-world examples of the interplay between gender, environmental change, and policymaking. The central argument running through the chapter is that considering the environment through gender lenses challenges our view of many of the pressing problems and solutions in the environmental realm in ways that are beneficial for policymaking.

Where are the connections between gender and the environment?

This chapter draws on feminist scholarship in order to explore how gender and environmental politics intersect. Feminist scholars have contributed important insights into the connections between gender and the environment, although they approach the issue from a variety of perspectives. Feminism is best thought of as a large umbrella term, which contains a range of subcategories. In other words, there are a great many feminisms. What holds them together is the goal of revealing and challenging widespread inequality in society (see Paterson, this book, on structural inequalities). Most feminists focus specifically on inequality in the form of women's subordination; however, this is typically part of a broader concern about how multiple forms of inequality and marginalization intersect. This can be marginalization not only based on gender but also based on race, class, ethnicity, or sexuality. Feminists will have different ways of understanding the sources of this inequality as well as different suggestions for how to deal with them, but at their root they all seek an international community in which people are not discriminated against or marginalized because they are identified as members of certain groups.

Gender is a central concept for feminist scholarship. It refers to a set of socially constructed ideas about what men and women *ought* to be. Most feminist scholarship stresses that sex and gender are two different concepts. Sex describes biological differences between people understood to be men and people understood to be women. Gender describes the socially constituted differences between these same groups. Gender roles and assumptions come directly out of a society's expectations about its members. They are not deterministic entities. Gender characteristics are cultural creations passed on through socialization. Society defines what is appropriate or acceptable behavior or qualities for men or women to demonstrate (Peterson and Runyan, 2010). Gender roles become so ingrained in society that they take on the appearance of being natural or 'normal'. This means when individuals act in ways that defy these gender norms they are seen as being unnatural (Sjoberg and Gentry, 2007).

Ecofeminism represents a widely discussed lens through which to view the combination of gender issues and the environment. The term 'ecofeminism' dates back to 1974 when French feminist Françoise d'Eaubonne used the word *ecoféminisme* to refer to the movement by women necessary to save the planet. The 1970s and 1980s saw the tendency for scholars and activists to use the term 'ecofeminist' to refer to their struggle to link feminism and ecology. The term 'ecofeminism' covers numerous approaches to connecting feminist concerns and environmental concerns (Mies and Shiva, 1993; Sandilands, 1999; Sturgeon, 1997; Warren, 1997). Heather Eaton and Lois Ann Lorentzen (2003: 3) explain:

> [W]hat makes ecological feminism *feminist* is the commitment to the recognition and elimination of male-gender bias and the development of practices, policies, and theories without this bias. What makes it *ecological* is a commitment to the valuing and preserving of ecosystems, broadly understood. Ecofeminism is a textured field of theoretical and experiential insights encompassing different forms of knowledge, embodied in the concrete.

They go on to explain that while ecofeminism covers a range of gender/environment connections, three are central – the empirical, the cultural, and the epistemological. The *empirical claim* refers to the fact that women are typically disproportionately impacted by the negative consequences of environmental damage (Stein, 2004). This is often because of their marginalization within society, or their gendered roles in household labor. The *cultural claim* is that cultures have established an idea of the world as dualistically and hierarchically divided. 'Dualisms such as reason/emotion, mind/body, culture/nature, heaven/earth, and man/woman give priority to the first over the second... Religion, philosophy, science, and cultural symbols reinforce this worldview, making male power over both women and nature appear "natural" and thus justified' (Eaton and Lorentzen, 2003: 2). The *epistemological claim* refers to the idea that women's experiences with environmental matters, due to socially constructed roles and responsibilities, make them potentially useful sources of knowledge and expertise for solutions to environmental change. These three claims have informed ecofeminist scholarship and critiques of issues like militarization, globalization, and status quo environmental policymaking (King, 1995; Seager, 1993). While some versions of ecofeminism are critiqued in turn for being essentialist, particularly versions of 'motherhood environmentalism' (Sandilands,

1999), the ecofeminist label is just one of a number used to describe scholarship on the connections between gender and the environment.

In addition to feminist scholars, several environmental organizations and states have recently called attention to the connections between gender and the environment. Non-governmental organizations (NGOs) like Women's Environment and Development Organization (WEDO), Gender CC – Women for Climate Justice, and Global Gender and Climate Alliance (GGCA) seek to highlight the unique environmental needs and connections of women and their role in environmental decision-making around the world. These organizations engage in a range of activities, including issuing reports on gender and environmental issues, lobbying at global environmental meetings and conferences, and pushing for gender-sensitive environmental policies at all levels of governance. Additionally, the United Nations has touched on gender–environment links for several decades.[1] In 2006, the United Nations Environment Programme (UNEP) launched the Gender Plan of Action, which outlined a framework for integrating a gender perspective within all the institution's divisions, branches, units, and activities. In particular, it sought to (1) promote human equality, equity, and rights across gender; (2) promote equality of opportunity and treatment of women and men in the environmental sector at multiple levels of governance; and (3) increase the quality and efficiency of the institution's work in environmental conservation and sustainable development promotion (UNEP, 2006). Finally, a number of states have also recognized a specific link between gender and environmental issues. For example, many African states in particular include a mention of gender and climate change impacts in their National Adaptation Programmes of Action (NAPAs) which are submitted as part of the United Nations Framework Convention on Climate Change (Detraz, 2012). In sum, scholars, NGOs, IGOs, and states have discussed gender as it relates to various environmental topics and problems. The following sections focus on the issues of sustainability/sustainable development, environmental justice, environment and security, and population growth/movement. The sections address feminist concerns and foci in these areas, and explain how incorporating gender into debates about these issues is helpful for just and effective policymaking.

Sustainability/sustainable development

Sustainable development is a concept with a relatively long and complex history in IEP (Happaerts and Bruyninckx, this book). The World

Commission on Environment and Development's (1987: 8) Brundtland Report *Our Common Future* defined sustainable development as 'development that meets the needs of the present without compromising the ability of future generations to meet their own needs'. It was introduced in the 1980s as a compromise position between those who wanted to draw attention to global environmental damage and those who wanted to ensure that Southern states could continue to develop their economies in ways consistent with the development trajectory of Northern states. The concept implies a sometimes delicate balance between the needs of humans to improve their well-being, and preserving ecosystems and natural resources in the interest of future generations (Nsiah-Gyabaah, 2010). Sustainable development has become an international buzzword that has come to mean almost everything and nothing. Actors use the term either with a focus on the 'sustainable' or the 'development' to mean very different things. Nonetheless, it remains an often-used concept within discussions of environmental change.

Feminism intersects with debates about sustainable development in various ways, including challenging expectations about where environmental expertise comes from, providing a critique of mainstream development politics, and providing a critique of notions of equity. One critique of sustainable development is that it has often been conceptualized in ways that allow change to come in the form of the current structures of society rather than calling for substantial change (Robinson, 2004; Sneddon et al., 2006; Worster, 1995). Feminists join other 'structuralists' who 'are critical of the belief that incremental changes, largely within the boundaries of the existing world order and global policy dynamics, will be able to bring about the social transformations required to come to a sustainable society' (Bruyninckx, 2006: 275). There is a fear that sustainable development policies are often focused on the 'development' side of the coin, with neoliberal development approaches ruling the day. Vandana Shiva has popularized the idea of 'maldevelopment' to refer to perspectives on development and modernization which are built on Western ideas of reductionism, duality, and linearity. She argues that maldevelopment is 'the violation of the integrity of organic, interconnected and interdependent systems, that sets in motion a process of exploitation, inequality, injustice and violence. It is blind to the fact that a recognition of nature's harmony and action to maintain it are preconditions for distributive justice' (Shiva, 1989: 5–6). Those concerned with women's frequent situation of being on the fringes of development are not satisfied with the traditional conceptualization of the term 'sustainable development'

(Harcourt, 1994; Mies and Shiva, 1993; Nsiah-Gyabaah, 2010). Since different paths to development often have survival implications for a population, a gendered approach to sustainable development that takes into account the needs of women, ecosystems, and future generations within a particular setting is necessary.

The Women's Environment and Development Organization (WEDO) is an NGO that promotes and protects human rights, gender equality, and the integrity of the environment. They claim that sustainable development is 'ecologically sound, economically viable, and socially just, and gender equality is prerequisite to it. Interdependent crises of food, fuel and climate – exacerbated by inequitable and fragile economies and social norms – need holistic attention and solutions' (WEDO, 2013). This perspective on sustainable development views gender equity as essential to both environmental sustainability and to a larger societal goal of justice. Environmental and development NGOs like WEDO were some of the earliest and most enthusiastic supporters of the ideas of sustainability and sustainable development. That being said, they specifically call for conceptualizations of sustainable development that recognize the harm that comes from the overlapping processes of patriarchy and globalization. For example, during the 57th session of the Commission on the Status of Women in March 2013, a side event was held, entitled 'Violence – Ecologies – Livelihoods: Feminists Confronting Unsustainable Development'. The event examined the various forms of violence resulting from unsustainable development and related environmental degradation. Female speakers from Guatemala, Colombia, Kazakhstan, Japan, and Fiji shared their experiences of land grabbing, mining, nuclear contamination, and nuclear disasters. These experiences of maldevelopment are illustrations of the necessity of reflecting on how sustainable our development approaches are for ecosystems and the humans that inhabit them. These gendered approaches to sustainable development have often come with a critique of both mainstream development models and the idea that sustainability goals require scientific environmental knowledge exclusively, with little space for local perspectives.

Environmental justice

Environmental justice is a concept that has been central to understanding the myriad ways that the issues of environmental change, inequality, vulnerability, and marginalization are intertwined (Okereke and Charlesworth, this book). The concept traces its roots to several topics in the international community, including environmental racism. Environmental racism is a term that originated in the United States, and was coined in 1982 by Benjamin Chavis, the head of the National Association for the Advancement of Colored People (NAACP) (Parks and Roberts, 2006). Environmental racism was the idea that degrading the environment was an additional component of the inequality that African American communities faced (Schlosberg and Carruthers, 2010). This early focus on environmental racism is still felt in discourses of environmental justice, with concern about race and racist dualisms as a prominent feature of many environmental justice ideas (Sturgeon, 2009). However, scholars and activists who use the concept of environmental justice utilize a variety of perceptions about the nature of justice. The early versions of environmental justice within the United States focused heavily on the notion of the distribution of environmental 'bads' (Schlosberg and Carruthers, 2010). This focus has been broadened to include concerns about unequal exposure by class and ethnicity alongside race, and also to include assessments of who benefits most from environmental degradation and solutions, and who suffers disproportionately from those processes (Newell, 2005; Park and Roberts, 2007). It has also become influential outside of the United States (Walker, 2012).

Feminist scholars have addressed the connections between environmental issues and justice issues for decades. In fact, the environmental justice movement has been described by some as a feminist movement for a number of reasons. To begin with, the larger environmental justice movement is typically concerned with issues and has goals viewed as important in the lives of many women (Stein, 2004). Robert Verchick (2004: 64) argues that 'to the extent that women remain the primary caregivers in their homes and communities, the responsibility over family health remains an immediate and primary goal for them'. This rationale may seem essentialist on the surface, but it does not have to be. Feminist scholars have long considered the regularized patterns of behavior of women and men in a society, but they trace these patterns to gender roles and assumptions that prevail in those societies. It is not that all women are caregivers or that it is 'natural' for them to undertake these care roles, but rather that societal expectations about what women *should* be or do result in many women serving as caregivers. These socially constructed, gendered roles have been used to explain the high proportion of women who take part in environmental justice activist campaigns (Newman, 1994; Stein, 2004; Sturgeon, 2009; Verchick, 2004).

The politics of food security is a topic with strong connections to gender and environmental justice debates. Debates about food security involve concerns about where the negative impacts of environmental damage are felt most. The Food and Agricultural Organization (FAO) defines food security as follows: 'food security exists when all people, at all times, have physical and economic access to sufficient, safe and nutritious food for a healthy and active life' (FAO, 2003). The concept contains four components: adequacy of food supply or availability, stability of supply without fluctuations or shortages from season to season, accessibility to food or affordability, and the quality and safety of food. Food insecurity is a gendered phenomenon. There is a great deal of evidence that during times of food scarcity it is women and children who suffer most (Davidson and Krull, 2011; Kerr, 2005; Steans, 2006; Uraguchi, 2010). Many instances of food shortage stem from distribution issues rather than total lack of food availability in a state or community. Around the world there are a disproportionate number of landless or land-poor rural households which are headed by women. This status has a direct link to food insecurity (FAO, 2011).

At the same time, several high-profile outlets have claimed that gender equity could help increase food security across the globe. United Nations Special Rapporteur on the right to food, Olivier De Schutter, claims that 'sharing power with women is a shortcut to reducing hunger and malnutrition, and is the single most effective step to realizing the right to food' (UNHR, 2013). He urged states to transform their approach to food, in ways that address cultural constraints and redistribute roles between women and men. Claims like these recognize the obstacles that many women face in gaining access to land, credit, and agricultural inputs (De Schutter, 2012). Focusing on women's agency is consistent with much feminist scholarship that seeks to avoid painting women as victims of the world's ills. This serves to essentialize the experiences of women, but it also potentially reinforces their marginalization. If women's insecurity is seen as part of the symptom of environmental degradation, then we can critically reflect on ways to address the larger societal processes that are responsible for these outcomes. Also, if women are seen as essential stakeholders and solutions to environmental degradation, then they will be given a place at the negotiating table, which addresses another component of gender inequality in the area of environmental politics – lack of representation of marginalized groups in environmental governance and decision-making.

Examining environmental justice through gender lenses includes assessing the areas that are currently incorporating gender, and the areas in which gender is missing. Ideas of environmental justice need to acknowledge the larger, structural contributions of power inequalities and marginalization. This includes forming a broad conceptualization of environmental justice that is concerned with issues of race, gender, sexuality, and class. It also requires understanding how these processes of domination appear in North–South relations. This must be done in ways that avoid essentialization, but point out relatively consistent patterns of behavior and experience.

Environment and security

Like environmental justice discourses, IEP scholars have examined proposed links between security and the environment for decades (Swatuk, this book). The terminology of 'food security' discussed above is an example of this tendency. The development of environmental security studies (ESS) was largely a response to two events: the end of the Cold War in the late 1980s to early 1990s, and the 1992 Rio Earth Summit. The first of these necessitated a rethinking of the concept of security by both security establishments and scholars, and the second mobilized evidence of global environmental change into a global policy agenda which became widely regarded as pressing and vital (Floyd and Matthew, 2013). While there have been examples of scholars making security–environment connections before the Cold War, since that time there has been a flood of academic scholarship on these issues. There has also been an increased tendency for policymakers and the media to increasingly use securitized language to talk about environmental issues. There are three main security–environment discourses that actors have used to connect these areas – the environmental conflict discourse, the environmental security discourse, and the ecological security discourse (Detraz, 2013).

The *environmental conflict* discourse is closely associated with traditional security concerns like state security and the potential for actors to engage in violent conflict. In this case, the central concern is that actors will engage in violent conflict over access to natural resources. These resource conflicts can spring either from scarcity of resources (Homer-Dixon, 1999) or from abundance of resources (de Soysa, 2013), and are understood to contribute to state instability and insecurity. Secondly, the *environmental security* discourse is rooted in a concern about the negative impacts that environmental change has for human communities. This discourse is closely associated with the idea of 'human security', which was popularized in the mid-1990s.[2] Human security rejects the militarized, state-centric nature of traditional security concerns and is concerned instead with the health and well-being of individuals and communities. When this concept is applied to security–environment discussions, it manifests in a concern that environmental change will cause human populations to experience insecurities like increased exposure to disease, increased experience with natural disasters, food insecurity, and livelihood insecurity, among others (Barnett, 2001). Finally, the *ecological security* discourse is concerned about the negative impacts of environmental change for ecosystems. This ecocentric discourse largely identifies human contributions to environmental degradation. Within this discourse, security is about ensuring environmental health and, by extension, human well-being. This holistic perspective assumes that humans are part of nature, not separate from it (Swatuk, this book). Narratives that relate to this theme include evaluating the relationship between humans and environment, and challenging 'traditional' conceptualizations of security.

Some feminist scholars have strongly critiqued ESS for failing to consider the implications of using securitized language to point to things like population growth, migration, or globalization as contributors to environmental change while ignoring the gendered aspects of these phenomenon (Hartmann, 2010; Urban, 2007). These scholars argue that securitizing these issues may result in their militarization, which has potentially negative implications for both the environment and its human inhabitants. Feminists have long called attention to the gendered and environmentally destructive consequences of militarization (Enloe, 2000; Seager, 1993, 2003). In response to these and other critiques, scholars have begun pushing for gender to become a central aspect of security–environment scholarship in ways that reject militarization and reflect on the gendered aspects of human insecurity (Detraz, 2013; Goldsworthy, 2010; Oswald Spring, 2008). For example, Úrsula Oswald Spring (2008) has advocated thinking about these issues in terms of 'human, gender and environmental security' (HUGE). This concept reflects on the ways that understanding gender helps us to re-evaluate priorities. Importantly, this means getting away from a focus on maintaining existing social orders, which often contribute to insecurity, and instead focusing on human vulnerabilities. More scholarship along these lines is needed, particularly work that speaks to the community of ESS scholars and those policymakers who utilize securitized environmental discourses.

While there is currently very little integration of gender in security–environment discourses, there is room for a *feminist environmental security discourse* which builds on and goes beyond aspects of existing discourses (Detraz, 2013). A feminist environmental security discourse seeks avenues toward (1) human security, (2) gender emancipation, and (3) environmental sustainability. The central focus of the discourse is revealing and removing the negative, gendered security threats that stem from environmental change. These threats include unequal exposure to environmental risk and high levels of vulnerability to environmental change. Both of these are directly related to societal processes that result in marginalization. By identifying these threats and vulnerabilities, a feminist environmental security discourse moves beyond state-centric and militarized notions of security, and instead centers on the various insecurities that people and communities face, for example, increased exposure to pollution or loss of livelihood security in the face of prolonged drought. The discourse requires critical assessment of the ways that these threats and vulnerabilities intersect with socially constructed assumptions about the appropriate roles and activities of men and women.

A feminist environmental security discourse represents a useful way for actors to highlight the range of negative implications of environmental change for people and ecosystems. It is a broad, critical perspective that includes space to reflect on the complexities of the causes, consequences, and solutions to environmental problems. Important components of a feminist environmental security discourse include adopting multilevel and multiple perspectives of security and the environment, employing broad and critical conceptualizations of key terms, engaging in critical reflection on potential causes of environmental insecurity, and working toward the adoption of inclusive and just solutions to environmental insecurity.

Population growth and movement

The issue of population growth and movement has had a central place in all three security–environment discourses outlined above, and within IEP more broadly. There has been a concern about the implications of population growth for the extraction or use of resources and species since the early days of IEP scholarship (Stevis, this book). The warnings of Thomas Malthus in the late 1700s about human population growth overwhelming our capacity to feed ourselves have consistently reappeared in the works of several prominent scholars like Garrett Hardin, Paul Ehrlich, and the Club of Rome. In *The Tragedy of the Commons*, Hardin (1968: 1248) claimed 'the most important aspect of necessity that we must now recognize is the necessity of abandoning the commons in breeding... Freedom to breed will bring ruin to all'. He argued that we cannot rely on technical fixes to the 'problem' of overpopulation, nor can we rely on appeals to conscience or morality to get people to freely give up their right or desire to have children. Instead, 'mutual coercion mutually agreed upon' is necessary.

These alarmist population growth narratives were given new life in the early 1990s with the rise of security–environment connections (Silliman and King, 1999; Sturgeon, 2009). For the most part, scholars, policymakers, and the media have discussed the connections between insecurity, population growth, and migration as if these are gender-neutral phenomena (Detraz, 2013). A lack of reflection on gender in discussions of population growth is a huge oversight. By identifying population increase as a contributor to environmental damage, these actors are automatically making women the potential target of policy 'solutions' because of their role as child bearers (Hartmann, 1995). This is particularly problematic when the issue of gender is completely ignored in many sources that address population specifically. Critical environmental scholars have called attention to negative examples like coalitions being built between environmental organizations and anti-immigration organizations, many of which rely on racist narratives to paint women of color as a root of environmental ills (Hultgren, 2012; Sturgeon, 2009).[3] When groups have petitioned states to undertake population control strategies, these measures have frequently had racist overtones in that they are rarely envisioned to reduce the numbers of children born in predominantly white countries where consumption levels tend to be disproportionately high (Gosine, 2005).

Many scholars criticize population limitation development strategies that otherwise ignore or exploit poor women, yet make them the main target of population programs (Hartmann, 1995, 1999; Seager, 2003; Sen, 2004; Silliman and King, 1999). They feel that population control should not be made a substitute for directly addressing the poor economic situation that many of the world's women face (Urban, 2007). Rather, population policies should be critically assessed in order to expose why they are introduced, and who benefits from them (Hartmann, 1995). Questions like this raise the point that in some cases the health of women may benefit from family-planning measures or other population-related policies, although an uncritical link between women's health and population control or between population reduction and development masks the potential problems that these policies raise for women. Some feminist scholars also highlight the unequal negative ramifications that population-reduction policies have on women and girls, including high levels of female child abandonment or abortion (Hartmann, 1995, 1999; Hudson and den Boer, 2005).

Critical approaches, like feminism, are also helpful in evaluating why the population storyline has continued to be so popular in mainstream environmental discourse. Andil Gosine (2005: 78) argues that in order to do this we must consider the composition of environmentalism in the Global North for much of its history.

> Given the preservationist-conservationist history of recognized environmental movements in the U.S., its white, middle-upper-class membership, and the hegemonic hold of neoliberalism, the issues that came to dominate popular green agendas tended to be uncritical of capitalist-industrial culture. Overpopulation, which reduced the complex social, political, economic, and cultural dimensions of environmental degradation, and, more important, posed little challenge to capitalist industrialization and consumption, thus became a highly preferred bogeyman for greens.

This assessment suggests that shining a light on population is often done without simultaneously (or alternatively) spotlighting the connections between economic processes, consumption levels, lifestyle choices, and environmental change. Attention has been paid to the issue of consumption by IEP scholars in recent years (Dauvergne, 2008; Princen et al., 2002), but the issue of population remains a central and consistent part of environmental debates.

Another facet of debates about population comes in the form of discussion of population movement. Environmental migration, or populations moving to escape environmental damage or processes, is a subject that has been discussed with increasing frequency within IEP. The concept of environmental migration views the environment as a push factor triggering migration (Warner, 2010). This concept has been given even greater attention due to concerns about climate change impacts necessitating migration as an adaptation strategy (Biermann and Boas, 2010; Trombetta, 2008). The Intergovernmental Panel on Climate Change (IPCC) discussed the phenomenon of climate migration in their 2007 Fourth Assessment reports. The synthesis report claims that population migration is likely to occur in areas affected by drought, as a result of increases in storms like tropical cyclones, and due to rising sea levels (IPCC, 2007). The International Organization for Migration (IOM) acknowledges the difficulty in predicting how many people will be forced to relocate due to climate change impacts in the future.[4] Despite these limitations in establishing definitive numbers, the most widely cited estimate is that there will be around 200 million climate migrants by the middle of the century (IOM, 2012). Much of the existing literature on climate migration has treated it primarily as a state security issue (White, 2011), meaning that the problems of and solutions for addressing climate migration are defined at the state level. This is despite the fact that there are several studies that suggest that climate migration is not a primary cause of instability or social unrest (Buhaug and Urdal, 2013).[5]

How are processes associated with environmental migration in general, and climate migration in particular, gendered? At present, the vast majority of climate migration debates do not take gender into consideration in a strong fashion. This is despite the fact that feminist scholars have long considered the various ways that decisions to migrate and experiences of migration are gendered (Piper, 2008). Characteristics like gender, age, and socioeconomic status affect patterns of migration in the case of environmental and climate migration. Vulnerable or marginalized populations, including the poor, very young, elderly, women, and so on, may be less likely to have the ability to use migration as an adaptation strategy. Factors like family roles, resources distribution, and property rights affect men and women's migration patterns when the environment becomes a strong push factor (Warner, 2010). Gendered patterns of labor also impact the ability of migrants to achieve livelihood security in their new destination communities (Detraz and Windsor, forthcoming). These factors need to feature more centrally in debates about environmental migration now and into the future.

Gendering debates about environmental and climate migration is an important component of shifting attention away from the idea of migration as a state security threat to recognizing the human security implications of this phenomenon, and the justice aspect of debates about how to address the problem of environmental migration (Detraz and Windsor, forthcoming). Environmental migration should be treated as an environmental justice and human security issue in order to justly and effectively address the individual-level security threats that face those most at risk of climate change impacts. For policymakers, this includes reflecting on environmental migration as forced migration in the framework of international law, as well as striving to ensure gender equity throughout the policymaking process (Warner, 2010). For scholars, future shifts include engaging in studies that reveal the experiences of environmental migrants in order to better understand the sources of insecurity that they face (Detraz and Windsor, forthcoming).

We need to utilize discourses which reflect the gendered nature of both population migration and overall population growth, and which call attention to the complex social, economic, and political processes which undergird decisions and choices about reproduction and migration. The next section reflects on debates about population growth and movement, along with sustainable development, environmental justice, and environment and security through the case of biodiversity in South America.

A case study: Gender and biodiversity

Biodiversity refers to diversity at all levels of life, including plants, animals, and microorganisms. Biodiversity loss can occur through 'natural' processes, like the extinction of the dinosaurs, but much recent biodiversity loss has been attributed to human behavior. The past several decades have witnessed humans changing their environments at a far greater scale and speed than previously (MEA, 2005). These actions alter ecosystems and are often detrimental to species that live within them. This section examines how issues like sustainable development, environment and security, and population intersect with debates about gender, environmental degradation, and biodiversity in the Amazon.

Reflecting on the connections between gender and the environment can be done through thinking about the gendered implications of certain development paths within the Amazon. This includes assessing the equity of benefits as well as consequences to development. A contribution that gender lenses make to debates about sustainable development in particular is in problematizing expectations about where environmental expertise comes from. There is a need to recognize that women often have specific knowledge about biodiversity due to socially constructed roles and responsibilities within society, while also avoiding essentializing the experiences of women. Policy documents from Ecuador in particular draw a clear line between gender equity and biodiversity protection in several places. A recent text identifies rural women as central to the development of knowledge about biodiversity maintenance and intergenerational transmission of that knowledge. It prioritizes research on the practices of rural women in its acknowledgement of the importance of research on the causes of biodiversity loss (Government of Ecuador, 2001). Along these same lines UNESCO (2010: 1) says,

> In particular, the gendered division of labour has resulted in women and men in many societies having different roles and knowledge

related to biodiversity within their communities. However, while women are increasingly seen as embodying specific biodiversity knowledge and although an increasing number of experiences highlight the sustainable manner in which women use biodiversity, their role in biodiversity management and decision-making process is often ignored.

This quote implies that gender often appears in biodiversity discussions through an acknowledgement of socially conditioned roles and responsibilities that men and women play in society. It also suggests that women are frequently excluded from biodiversity policymaking and management.

Women are frequently excluded from policymaking which is securitized (Enloe, 2000). Recently there has been a tendency for states and other actors to use securitized discourses to discuss biodiversity in policy circles. There are examples of actors using an environmental conflict discourse to focus on the connections between biodiversity as natural resource and the state. In this case, this means a state is regarding biodiversity as a sovereign matter of national security, or more directly, a potential source of violent conflict over access to resources. In the case of the Amazon, states like Brazil have used the biopiracy frame when governing the Amazon to legitimize tight governmental controls on visas for academic research and other forms of control over the region.

A feminist environmental security discourse focuses on human security and the security of ecosystems. It is critical of the tendency for the value of biodiversity to be translated into militarized policies. This includes militarization of Amazonian borders. There have been several academic works recently published which criticize this militarized approach to the Amazon (Salisbury et al., 2010). States have used discourses of sustainable development and national security to justify the creation and maintenance of military outposts in the Amazon, and several scholars have feared that securitized language in the environmental realm could lead to the militarization of environmental policy. An environmental conflict discourse is consistent with marrying the narratives of state security and environmental protection. A feminist environmental security discourse, on the other hand, includes narratives of human security while also privileging the health and well-being of nonhuman species. This set of narratives also shifts attention away from the security of the state in considering resource conflicts, and focuses on the communities and ecosystems that are adversely impacted by those conflicts. It acknowledges that sources of insecurity are gendered. Men and women experience phenomena like land use conflicts or food insecurity – both of which are tied to biodiversity policies – based on the socially constructed roles and responsibilities they perform. A feminist environmental security discourse envisions a non-militarized approach to biodiversity conservation that shifts the policymaking discourse toward a concern about gendered instances of human insecurity.

One example of an important shift is how we think about issues of population. Population growth has been one of the central narratives in debates about both environmental security and environmental justice in the Amazon. There have been several examples of actors linking population growth or population migration to degradation of the Amazon (Barbieri and Carr, 2005; Caldas et al., 2010; Marcoux, 2000).[6] Brazil in particular has been the subject of debate over the trends of population migration to urban centers in the Amazon. According to recent government statistics, ten Brazilian cities in the Amazon have seen their population double over the past decade. This region's population climbed 23 percent between 2000 and 2010, compared to 12 percent growth for Brazil as a whole. Brazil has one of the lowest birthrates in South America at 1.86 children per woman. The Amazon region has the highest birthrate for the country at 2.42 children per woman (Romero, 2012).

It is also important to consider how migration in this region is gendered. Several studies have found that men and women tend to follow different migration patterns, with women being more likely to migrate to urban areas and men being more likely to migrate to rural areas and engage in farm work (Barbieri and Carr, 2005; Radcliffe, 1992).[7] Alisson Barbieri and David Carr (2005: 99) have studied migration patterns in the Ecuadorian Amazon and found:

> Economic factors such as engagement in on-farm work, increasing resources scarcity- measured by higher population density at the farm and reduction in farm land on forest and crops and increase in pasture land are more associated with male out-migration to rural areas. On the other hand, increasing resource scarcity, higher population density and weaker migration networks are more associated with female out-migration to urban areas.

The gendered nature of migration is thus an important component to understanding whether and how people migrate in the region. This is important for policymaking because if Amazon states choose policies to address migration in the region, they need to be grounded in an understanding of the socially conditioned factors which influence why people choose to migrate and where they go.

This discussion of biodiversity in the Amazon problematizes existing mainstream perspectives on the issue which are largely missing gender. It reveals that incorporating gender into discussions of biodiversity necessitates reflecting on questions like who suffers from biodiversity loss, who has specific expertise on biodiversity in the Amazon and where does this expertise come from, and how can we ensure that biodiversity is preserved while simultaneously protecting human security? Gender lenses contribute to raising questions about the causes of biodiversity loss and possible avenues toward its protection, which seek the health and well-being of both ecosystems and humans that inhabit many of them.

Conclusion

This chapter highlights the important connections that exist between gender and several topics/concepts that have been foundational to IEP. It is intended to contribute to the ongoing conversation between feminist environmental scholars and activists, and in general IEP scholars. Gender lenses contribute to our understanding of sustainable development, environmental justice, environmental security, and population growth/movement by problematizing existing ways of understanding each of these topics and revealing various ways that each intersect with gender. It is impossible to treat these topics as gender-neutral, just as it is impossible to treat them as separate categories for discussion and policymaking.

Some may claim that calling for broad, critical definitions of environmental concepts contributes to muddying the waters of IEP. Environmental issues are often incredibly complex and involve a large number of factors and actors. While I acknowledge this, I argue that it is a necessary component of identifying the larger processes that contribute to the related social ills of environmental degradation and the discrimination of marginalized populations. This can be related to environmental and ecological injustice, as well as to environmental and ecological insecurity. Feminist scholars have a great deal to contribute to the ways in which we understand and act on environmental issues. The task at hand encourages a dialog between scholars who have corresponding goals of preserving ecosystems and their inhabitants, including the humans within those ecosystems.

Notes

1. UN initiatives include specific references to gender in Agenda 21 (1992 United Nations Conference on Environment and Development); the Johannesburg Plan of Implementation (2002 World Summit on Sustainable Development); paragraph K of the Beijing Platform for Action (1995 Fourth World Conference on Women); the 1994 International Conference on Population and Development; the 2000 Millennium Declaration; and the requirements and agreements set out in the Convention on the Elimination of All Forms of Discrimination Against Women (UNEP, 2006).
2. The 1994 United Nations Human Development report outlined seven areas of human security: (1) economic security, (2) food security, (3) health security, (4) environmental security, (5) personal security, (6) community security, and (7) political security (UNDP, 1994).
3. Sturgeon (2009: 143) explains that

    a former board member of the Sierra Club, John Tanton, has since the 1970s built a set of anti-immigration organizations (including founding the Federation for American Immigration Reform) that are virulently racist while he simultaneously supported organizations (such as Population-Environment Balance) that claim that immigrants and overpopulation are the source of environmental problems.

4. Some of the difficulty in predicting the number of people who will become climate migrants stems from the differing definitions used to distinguish climate migrants from other types of migrants, along with the variability in climate models (Biermann and Boas, 2010).
5. Buhaug and Urdal (2013: 1) conducted a study which examined how population growth impacts patterns of public unrest in urban centers, and found 'little support for the notion that high and increasing urban population pressure leads to a higher risk or frequency of social disorder'. 'Instead, we find that urban disorder is primarily associated with a lack of consistent political institutions, economic shocks, and ongoing civil conflict.'
6. The population growth–deforestation linkage narrative has been applied to areas outside of the Amazon as well. Several studies find little support for this connection. For example, in a study of Basho Valley in the Western Himalayas of Pakistan, Ali et al. (2005) find little support that population growth was responsible for deforestation. Instead, they claim that mismanagement and illegal commercial harvesting endorsed by the Forest Department are primarily to blame for deforestation in the region.
7. While patterns of migration exist, there can be exceptions to these. For example, men sometimes move to rural areas of the Amazon to help build large hydroelectric dams.

Annotated bibliography

Arora-Jonsson, S. (2012) *Gender, Development, and Environmental Management: Theorizing Connections* (New York: Routledge).

This book is a recent exploration of the specific issue of gender in environmental governance. Bringing in examples and issues from India and Sweden, the book problematizes simplistic assumptions about development, gender equity, and environmental management.

Bretherton, C. (1998) 'Global Environmental Politics: Putting Gender on the Agenda?' *Review of International Studies*, 24, 1, 85–100.

This article, published in a widely read outlet for international relations scholarship, is an early example of a gendered perspective on the politics of environmental change. Bretherton challenges the effectiveness of attempts to include gender in environmental politics if we are just 'adding women in'. Instead she advocates a more holistic approach to gender.

Seager, J. (1993) *Earth Follies: Coming to Feminist Terms with the Global Environmental Crisis* (New York: Routledge).

This book is an early and foundational example of a broad, feminist environmentalism. Seager provides a thorough critique of many social institutions which she links to both patriarchy and environmental damage. These include the military (and processes of militarization), multinational corporations (and global capitalism), states, and elements of the environmental movement.

Silliman, J. and Y. King. (eds) (1999) *Dangerous Intersections: Feminist Perspectives on Population, Environment, and Development* (Cambridge, MA: South End Press).

This edited volume contains chapters by well-known feminist environmental scholars and covers an array of topics, such as population, development, militarism, consumption, and others. Overall, its chapters challenge simplistic assumptions about the links between population and environmental destruction, and call for treating the world's women as beings with agency.

Sturgeon, N. (1997) *Ecofeminist Natures: Race, Gender, Feminist Theory and Political Action* (New York: Routledge).

This book traces the history of ecofeminism, which has been an influential way that scholars have combined the areas of gender and the environment. Sturgeon provides a useful historical perspective on the early days of ecofeminist though, as well as a necessary account of the various ways that ecofeminism incorporates race within its understandings of environmental change and promise.

Works cited

Ali, J., T. A. Benjaminsen, A. A. Hammad and Ø. B. Dick (2005) 'The Road to Deforestation: An Assessment of Forest Loss and Its Causes in Basho Valley, North Pakistan.' *Global Environmental Change*, 15, 370–80.

Barbieri, A. F. and D. L. Carr (2005) 'Gender-specific Out-migration, Deforestation and Urbanization in the Ecuadorian Amazon.' *Global and Planetary Change*, 47, 99–110.

Barnett, J. (2001) *The Meaning of Environmental Security: Ecological Politics and Policy in the New Security Era* (New York: Zed Books).

Biermann, F. and I. Boas (2010) 'Preparing for a Warmer World: Towards a Global Governance System to Protect Climate Refugees.' *Global Environmental Politics,* 10, 1, 60–88.

Boehm, D. A. (2008) 'Now I Am a Man and a Woman! Gendered Moves and Migrations in a Transnational Mexican Community.' *Latin American Perspectives,* 35, 1, 16–30.

Bruyninckx, H. (2006) 'Sustainable Development: The Institutionalization of a Contested Policy Concept.' In *Palgrave Advances in International Environmental Politics,* ed. M. M. Betsill, K. Hochstetler and D. Stevis (New York: Palgrave Macmillan), 265–98.

Buhaug, H. and H. Urdal (2013) 'An Urbanization Bomb? Population Growth and Social Disorder in Cities.' *Global Environmental Change,* 23, 1–10.

Caldas, M. M., C. Simmons, R. Walker, S. Perz, S. Aldrich, R. Pereira, F. Leite and E. Arima (2010) 'Settlement Formation and Land Cover and Land Use Change: A Case Study in the Brazilian Amazon.' *Journal of Latin American Geography,* 9, 1, 125–44.

Convention on Biological Diversity and Women's Environment and Development Organization (2012) 'Gender Equality and the Convention on Biological Diversity: A Compilation of Decision Text.' http://www.wedo.org/wp-content/uploads/folletofinalCBD-WEB.pdf

Crowley, T. (2010) 'From "Natural" to "Ecosocial Flourishing": Evaluating Evaluative Frameworks.' *Ethics and the Environment,* 15, 1, 69–100.

Dauvergne, P. (2008) *The Shadows of Consumption: Consequences for the Global Environment* (Cambridge, MA: The MIT Press).

Davidson, M. J. and C. Krull (2011) 'Adapting to Cuba's Shifting Food Landscapes: Women's Strategies of Resistance.' *Cuban Studies,* 42, 59–77.

De Schutter, O. (2012) 'Women's Rights and the Right to Food.' http://www.srfood.org/images/stories/pdf/officialreports/20130304_gender_en.pdf

De Soysa, I. 2013. 'Environmental Security and the Resource Curse.' In *Environmental Security: Approaches and Issues,* ed. R. Floyd and R. A. Matthew (New York: Routledge), 64–81.

Detraz, N. (2012) *International Security and Gender* (Malden, MA: Polity Press).

Detraz, N. (2013) 'Gender and Environmental Security.' In *Environmental Security: Approaches and Issues,* ed. R. Floyd and R. A. Matthew (New York: Routledge), 154–68.

Detraz, N. and L. Windsor (2014) 'Evaluating Climate Migration: Population Movement, Insecurity and Gender.' *International Feminist Journal of Politics,* 16(1): 127–146.

Eaton, H. and L. A. Lorentzen (eds) (2003) *Ecofeminism and Globalization: Exploring Culture, Context, and Religion* (New York: Rowman & Littlefield Publishers, Inc).

Enloe, C. (2000) *Maneuvers: The International Politics of Militarizing Women's Lives* (Berkeley: University of California Press).

FAO (2011) 'The State of Food Insecurity in the World: How Does International Price Volatility Affect Domestic Economies and Food Security?' http://www.wedo.org/wp-content/uploads/The-State-of-Food-Insecurity-in-the-World-2011.pdf

Floyd, R. and R. A. Matthew (eds) (2013) *Environmental Security: Approaches and Issues* (New York: Routledge).

Goldsworthy, H. (2010) 'Women, Global Environmental Change, and Human Security.' In *Global Environmental Change and Human Security*, ed. R. A. Matthew, J. Barnett, B. McDonald and K. L. O'Brien (Cambridge, MA: The MIT Press), 215–36.

Gosine, A. (2005) 'Dying Planet, Deadly People: "Race"-Sex Anxieties and Alternative Globalizations.' *Social Justice*, 32, 4, 69–86.

Hardin, G. (1968) 'The Tragedy of the Commons.' *Science* December, 13, 127–51.

Hartmann, B. (1995) *Reproductive Rights and Wrongs: The Global Politics of Population Control* (Boston: South End Press).

Hartmann, B. (1999) 'Population, Environment, and Security: A New Trinity.' In *Dangerous Intersections: Feminist Perspectives on Population, Environment, and Development*, ed. J. Silliman and Y. King (Cambridge, MA: South End Press), 1–23.

Hartmann, B. (2010) 'Rethinking the Role of Population in Human Security.' In *Global Environmental Change and Human Security*, ed. R. A. Matthew, J. Barnett, B. McDonald and K. L. O'Brien (Cambridge, MA: The MIT Press), 193–214.

Homer-Dixon, T. (1999) *Environment, Scarcity, and Violence* (Princeton: Princeton University Press).

Hultgren, J. (2012) 'Natural Exceptions to Green Sovereignty? American Environmentalism and the "Immigration Problem."' *Alternatives: Global, Local, Political*, 37, 4, 300–316.

Hunter, L. M. and E. David (2011) 'Displacement, Climate Change and Gender.' In *Migration and Climate Change*, ed. É. Piguet, A. Pécoud and P. de Guchteneire (New York: Cambridge University Press), 306–30.

International Organization for Migration (2012) 'Migration and Climate Change.' *International Organization for Migration: Migration for the Benefit of All*. http://www.iom.int/jahia/Jahia/activities/by-theme/migration-climate-change-and-environment/copenhagen-and-beyond

IPCC (2007) 'Summary for Policymakers.' In ed. S. Solomon, D. Qin, M. Manning, Z. Chen, M. Marquis, K. B. Averyt, M. Tignor and H. L. Miller. http://www.ipcc.ch/pdf/assessment-report/ar4/wg1/ar4-wg1-spm.pdf

Kerr, R. B. (2005) 'Food Security in Northern Malawi: Gender, Kinship Relations and Entitlements in Historical Context.' *Journal of Southern African Studies*, 31, 1, 53–74.

King, Y. (1995) 'Engendering a Peaceful Planet: Ecology, Economy, and Ecofeminism in Contemporary Context.' *Women's Studies Quarterly*, 23, 15–21.

MacQuarrie, P. and A. T. Wolf. 2013. 'Understanding Water Security.' In *Environmental Security: Approaches and Issues*, ed. R. Floyd and R. A. Matthew (New York: Routledge), 169–86.

Marcoux, A. (2000) *Population and Deforestation*. Sustainable Development Department, FAO. http://www.fao.org/sd/wpdirect/WPan0050.htm

Merchant, C. (1996) *Earthcare: Women and the Environment* (New York: Routledge).

Mies, M. and V. Shiva (1993) *Ecofeminism* (Halifax, Nova Scotia: Fernwood Publications).

Newell, P. (2005) 'Race, Class and the Global Politics of Environmental Inequality.' *Global Environmental Politics*, 5, 3, 70–94.

Newman, P. (1994) 'Killing Legally with Toxic Waste: Women and the Environment in the United States.' In *Close to Home: Women Reconnect Ecology, Health and Development*, ed. V. Shiva (London: Earthscan Publications Ltd), 43–59.

Nsiah-Gyabaah, K. (2010) 'Human Security as a Prerequisite for Development.' In *Global Environmental Change and Human Security*, ed. R. A. Matthew, J. Barnett, B. McDonald and K. L. O'Brien (Cambridge, MA: The MIT Press), 237–60.
Oswald Spring, Ú. (2008) *Human, Gender and Environmental Security: A HUGE Challenge* (Bonn, Germany: UNU Institute for Environment and Human Security).
Parks, B. C. and J. T. Roberts (2006) 'Environmental and Ecological Justice.' In *Palgrave Advances in International Environmental Politics*, ed. M. M. Betsill, K. Hochstetler and D. Stevis (New York: Palgrave), 329–60.
Peterson, V. S. and A. S. Runyan (2010) *Global Gender Issues in the New Millennium* (Boulder: Westview Press).
Piper, N. (ed.) (2008) *New Perspectives on Gender and Migration: Livelihood, Rights and Entitlements* (New York: Routledge).
Princen, T., M. Maniates and K. Conca (2002) *Confronting Consumption* (Cambridge, MA: The MIT Press).
Radcliffe, S. (1992) 'Mountains, Maidens and Migration: Gender and Mobility in Peru.' In *Gender and Migration in Developing Countries*, ed. S. Chant (London: Belhaven Press), 30–48.
Rolston, H. III (2002) 'Justifying Sustainable Development: A Continuing Ethical Search.' *Global Dialogue*, 4, 1, 103–13.
Romero, S. (2012) 'Swallowing Rain Forest, Cities Surge in Amazon.' *The New York Times*. http://www.nytimes.com/2012/11/25/world/americas/swallowing-rain-forest-brazilian-cities-surge-in-amazon.html
Sandilands, C. (1999) *The Good-Natured Feminist: Ecofeminism and the Quest for Democracy* (Minneapolis: University of Minnesota Press).
Schlosberg, D. and D. Carruthers (2010) 'Indigenous Struggles, Environmental Justice, and Community Capabilities.' *Global Environmental Politics*, 10, 4, 12–35.
Seager, J. (2003) 'Rachel Carson Died of Breat Cancer: The Coming of Age of Feminist Environmentalism.' *Signs: Journal of Women in Culture and Society*, 28, 3, 945–72.
Shiva, V. (1989) *Staying Alive: Women, Ecology, and Development* (Atlantic Heights, NJ: Zed Books Ltd).
Silliman, J. and Y. King (1999) *Dangerous Intersections: Feminist Perspectives on Population, Environment, and Development* (Cambridge, MA: South End Press).
Simmons, C. S., M. M. Caldas, S. P. Aldrich, R. T. Walker and S. G. Perz (2007) 'Spatial Processes in Scalar Context: Development and Security in the Brazilian Amazon.' *Journal of Latin American Geography*, 6, 1, 125–48.
Sjoberg, L. and C. E. Gentry (2007) *Mothers, Monsters, Whores: Women's Violence in Global Politics* (London: Zed Books).
Stein, R. (ed.) (2004) *New Perspectives on Environmental Justice: Gender, Sexuality, and Activism* (New Brunswick, NJ: Rutgers University Press).
Sturgeon, N. (1997) *Ecofeminist Natures: Race, Gender, Feminist Theory and Political Action* (New York: Routledge).
Sturgeon, N. (2009) *Environmentalism in Popular Culture: Gender, Race, Sexuality, and the Politics of the Natural* (Tucson, AZ: University of Arizona Press).
Terry, G. (2009) 'No Climate Justice Without Gender Justice: An Overview of the Issues.' *Gender and Development*, 17, 1, 5–18.

Trombetta, M. J. (2008) 'Environmental Security and Climate Change: Analysing the Discourse.' *Cambridge Review of International Affairs,* 21, 4, 585–602.

UNDP (1994) 'New Dimensions of Human Security: Human Development Report 1994.' http://hdr.undp.org/en/reports/global/hdr1994/chapters/

United Nations Human Rights (2013) 'Empowering Women Is Shortcut to Tackling Hunger – UN Expert on Right to Food.' http://www.ohchr.org/EN/NewsEvents/Pages/DisplayNews.aspx?NewsID=13069&LangID=E

Uraguchi, Z. B. (2010) 'Food Price Hikes, Food Security, and Gender Equality: Assessing the Roles and Vulnerability of Women in Households of Bangladesh and Ethiopia.' *Gender and Development,* 18, 3, 491–501.

Urban, J. L. (2007) 'Interrogating Privilege/Challenging the "Greening of Hate".' *International Feminist Journal of Politics,* 9, 2, 251–64.

Verchick, R. R. M. (2004) 'Feminist Theory and Environmental Justice.' In *New Perspectives on Environmental Justice: Gender, Sexuality, and Activism,* ed. Rachel Stein (New Brunswick, NJ: Rutgers University Press), 63–77.

Walker, G. (2012) *Environmental Justice: Concepts, Evidence and Politics* (New York: Routledge).

Warner, K. (2010) 'Global Environmental Change and Migration: Governance Challenges.' *Global Environmental Change,* 20, 402–13.

Warren, K. J. (1997) *Ecofeminism: Women, Culture, Nature* (Bloomington: Indiana University Press).

Warren, K. J. (2000) *Ecofeminist Philosophy: A Western Perspective on What It Is and Why It Matters* (Boulder: Rowman & Littlefield).

7
Knowledge and the Environment
Eva Lövbrand

Introduction

Environmental policy debates are often permeated by claims to knowledge and expertise, perceptions of environmental risk, and scientific uncertainty. As a consequence, a growing scholarship has asked questions about how knowledge interacts with power and gains political effect in environmental affairs. Students of International Relations (IR) have primarily contributed to this analytical inquiry by examining the conditions under which scientific expert networks influence state behavior and help to foster interstate coordination on collective action problems. Speaking to liberal institutionalist analyses of environmental regime formation and implementation, this is a literature that has sought to explain how scientific expert advice informs states' calculations of costs and benefits of different courses of action and hereby affects patterns of international behavior. Some work in this field remains informed by the rationalist ontology of neoliberalism and treats scientific knowledge as a causal factor that is exchanged by fully informed actors. While scientific information may affect outcomes of strategic situations by increasing states' clarity about goals or ends-means relationships (Goldstein and Keohane, 1983: 3), it does not challenge the material foundations of global environmental politics.

For social constructivists, by contrast, the study of scientific knowledge in global environmental affairs has turned into a potent analytical site for exploring the constitutive power of ideas and discourses of the environment. Located at the intersection of constructivist IR theory, science and technology studies (STS), and postmodern social theory, this is a diverse literature with different analytical origins and trajectories. In this chapter I draw upon Fierke's (2013) distinction between conventional and critical constructivist approaches to outline some of these differences. Whereas both variations of constructivist theorizing are grounded in a social ontology that challenges the rationalist image of actors as atomistic egoists whose interests are formed prior to social interaction, 'conventional' constructivist IR theory maintains an interest in 'the standard international relations repertoire of interstate cooperation and diplomacy' (O'Neill, 2009: 6) and holds on to a positivist epistemology which includes hypothesis testing, causality, and explanation (Fierke, 2013). The social ontology is here advanced to explain how states respond to, and how international environmental cooperation is shaped by, the introduction of scientific information and ideas. The critical constructivist project, by contrast, understands knowledge as a political category that is situated in, and conditioned by, a particular social and historical context. As such it challenges the notion of value-neutral theorizing and instead calls for interpretative modes of understanding (Fierke, 2013). Informed by the postmodern ambition to destabilize and rethink the assumptions that have made things as they are, the critical social constructivist scholarship addressed in this chapter has critically interrogated the international as a natural or necessary starting point for a politics of the environment, and the nation-state as the primary mode of subjectivity in global environmental affairs.

In the following I seek to offer an overview of these scholarly debates and outline ontological differences that inform how the links between science and politics, knowledge and power, are conceptualized in the study of global environmental politics. To that end it is useful to make a distinction between the concepts of science, knowledge, and expertise. Although these concepts sometimes are used interchangeably, the social constructivist literature drawn upon in this chapter has devoted considerable effort to problematizing science as the only authoritative source of knowledge on the environment, and scientists as the primary environmental experts. By asking who counts as an expert and whose ways of knowing nature inform environmental decision-making, critical constructivists have fostered a normative research agenda that explores how marginalized forms of knowledge (for example, lay, citizen, local) may bring new ideas into environmental policy debates and hereby render nature governable in new ways. In order to reflect on these debates, this chapter navigates a diverse literature that extends well beyond the disciplinary parameters of IR theory and includes work from the fields of STS, political geography, and Foucauldian analyses of discourse and government.

The chapter is divided into four main sections. In the first I revisit Peter M. Haas' epistemic community theory and situate it in relation to environmental regime analysis and conventional constructivist theorizing. The second section reviews how the more recent literature on global environmental assessments has shifted the study of scientific influence in a more critical-interpretative direction. The making and validation of authoritative expertise here emerges as a site of politics in its own right that may redirect the patterns and dynamics of international behavior. The third section is concerned with postmodern efforts to deconstruct the constitutive effects of scientific representations of nature, and the fourth asks how scholarly efforts to bring non-scientific forms of knowledge to the fore may render nature governable in new ways. The chapter ends with a brief case study of the Intergovernmental Panel on Climate Change (IPCC). Through this case study I primarily illustrate how constructivist scholars, both conventional and critical, have interrogated the links between the politics of representation and representativeness in the IPCC.

Tracing the influence of scientific expert networks

The emergence of the environment as an issue area for scholars of IR has a distinct spatial history (Stevis, this book). Early works in the field appeared in the 1970s when environmental problems were increasingly understood as trans-boundary or global in character, traversing the political geography of the modern nation-state. As a consequence the environmental problématique was initially placed within the liberal institutionalist debate on interstate interdependence and coordination. Inspired by the extensive institutional development following the UN Conference on the Human Environment in Stockholm in 1972, neoliberal scholars used the environmental example to question the dominant realist assumption that cooperation among rival states was purely transitory in an anarchic international system (O'Neill, 2009: 9). Focus was instead turned to the conditions under which the environment as a collective action problem may foster 'cooperation under anarchy' (Oye, 1986) and produce durable international regimes around which state behavior converges (Young, 1987; Vogler, 1995). In the years to come, international environmental politics turned into a rich site for regime analysis that contributed greatly to empirical and theoretical refinement (Young, 1999; Paterson in this book).

Early work on the role of knowledge in environmental affairs emerged in relation to this neoliberal scholarship and raised questions about the conditions under which scientific expert networks may spur interstate coordination on environmental issues. Peter M. Haas' epistemic community theory represents the most influential contribution to this debate. Drawing upon a study of the Mediterranean Action Plan, Haas proposed a research agenda that 'examines the role that networks of knowledge-based experts – epistemic communities – play in articulating the cause-and-effect relationships of complex problems, helping states identify their interests, framing the issues for collective debate, proposing specific policies, and identifying salient points for negotiation' (Haas, 1992: 2, 1989). Haas defined epistemic communities as a knowledge-based transnational network of professionals with an authoritative claim to policy-relevant knowledge within a particular issue area. Members may come from various disciplines but share the same causal and normative beliefs derived from a basic body of knowledge within their domain of expertise. They also share the same normative agenda, that is, a conviction that certain policy actions will increase human welfare (Haas, 1992).

Haas' epistemic community theory gained ground in the post-Cold War era when social constructivists IR scholars were exploring how to combine material and ideational factors in the explanation of world politics. Rather than accepting the rationalist assumption that states are ontologically prior actors who can be studied independently of social interaction, the constructivist scholarship claimed that states are social beings whose identities and interests are formed *through* social interaction (Wendt, 1998). This social ontology opened up a new research agenda on 'how international norms evolve, how ideas and values come to shape political action, how argument and discourse condition outcomes and how identity constitutes agents and agency, all in ways that contradict materialist and rationalist theories' (Reus-Smith, 2005: 207). The epistemic community theory resonates with an actor-centered strand of constructivist theorizing that has explored how networks of domestic and transnational actors can bring new ideas and norms into policy debates and hereby influence the patterns of international behavior (see, for instance, Finnemore, 1996; Keck and Sikkink, 1998). This 'conventional' (Fierke, 2013) or 'first wave' of (Bernstein, 2013) constructivist scholarship does not denounce the importance of material factors in world affairs, nor strategic rationality. It does, however, recognize that state interests are bound up with a dense web of social interrelations which both reflect and help to sustain shared values, beliefs, and projects (Keck and Sikkink, 1998: 213).

In his early writings, Haas proposed that epistemic communities are most likely to shape international environmental cooperation under conditions of uncertainty, when decision-makers do not automatically know what course of action is in their political interest. When sorting out cause-and-effect relationships of complex environmental problems, epistemic communities may create a focal point around which international cooperation might converge (Haas, 1992). More recently, however, Haas has revisited this assumption and asked IR scholars not to overestimate the instances when ideational factors trump material capabilities. Leaning toward the rationalist ontology of neoliberalism, Haas has suggested that constructivist approaches are warranted only 'in cases with greater uncertainty about material national interests and national effects of environmental harm' (Haas and Stevens, 2011: 150). Against this 'thin constructivist' (Reus-Smith, 2005) backdrop, environmental scholars have over the years sought to specify when scientific information has causal weight in explaining international environmental policy coordination.

Considering the complexity and uncertainty of most global environmental problems, many neoliberals and constructivists agree that science is a central agenda-setter for international environmental politics. Stratospheric ozone depletion, long-range transboundary air pollution, and forest conservation are all examples of environmental issue areas where scientists are recognized as authoritative experts and have served to precipitate some kind of policy response (Litfin, 1994; Gulbrandsen, 2008; Lidskog and Sundqvist, 2011). However, the extent to which scientific expert advisors also have managed to shape the content of that response and hereby influence the 'principles, norms, rules and decision-making procedures' (Krasner, 1983: 2) designed by states to collectively solve environmental problems has been subject to considerable debate. Summarizing findings from a comparative study of interstate cooperation across environmental issue areas, Underdal (2000) and colleagues, for instance, found that scientific knowledge about environmental problems and available response options is a necessary – although by no means sufficient – condition for designing and operating effective international regimes. In order to shape policy outcomes, scientific expert networks need to perform a difficult balancing act between scientific autonomy and integrity, on the one hand, and policy responsiveness and involvement, on the other. Influential expert networks are those that maintain their privileged position as nature's objective 'ombudsman', and yet present knowledge in ways that are deemed relevant and useful to decision-makers (Underdal, 2000).

Interestingly, the comparative analyses conducted by Underdal et al. demonstrate that scientific autonomy serves different functions for developed and developing states. For policymakers in the North, the autonomy of scientific expert networks to operate according to their own professional standards is often a major source of confidence in their findings. However, for developing states that play a more marginal role in international research and have political interests that diverge from those in the North, the autonomy of science has proved to be a less reassuring arrangement (Underdal 2000:198). The finding that states' interpretations of scientific ideas can vary and clash has been developed by the more recent literature on global environmental assessments, and has raised important questions about the politics of scientific expert advice.

Scientific expertise as a site of politics and contestation

Global environmental assessments have been defined as organized efforts to assemble scientific knowledge with a view toward making it publicly available in a form intended to be useful for decision-making (Mitchell et al., 2006: 3). The Millennium Ecosystem Assessment (MA), the IPCC, and the more recent Intergovernmental Platform on Biodiversity and Ecosystem Services (IPBES) are renowned examples of such efforts that operate at the boundary of science and politics. According to Biermann (2002), environmental assessments of this kind have since the 1990s engaged large-enough networks of scientists, experts, government officials, and private bodies to be considered as distinct international institutions in their own right, consisting of internationally accepted principles, norms, and decision-making procedures for producing, synthesizing, evaluating, and legitimizing expert knowledge. When summarizing the political influence of this growing set of international expert institutions, Mitchell et al. (2006) identify scientific credibility, salience, and legitimacy as important procedural principles that determine their ability to speak authoritatively to governments.

Whereas the *credibility* of global environmental assessments hinges on the conviction that the included facts, theories, models, and causal beliefs are true or at least better than competing information, their *salience* has to do with expert advisors' ability to deliver information that meets the needs of decision-makers. The *legitimacy* of global environmental assessments, in turn, implies that the assessment process is viewed as fair, inclusive, and unpartisan by potential users

(Mitchell et al., 2006). The latter principle primarily speaks to the Southern distrust in the cognitive and normative homogeneity of dominant epistemic communities (Lahsen, 2004). Considering the well-documented disparity in the production of science among developed and developing countries (Karlsson et al., 2007), international expert institutions such as the IPCC have been criticized for feeding Northern research agendas and norms into global decision-making while neglecting the environmental concerns of the developing world (Jasanoff and Martello, 2004). This attention to the contextual and relational character of expert credibility, saliency, and legitimacy has challenged conventional constructivist efforts to develop a general theory that explains the causal weight of scientific ideas in global environmental politics. It has also gradually redrawn the boundaries of the political in the study of international environmental cooperation.

In contrast to Haas' early description of epistemic communities as a source of consensual knowledge exterior to the world of politics, the more recent literature on global environmental assessments approaches the making, validation, and legitimation of scientific expertise as an important site of environmental politics in its own right where shared causal beliefs and norms are negotiated, contested, and transformed through social interaction and learning (Litfin, 1994; Miller, 2004). Inspired by STS, this is a critical constructivist literature that does not take the cognitive authority of science for granted and hereby refutes the idea that scientific expert networks act as value-neutral or detached mediators of political disputes. Recognizing that knowledge often works to reinforce rather than resolve political conflict (Sarewitz, 2004), focus has instead turned toward the social processes by which authoritative knowledge gets framed, negotiated, and acted upon in distinct political settings. Informed by rich empirical case studies across environmental issue areas, work in this field has demonstrated how the boundary between science and politics, knowledge and power, is far from given but constantly redrawn in light of specific interests, ideas, and discourses.

While this politically entrenched understanding of scientific expertise has moved the study of global environmental assessments in a more critical-interpretative direction, an important dividing line can be drawn between social constructivist and structural forms of critique. The former is embedded in the STS scholarship's interest in the micro-social contexts of meaning-making, in which state and non-state actors translate the production and validation of knowledge claims into political authority (Miller, 2007). By examining the fluid and provisional boundaries that separate good science from bad, expertise from non-expertise, work in this field has sought to deconstruct the social practices that constitute 'certified knowledge' in global environmental affairs. The structural critique is more concerned with global knowledge asymmetries and the resulting North–South trust deficit in international environmental diplomacy. Resonating with neo-Marxist world system theory (for example, Roberts and Parks, 2007), this is a critique of Northern elitism and unjust rules of the game reproduced by international expert institutions at the expense of marginalized majorities in the South. While different in their analytical origin and trajectories, both forms of critique have recast global environmental assessments as sites of politics and contestation and hereby added new layers of analysis to Haas' original epistemic community theory. This critical literature does, however, retain an interest in 'the standard international relations repertoire of interstate cooperation and diplomacy' (O'Neill, 2009: 6). As such it differs from the critical constructivism advanced by postmodern IR scholars.

The representational politics of the environment

Environmental interdependence has since the 1970s been the starting point for studies of global environmental politics. The representation of the environment as a 'beyond-the-borders-issue' was conducive to its establishment within IR and formed the ground for international environmental policy coordination in the years to come. Postmodern IR theorists have used global environmental interdependence as a vantage point to question the nation-state as the organizing principle for world politics (Lipschutz and Conca, 1993; Litfin, 1998). In this literature 'the international' is not a natural or necessary starting point for a politics of the environment, nor is the nation-state the only possible mode of subjectivity. Global environmental politics, as we know it, is instead approached as a historical and conceptual achievement closely tied to particular representations of nature. Driven by the postmodern ambition to 'denaturalize the ways in which we think about the world, and relate to each other, state to state, person to person, people to nature' (Lipschutz, 1998: 132), work in this field has sought to historicize how the state has been given such centrality in environmental politics and hereby critically interrogate how things could be otherwise.

One analytical theme emerging from this critical constructivist scholarship focuses on the linkages between the micro-social contexts in which knowledge about the environment are produced and the macropolitical institutions that shape social and environmental change on global scales (Miller and Edwards, 2001: 12). Under the analytical label of 'co-production' (Jasanoff, 2004) STS scholars have asked how ways of seeing, knowing, and representing nature, often originating from the domains of science and technology, shape how the environment is construed and acted upon in political life. Sheila Jasanoff (2001) has, for instance, traced the complex pathways by which the image of planet Earth, first captured by the US space program in the 1960s, was established as an icon of global environmentalism. Rather than approaching satellite images as neutral optical tools that simply mirror reality, Earth observation technologies here emerge as powerful representational practices that are integral to the constitution of reality. Whereas Jasanoff's (2001) work effectively illustrates how the view of Earth from space was conducive to international environmental cooperation prior to and after the UN Conference on the Human Environment in 1972, critical constructivists are eager to highlight the historically bounded effects of representational practices. Since knowledge is never unconditioned, there can be no single Archimedean perspective that trumps all others. Accordingly, Karen Litfin (1998) has demonstrated how the global imagery produced by earth observation satellites has worked to both reinforce and undercut states' territorial control and authority in environmental politics.

The spatial work performed by more mundane scientific methods and techniques for collecting, measuring, aggregating, and projecting environmental data has also been subject to extensive postmodern scrutiny. The representational power of computer modeling is one such example. In Paul Edwards' (2001, 2010) historical analyses of climate science, for instance, the introduction of dynamic and coupled computer models during the 1980s and 1990s was central to the imagining of climate change as a global problem. By making 'inaccurate, incomplete, inconsistent, poorly calibrated, and temporally brief data *function as* global by correcting, interpolating, completing and gridding them' (Edwards, 2001: 62, italics in original), the practice of computer modeling shifted the very meaning of climate from a local to global understanding. Whereas several STS scholars have pointed at the close links between these scientific representations of the global climate system and the emergence of a global politics of the climate (Miller, 2004; Boyd, 2010), work in this field has also illustrated how international climate diplomacy has re-territorialized the climate problem and produced 'national natures'. In Lövbrand's and Stripple's (2006) study of the Kyoto negotiations on land use change and forestry, for instance, the introduction of standardized monitoring and reporting guidelines under the UN Framework Convention on Climate Change (UNFCCC) in the 1990s worked to rearticulate global flows of carbon as 'national sinks' and hereby adjust climate change to the territorial grammar of the nation-state.

This latter finding speaks to earlier 'histories of nature's production' (Braun, 2000: 14) offered by political geographers and historians of science. Inspired by Michel Foucault's rethinking of government and statehood in the late 1970s (for an overview, see Walters, 2012), this is a postmodern literature that has illustrated how graphical inscriptions such as surveys, inventories, maps, and accounting schemes are bound up with modern statecraft and the production of national territory as a political category. Scott (1998) and Agrawal (2005) have, for instance, detailed how the introduction of scientific forestry in early modern Europe and colonial India constituted forests as a natural resource to be systematized, ordered, and controlled on national scales. New procedures for measuring, aggregating, differentiating, and analyzing landscapes, vegetation types, and species created a legible national terrain that made possible the reworking of existing vegetation in terms of sustainable yields and profit maximization. Murdoch and Ward (1997) and Braun (2000) have, in turn, illustrated how mundane spatial practices such as the invention of agricultural statistics or geological surveys in 18th- and 19th-century Britain and Canada helped to make agricultural and geological territories thinkable as discrete sectors of the British and Canadian national economies.

Central to these critical constructivist analyses of modern statecraft is the assumption that natures do not come ready-made. As suggested by Baldwin (2003), the environment must first and foremost be understood as a political space, or a technological artifact, that is brought into being and gains meaning through representational practices and technologies. Hence, rather than looking for any presumed essence – a pristine, absolute nature – work in this field has drawn attention to the historically situated knowledge practices that render the environment intelligible as an object of government amenable to state intervention. In the 1990s, Foucauldian scholars such as Luke (1995) and Rutherford (1999) interpreted scientific efforts to quantify, map, monitor, and visualize the Earth's biogeochemical processes on national to global scales as expressions of a particular rationality of government – a 'green governmentality' – that reinforces the power of the administrative state in the name of responsible environmental stewardship. To save the planet, suggests Luke (1995: 77), 'it becomes necessary to environmentalize it, enveloping its system of systems in new disciplinary discourses to regulate population growth, economic development, and resource exploitation on a global scale with continual managerial intervention'.

These analytical efforts to trace the systems of thought that inform modern environmentalism represent another distinct theme in the postmodern study of global environmental politics. In recent years numerous studies have outlined the diverse, and often competing, set of ideas, discourses, and norms that underpin how the environment is represented and acted upon in political life (Litfin, 1994; Death, 2010). Through work in this field we have learned that the eco-managerial ambitions of green governmentality in recent years have been challenged by more decentralized discourses or norm-complexes such as 'liberal environmentalism' (Bernstein, 2001) or 'advanced liberal government' (Oels, 2005; Death, 2010). Rather than reproducing the administrative state as the only mode of subjectivity in environmental politics, these parallel ideational frameworks draw upon a new set of representational practices (for example, voluntary codes of conduct, private certification schemes, public–private partnerships) to govern environmental problems *through* the market or *through* responsible environmental citizenship (Biermann, this book). These efforts to explore how ideas and material practices interact in the constitution of the environment as a site of political intervention have often emerged outside the disciplinary boundaries of IR theory. Although critical constructivist IR scholars maintain an interest in the normative underpinnings of environmental regimes, and shifts in the identities of states participating in them (see, for instance, Bernstein, 2013), the nation-state is not approached as a *natural* starting point for work in this field. Postmodern contributions to this analytical inquiry have instead used the environment as a potent site to raise critical questions about *how* the world is spatially and temporally produced and what subject formations and modes of engagement such productions make possible (Litfin, 1998).

By recognizing that environmental politics is not a product of nature, but constituted through distinct representational practices, this postmodern scholarship has paved the way for a normative research agenda that seeks to open up environmental policy debates to alternative ways of knowing nature. By asking whose nature is being represented and what the material effects of such representations are, this is a literature that seeks to 'democratize' the tools that mark the world (Baldwin, 2003: 425) and hereby bring 'the local back in' (Jasanoff and Martello, 2004) to the study and practice of global environmental politics.

Bringing the local back in

Since the UN Conference on Environment and Development (UNCED) was held in Rio de Janeiro in 1992, new voices have been brought to the fore in global environmental politics. As recognized in Agenda 21, indigenous peoples, environmental groups, and youth all have a role to play in the quest for sustainable development. As a consequence, there is today a well-documented participation of transnational actors and interest groups in international environmental fora (Betsill and Corell, 2008; Betsill, this book). This participatory or deliberative turn in global environmental politics (Bäckstrand et al., 2010) speaks to scholarly efforts to refashion environmental expertise in more transparent, accountable, and democratic directions. Rather than approaching 'science' as the primary cognitive resource for addressing environmental problems, a growing number of STS scholars have promoted the broader category of 'knowledge' to re-embed expert claims in local experiences of nature. As argued by Jasanoff and Martello (2004), environmental politics has historically been a politics of the local. 'It derives emotional force from people's attachment to particular places, landscapes, and livelihoods, and to an ethic of communal living that can sustain stable, long-term regimes for the protection of shared resources' (Jasanoff and Martello, 2004: 7). However, through the work of modern science and technology, these ties to the manifold local and subjective experiences of nature have been cut loose so that claims may float freely and persuade people as objective facts. As a consequence, it is argued, the making and mobilizing of knowledge about global environmental problems has lost its geographical sensibility and meaning (Hulme, 2010a: 3).

Global representations of climate change are often invoked as typical examples of the detached and impoverished gaze of science. ' "Climate change" – the scientific phenomenon – employs techniques of aggregation and deletion, calculation, and comparison that exhaust the capacities of even the most meticulously recorded communal memories. Indeed, climate change arguably displaces the very notion of community by displacing human beings, both as a species and as a source of norms, in favor of an impersonal, but naturalized, object of concern' (Jasanoff, 2010: 237). From this critical constructivist vantage point, scientific ways of knowing and seeing the climate constitute an immensely powerful technology of simplification and legibility that decouples the various components of the climate system (for example, forests, agricultural lands) from their ecological contexts and paves the way for technical fixes as the only imagined response options (Boyd, 2010). Derived from a Western epistemology that constructs the environment as an exterior object to be monitored and controlled, the representational practices of climate science pay little attention to alternative ways of experiencing the weather, embedded in local cultural practices and knowledge-making traditions (Long-Martello, 2008; Hulme, 2010a: 3).

Scholarly efforts to escape this detached epistemic frame of environmental politics, and to introduce broader and more multiple meanings to the practice of environmental governance, have in recent years drawn renewed attention to the institutional processes for producing, validating, and disseminating expert advice in global environmental assessments such as the IPCC and the emerging IPBES. When recognizing the power of these international expert institutions to shape the epistemic foundations for global policy decisions, critical constructivist STS scholars have brought questions of representativeness, transparency, and accountability to the fore. 'Who is participating and in what ways in deliberating policy ideas? How are policy problems framed? What rules of evidence and evidentiary standards are adopted? How is expertise defined? How are the credibility and trustworthiness of epistemic claims certified and warranted?' (Miller, 2009: 150). Most work in this field shares the assumption that the doors to previously closed expert fora need to be opened up to public scrutiny, debate, and non-scientific ways of knowing nature. However, the normative rationales for such epistemic pluralism are not always the same.

Some writers in this field see more transparent and participatory procedures for producing, validating, and disseminating expert knowledge as a way to bridge the increasing gulf between science and society and hereby securing public trust in global environmental assessments (Lidskog and Sundqvist, 2011). In the wake of scientific controversies such as the 'Climategate' scandal in 2009, it is no longer enough for global environmental assessments to be transparent toward their scientific peers. Expert credibility and authority instead hinge on inclusive public debate and accountability (Jasanoff, 2010; Beck, 2012). Others have taken the argument one step further and suggested that more inclusive expert practices represent a democratic value in its own right. If we accept that the representational practices of international expert institutions have ramifications on the every-day life of people around the world, they must be subject to democratic scrutiny and control. While some scholars have invoked deliberative or cosmopolitan ideals of democracy to imagine new arenas for civic deliberation, dialog, and knowledge exchange (Hulme, 2010b), others have focused on how to strengthen democratic values within existing intergovernmental decision processes. As suggested by Miller (2009: 159), '[m]any of the same principles that are routinely required of policymaking processes at national scales – openness and transparency, critique and appeal, accountability – can also be required of institutional processes that contribute to the creation, deliberation and certification of policy-relevant knowledge and ideas in international governance'.

The third, and closely related, rationale for such 'epistemic constitutionalism' (Miller, 2009) is grounded in the recognition that complex environmental problems cannot be resolved by the uniform and reductionist ways of seeing and knowing nature offered by science. To rethink '[w]ho counts as expert, whose knowledge is deemed relevant, and who participates in advisory and negotiating bodies' (Jasanoff and Martello, 2004: 344) here emerges as an important avenue to increased problem-solving capacity. By recognizing the validity of traditional, indigenous, and lay knowledge ways, international expert institutions may foster a healthy epistemic pluralism that can help to render the environment governable in new and productive ways (Long-Martello, 2008; Turnhout et al., 2012). It is worth emphasizing that these efforts to re-embed international environmental expertise in 'the local' do not necessarily imply a rethinking of global environmental politics in more decentralized directions. The critical constructivist STS scholarship addressed in this chapter does not primarily speak to green political ideals such as bioregionalism or post-sovereign politics (for an overview, see Paterson, this book), neither does it represent a critique of the nation-state *per se*. 'The local' is instead mobilized to displace the modern quest for a universal point of reference with the postmodern recognition that there is always more than one perspective and that each perspective embodies a particular set of values. As suggested by Jasanoff and Martello (2004), all knowledge, including the 'hardest' scientific facts, is situated in particular social and cultural contexts and hereby brings with it locally embedded ways of seeing and knowing nature. To develop geographically sensitive expert institutions thus implies fostering environmental knowledge that recognizes 'its place' and, in the words of Hulme (2010a: 1), turns 'attention away from the globalizing instincts that so easily erase difference and collapse meaning'.

The Intergovernmental Panel on Climate Change

The IPCC is at the center of most debates on the power of knowledge and expertise in global environmental affairs. Acting as one of the largest and most recognized global environmental assessments, the IPCC has since its inception been subject to extensive scholarly scrutiny and critique. In the following section I will exemplify how conventional and critical constructivist IR and STS scholars have interrogated the links between the politics of *representation* and *representativeness* in the IPCC. While much work in this field has highlighted the constitutive role that climate science has played in the making of climate change as a coherent object of interstate cooperation and regulation, critics have also questioned the selective representation of climate change produced by the knowledge practices and supporting infrastructures of the IPCC. By disclosing patterns of inclusion and exclusion generated by the predominant Northern gaze of the IPCC's assessment reports, much of this literature has sought to denaturalize how the climate problem has been rendered amenable to political intervention and hereby make more participatory forms of climate governance possible.

Brief background to the IPCC

The IPCC was founded in 1988 by the World Meteorological Organization (WMO) and the United Nations Environment Programme (UNEP). Following a series of expert workshops and conferences on natural and human-induced climate variations in the 1980s (for a detailed historical account, see Agrawala, 1998b), the IPCC responded to a mounting scientific and political concern for the potential risks of rising concentrations of greenhouse gases in the atmosphere. The official mandate of the IPCC is outlined in the UN General Assembly Resolution 43/53 of 6 December 1988 which asks the WMO and UNEP to conduct a comprehensive review with respect to the state of scientific knowledge on climate change and available legal instruments, and to give advice on elements to be included in a future international convention on climate (UNGA, 1988: article 10). At the first IPCC plenary session in November 1988, three working groups were set up to fulfill this mandate. Working Group (WG) I was given the task to assess the state of scientific knowledge on climate change; WG II to assess the environmental and socioeconomic impacts of climate change; and WG III to formulate response strategies (Skodvin, 2000).

The first chairman of the IPCC, the Swedish meteorologist Bert Bolin, has reflected upon this division of labor and has suggested that it was central for the scientific credibility of the IPCC. Acting as an intergovernmental body, governed by the member states of the WMO and UNEP, the IPCC had to adopt an institutional design that secured the autonomy of its more than 2000 participating scientists. The separation of the science of climate change assessed by WG I, from the more politically charged policy recommendations produced in WG III, was thought to be a necessary step in that direction (Bolin, 2010). At the same time, the intergovernmental design of the IPCC has been central for its political salience and legitimacy. By granting government delegates the power to determine the mandate of the panel, to elect chairs for its subsidiary bodies, and to review and approve the assessment reports before they are published, the IPCC was designed to foster results of relevance to its principals, that is, the UN member states (Agrawala, 1998b).

Since the publication of the IPCC's First Assessment Report at the Second World Climate Conference in 1990, the IPCC has in total produced five assessment reports. During this time the IPCC has also published a wide range of technical papers and nine special reports on topics such as emission scenarios, carbon capture and storage, renewable energy, and extreme weather events. All reports have been produced in tandem with the interstate negotiations on climate change. Although the IPCC is formally and operationally autonomous from the UN climate regime, many of the special reports and technical papers have been prepared in direct response to requests by the Conference of the Parties to the UNFCCC.

Representation and representativeness in the IPCC

Many STS scholars have interpreted this close link between the IPCC and the UN climate regime as the prime example of how scientific knowledge practices are co-constitutive with the patterns and practice of global environmental politics. Since its inception, the IPCC has continuously informed states on how to understand and respond to the problem of climate change and hereby defined the epistemic framework for global climate policy. Over the course of two decades, argues Miller (2004: 339), 'scientists and other advocates persuaded the global community not only of the existence of a new global kind – the climate system – but also that threats to that system demanded concerted, worldwide cooperation and the creation of a major new suite of international institutions'. Climate change, which previously had been framed in terms of local changes in weather, was through the IPCC's assessment reports re-defined as a global threat that required global solutions (Miller, 2004).

Conventional constructivist IR scholars have used the epistemic community framework to analyze this representational power of the IPCC. According to Sonja Boehmer-Christiansen (1996: 190), for instance,

[t]he climate-change threat rose to the top of the agenda in international and diplomatic debates not so much because science meekly responded to public concern but more because a global network of organisations, devoted to research into atmospheric sciences, climate forecasting, ecosystems, energy-demand forecasting and new energy and information-gathering technologies, actively helped to create and disseminate concern.

The IPCC here emerges as a powerful epistemic community that has used the narrative structures of its assessment reports to boost credibility for an understanding of climate change that supports the policy agendas of particular interest groups (Miller, 2004). Haas and Stevens (2011), by contrast, are more skeptical about IPCC's ability to influence international climate politics and policymaking. Since the IPCC was designed to 'keep science on a tight leash', they find that this particular epistemic community has been unable to exercise sufficient discretion to develop politically tractable advice (Haas and Stevenson, 2011: 147). Rather than offering a focal point around which state interests can converge, several case studies have illustrated how the IPCC assessment process has functioned as a site of politics where struggles over what counts as credible, salient, and legitimate knowledge are played out (Fogel, 2005; Beck, 2012).

For critical constructivists, these struggles foreground the constitutive effects of representation. Rather than offering passive reflections of reality, visual images of the world – such as the global climate system – favor and promote particular forms of social relations. The extensive social constructivist critique of the IPCC should be read against this background. Many scholars have over the years brought the question of representativeness to the fore in order to problematize the selective politics of the IPCC's knowledge practices and supporting infrastructures. Already prior to the publication of the IPCC's first assessment report, critics raised questions about Northern bias (Agrawala, 1998a; Bolin, 2010). In order to counter the unbalanced geographical representation of the participating scientists, the IPCC Bureau at its very first session in February 1989 decided to establish an 'Ad-hoc Sub-group on Ways to Increase Participation of the Developing Countries in IPCC Activities'. Recognizing that the Northern dominance in the IPCC assessment process could undermine the credibility of the resulting reports, the Bureau has since then worked to increase the proportion of developing country experts in the IPCC network (Biermann, 2002; Bolin, 2010).

While the geographical balance has improved over time, critical constructivists maintain that the IPCC continues to represent climate change in ways that primarily speak to Northern interests and concerns (Jasanoff, 2010; Hulme, 2010a). This critique has in recent years resulted in suggestions for IPCC reform. In the wake of the 'Climategate' controversy that erupted in November 2009 after compromising emails by leading IPCC scientists had leaked to the blogosphere, many STS scholars have taken the opportunity to rethink the IPCC's procedures for producing and validating climate expertise. Interpreting the widespread critique in media and the blogosphere as an example of the limited public trust in the elite procedures of the IPCC, the need for new forms of representation have been at the forefront of scholarly concern (for an overview, see Beck, 2012). In October 2010 the IPCC plenary adopted new rules of procedure that require all working groups to better document and communicate how they respond to external review comments throughout the assessment process. While this institutional reform was set in motion to ease the public critique against the panel, Beck (2012) notes that it does not reflect any major rethinking of the IPCC's role and place in global climate politics. On the contrary, the current IPCC chairman Rajendra Pachauri continues to portray 'the panel as untainted, distinct from politics, and as a provider of neutral, sound scientific expertise' (Beck, 2012: 157).

Conclusions

The power of scientific knowledge and ideas is today an established analytical theme in the study of global environmental politics. Efforts to understand how environmental problems are construed and acted upon in world politics can seldom escape the centrality of science and international expert institutions. As a consequence, the making, validation, and dissemination of legitimate environmental expertise has turned into an important analytical site in its own right that now engages a broad scholarship well beyond the disciplinary parameters of IR theory. In this chapter I have investigated the different ways in which the links between science and politics, knowledge and power, are conceptualized and studied in this literature. Most studies emerging from this analytical domain have pointed at the close links between scientific representations of environmental problems and the political response options devised in national and international settings. When examining in detail how expert knowledge is mobilized and deployed in international environmental diplomacy, it seems difficult to retain a strict demarcation between science and politics, knowledge and power.

Peter M. Haas' epistemic community theory remains an important reference point for work in this field. Rather than explaining international environmental policy coordination as the result of states' material interests and capabilities alone, the epistemic community model has foregrounded scientific expertise and epistemic consensus as important causal variables for environmental regime formation and implementation. While Haas' early constructivist theorizing has offered IR scholars an analytical platform for the study of ideational factors in global environmental politics, recent work in this field has taken a more critical direction and today seldom approaches scientific expertise as a source of consensual knowledge around which state interests may converge. Grounded in the postmodern assumption that knowledge always is situated and conditioned, much of the critical literature addressed in this chapter has challenged the positivist language of conventional constructivism and instead turned attention to the situated modes of representation that give rise to the objects of environmental politics to which IR theory is responding.

Whereas some critical constructivist scholars maintain an interest in international environmental regimes and the role international expert institutions plays therein, important work in this field draws upon postmodern social theory to question the naturalness or necessity of existing ways of organizing, occupying, and administrating political space. By asking how conventional ways of representing nature are bound up with the states' historical re-constitution, postmodern IR scholars have begun to problematize the nature and location of environmental politics. The postmodern STS scholarship, by contrast, has to date directed most critical attention to the cognitive power of science and critically interrogated how Northern expert claims condition the epistemic foundations of global environmental politics. By asking who counts as an expert, and whose understandings of nature inform environmental decision-making, work in this field has sought to transcend the 'global gaze' of international expert institutions and, in the words of Jasanoff (2010: 49), 'make us more aware, less comfortable, and hence more reflective about how we intervene, in word or deed, in the changing order of things'.

The extent to which contemporary efforts to reform the procedures of global environmental assessments, such as the IPCC, will harness the critical potential of postmodern social theory and hereby foster political imaginaries that cut across the conventional spatial grammar of IR remains to be seen. To invite new actors to debate the epistemic frameworks underlying environmental policy decisions may indeed bring different styles of reasoning in productive dialog and exchange with another (Miller, 2009). However, how such epistemic plurality will transform the environment as a site of political intervention remains a central analytical question that should engage critical STS and IR scholars in the years to come.

Annotated bibliography

Haas, P. M. (1989) *Saving the Mediterranean. The Politics of International Environmental Cooperation* (New York: Columbia University Press).

This is the volume in which Haas' first developed his 'epistemic community' theory. Although his depiction of science as a source of consensual knowledge has been widely challenged, his study remains a key reference point in the field.

Litfin, K. (1994) *Ozone Discourses, Science and Politics in Global Environmental Cooperation* (New York: Columbia University Press).

In this book Litfin offers an early critical constructivist analysis of the role of knowledge and expertise in international environmental cooperation. Informed by Michel Foucault, Litfin develops a productive power concept to examine the constitutive effects of scientific expert practices.

Jasanoff, S. and M. Long-Martello (eds) (2004) *Earthly Politics. Local and Global in Environmental Governance* (Cambridge, MA: The MIT Press).

This edited volume draws upon STS to critically interrogate the close links between scientific representations of nature and the global governance of environmental problems. The volume offers a critique of the global gaze of scientific expert institutions and calls for more locally grounded and humble forms of environmental expertise.

Bernstein, S. (2001) *The Compromise of Liberal Environmentalism* (New York: Columbia University Press).

In this important volume, Bernstein effectively illustrates how constructivist IR scholars can add to neoliberal analyses of environmental regime formation and implementation. In focus are the ideas and norms underpinning environmental regimes, and how these ideational structures may affect the patterns of international behavior.

Works cited

Agrawal, A. (2005) *Environmentality. Technologies of Government and the Making of Subjects* (Durham and London: Duke University Press).

Agrawala, S. (1998a) 'Structural and Process History of the Intergovernmental Panel on Climate Change', *Climatic Change* 39, 4, 621–642.

Agrawala, S. (1998b) 'Context and Early Origins of the Intergovernmental Panel on Climate Change', *Climatic Change* 39, 4, 605–620.

Bäckstrand, K., J. Khan, A. Kronsell and E. Lövbrand (eds) (2010) *Environmental Politics and Deliberative Democracy: Examining the Promise of New Modes of Governance* (Cheltenham: Edward Elgar).

Baldwin, A. (2003) 'The Nature of the Boreal Forest: Governmentality and Forest-Nature', *Space and Culture* 6, 415–428.

Barnett, M. and R. Duvall (eds) (2005) *Power in Global Governance* (Cambridge: Cambridge University Press).

Beck, S. (2012) 'Between Tribalism and Trust: The IPCC Under the "Public Microscope"', *Nature and Culture* 7, 2, 151–173.

Bernstein, S. (2001) *The Compromise of Liberal Environmentalism* (New York: Columbia University Press).

Bernstein, S. (2013) 'Global Environmental Norms', in Falkner, R. (ed.) *The Handbook of Global Climate and Environment Policy* (Chichester: Wiley-Blackwell).

Betsill, M. and E. Corell (eds) (2008) *NGO Diplomacy: The Influence of Non-Governmental Organisations in International Environmental Negotiations* (Cambridge, MA and London: The MIT Press).

Biermann, F. (2002) 'Institutions for Scientific Advice: Global Environmental Assessments and their Influence in Developing Countries', *Global Governance* 8, 195–219.

Boehmer-Christiansen, S. (1996) 'The International Research Enterprise and Global Environmental Change. Climate-Change Policy as Research Process', in Vogler, J. and M. Imber (eds) *The Environment and International Relations* (London and New York: Routledge).

Bolin, B. (2010) *A History of the Science and Politics of Climate Change. The Role of the Intergovernmental Panel on Climate Change* (Cambridge: Cambridge University Press).

Boyd, W. (2010) 'Ways of Seeing in Environmental Law: How Deforestation Became an Object of Climate Governance', *Ecology Law Quarterly*, 37, 843–914.

Braun, B. (2000) 'Producing Vertical Territory: Geology and Governmentality in Late Victorian Canada', *Ecumene*, 7, 1, 7–46.

Death, C. (2010) *Governing Sustainable Development. Partnerships, Protests and Power at the World Summit* (New York: Routledge).

Edwards, P. N. (2010) *A Vast Machine. Computer Models, Climate Data and the Politics of Global Warming* (Cambridge, MA and London: The MIT Press).

Edwards, P. N. and S. H. Schneider (2001) 'Self-Governance and Peer Review in Science-for-Policy: The Case of the IPCC Second Assessment Report', in Miller, C. A. and P. N. Edwards (eds) *Changing the Atmosphere* (Cambridge, MA and London: The MIT Press).

Epstein, C. (2006) 'The Making of Global Environmental Norms: Endangered Species Protection', *Global Environmental Politics* 6, 2, 32–54.

Fierke, K. M. (2013) 'Constructivism', in Dunne, T., M. Kurki and S. Smith (eds) *International Relations Theory. Discipline and Diversity*, Third Edition (Oxford: Oxford University Press).

Fogel, C. (2005) 'Biotic carbon sequestration and the Kyoto Protocol: The Construction of Global Knowledge by the Intergovernmental Panel on Climate Change', *International Environmental Agreements* 5,191–210.

Goldstein, J. and R. O. Keohane (eds) (1993) *Ideas and Foreign Policy: Beliefs, Institutions, and Political Change* (USA: Cornell University Press).

Gulbrandsen, L. (2008) 'The Role of Science in Environmental Governance: Competing Knowledge Producers in Swedish and Norwegian Forestry', *Global Environmental Politics* 8, 2, 99–122.

Haas, P. M. (1989) *Saving the Mediterranean. The Politics of International Environmental Cooperation* (New York: Colombia University Press).

Haas, P. M. (1992) 'Introduction: Epistemic Communities and International Policy Coordination', *International Organization* 46, 1, 1–35.

Haas, P. M. and C. Stevens (2011) 'Organized Science, Usable Knowledge, and Multilateral Environmental Governance', in Lidskog, R. and G. Sundqvist (eds) *Governing the Air. The Dynamics of Science, Policy, and Citizen Interaction* (Cambridge, MA and London: The MIT Press).

Hulme, M. (2010a) 'Problems with Making and Governing Global Kinds of Knowledge', *Global Environmental Change* 20, 558–564.

Hulme, M. (2010b) 'Cosmopolitan Climates: Hybridity, Foresight and Meaning', *Theory, Culture and Society* 27, 2–3, 267–276.

Jasanoff, S. (2001) 'Image and Imagination', in Miller, C. and P. Edwards (eds) *Changing the Atmosphere. Expert Knowledge and Environmental Governance* (Cambridge, MA: MIT Press).

Jasanoff, S. (ed.) (2004) *States of Knowledge: The Co-Production of Science and Social Order* (London: Routledge).

Jasanoff, S. (2010) 'A New Climate for Society', *Theory, Culture and Society* 27, 2–3, 233–253.

Jasanoff, S. and M. Long Martello (eds) (2004) *Earthly Politics. Local and Global in Environmental Governance* (Cambridge, MA: The MIT Press).

Keck, M. and K. Sikkink (1998) *Activists Beyond Boarders* (Ithaca and London: Cornell University Press).

Krasner, S. D. (ed.) (1989) *International Regimes* (Ithaca: Cornell University Press).

Lahsen, M. (2004) 'Transnational Locals: Brazilian Experiences of the Climate Regime', in Jasanoff, S. and M. Long-Martello (eds) *Earthly Politics. Local and Global in Environmental Governance* (Cambridge, MA: The MIT Press).

Lidskog, R. and G. Sundqvist (eds) (2011) *Governing the Air. The Dynamics of Science, Policy, and Citizen Interaction* (Cambridge, MA and London: The MIT Press).

Lipschutz, R. D. (1998) 'The Nature of Sovereignty, and the Sovereignty of Nature', in Litfin, K. (ed.) *The Greening of Sovereignty in World Politics* (Cambridge and London: The MIT Press).

Lipschutz, R. D. and K. Conca (eds) (1993) *The State and Social Power in Global Environmental Politics* (New York: Columbia University Press).

Litfin, K. (1994) *Ozone Discourses, Science and Politics in Global Environmental Cooperation* (New York: Columbia University Press).

Litfin, K. (1998) 'Satellites and Sovereign Knowledge', in Litfin, K. (ed.) *The Greening of Sovereignty in World Politics* (Cambridge and London: The MIT Press).

Long-Martello, M. (2008) 'Arctic Indigenous Peoples as Representations and Representatives of Climate Change', *Social Studies of Science* 38, 3, 351–376.

Lövbrand, E. and J. Stripple (2006) 'The Climate as Political Space: On the Territorialisation of the Global Carbon Cycle', *Review of International Studies* 32, 217–235.

Luke, T. (1995) 'On Environmentality: Geo-Power and Eco-Knowledge in the Discourses of Contemporary Environmentalism', *Cultural Critique* 31, 57–81.

Miller, C. (2004) 'Climate Science and the Making of a Global Political Order', in Jasanoff, S. (ed.) *States of Knowledge: The Co-Production of Science and Social Order* (London, New York: Routledge).

Miller, C. (2007) 'Democratization, International Knowledge Institutions, and Global Governance', *Governance* 20, 2, 325–357.

Miller, C. (2009) 'Epistemic Constitutionalism in International Governance: The Case of Climate Change', in Heazle, M. G. and T. Conley (eds) *Foreign Policy Challenges in the 21st Century* (London: Edward Elgar Press).

Miller, C. and P. Edwards (2001) *Changing the Atmosphere. Expert Knowledge and Environmental Governance* (Cambridge, MA: MIT Press).

Mitchell, R. B., W. C. Clark, D. W. Cash and N. M. Dickson (eds) (2006) *Global Environmental Assessments. Information and Influence* (Cambridge MA and London: The MIT Press).

Murdoch, J. and N. Ward (1997) 'Governmentality and Territoriality: The Statistical Manufacture of Britain's "National Farm"', *Political Geography* 16, 4, 307–324.

Oels, A. (2005) 'Rendering Climate Change Governable: From Biopower to Advanced Liberal Government', *Journal of Environmental Policy and Planning* 7, 3, 185–207.

O'Neill, K. (2009) *The Environment and International Relations* (Cambridge: Cambridge University Press).

Oye, K. (ed.) (1986) *Cooperation under Anarchy* (Princeton: Princeton University Press).

Reus-Smith, C. (2005) 'Constructivism', in Burchill, S., Linklater, A., Devetak, R., Donnelly, J., Paterson, M., Reus-Smith, C. and True, J. (eds) *Theories of International Relations*, 3rd Edition (Hampshire and New York: Palgrave Macmillan).

Roberts, J. T. and B. C. Parks (2007) *A Climate of Injustice. Global Inequality, North-South Politics, and Climate Policy* (Cambridge MA and London: The MIT Press).

Rutherford, P. (1999) 'The Entry of Life into History', in Darier, E. (ed.) *Discourses of the Environment* (Massachusetts: Blackwell Publishers).

Sarewitz, D. (2004) 'How Science Makes Environmental Controversies Worse', *Environmental Science and Policy* 7, 385–403.

Scott, J. (1998) *Seeing Like a State: How Certain Schemes to Improve the Human Condition Have Failed* (Yale University Press).

Skodvin, T. (2000) 'The Intergovernmental Panel on Climate Change', in Andresen, S., T. Skodvin, A. Underdal and J. Wettestad (eds) *Science and Politics in International Environmental Regimes. Between Integrity and Involvement* (Manchester and New York: Manchester University Press).

Underdal, A. (2000) 'Science and Politics: The Anatomy of an Uneasy Partnership', in Andresen, S., T. Skodvin, A. Underdal and J. Wettestad (eds) *Science and Politics in International Environmental Regimes. Between Integrity and Involvement* (Manchester and New York: Manchester University Press).

Vogler, J. (1995) *The Global Commons. Environmental and Technological Governance* (Chichester, New York, Weinheim, Brisbane, Singapore and Toronto: John Wiley and Sons Ltd).

Walters, W. (2012) *Governmentality: Critical Encounters* (New York: Routledge).

Wendt, A. (1999) *Social Theory of International Politics* (Cambridge: Cambridge University Press).

Wynne, B. (2010) 'Strange Weather, Again: Climate Science as Political Art', *Theory, Culture and Society* 27, 2–3, 289–305.

Young, O. (ed.) (1997) *Global Governance: Drawing Insights from the Environmental Experience* (Cambridge, MA: The MIT Press).

Young O. (ed.) (1999) *The Effectiveness of International Environmental Regimes: Causal Connections and Behavioral Mechanisms* (Cambridge, MA: The MIT Press).

8
Transnational Actors in International Environmental Politics

Michele M. Betsill

Over the last decade, there has been considerable growth in academic research on actors 'beyond the state' as part of a broader 'rescaling' of global environmental politics whereby non-state actors and institutions operating at multiple levels of social and political organization are seen to shape the politics and governance of global environmental issues (Andonova and Mitchell, 2010; Biermann and Pattberg, 2008; Bulkeley, 2005; Newell et al., 2012). Non-state actors include grassroots organizations, scientific associations, special interest groups (national and international), universities, businesses, trade associations, environmentalists, individuals, the media, churches and religious organizations, independence movements, sub-national governments, political parties, government bureaucrats, foundations, think tanks, social entrepreneurs, and consumer groups. While realists dismiss claims about the significance of these actors in world politics, scholars of international environmental politics (IEP) have long recognized their importance, particularly in processes of global governance, and have shaped discussions in the wider discipline of international relations. This largely reflects the fact that non-state actors have had a stronger presence in the environmental issue area than in many other areas of concern to international relations scholars, such as security, trade, and health.

This chapter reviews IEP scholarship on transnational actors, a subcategory of non-state actors defined here as collectivities of individuals and organizations that interact across national borders but do not

---

I gratefully acknowledge the research assistance of Abigail Harder in updating this chapter.

formally represent the foreign policy interests of a nation-state (Risse-Kappen, 1995). Three general categories of transnational actors are most prominent in the literature: international non-governmental organizations (INGOs), multinational corporations (MNCs), and transnational networks. INGOs typically are defined as non-profit organizations with operations in two or more countries that have not been established by a government. While INGOs with many different substantive interests engage in global environmental politics (for example, human rights and development organizations), IEP scholars often concentrate their research on environmental INGOs, such as the World Wide Fund for Nature, Greenpeace, and Friends of the Earth International (for example, Doherty and Doyle, 2012; Mason, 2007; Wapner, 1996), and business/industry INGOs, such as the International Chamber of Commerce and the World Business Council on Sustainable Development (for example, Burgiel, 2008; Clapp, 2003; Vormedal, 2008). There is some debate about whether non-profit associations representing commercial interests should be treated the same as other INGOs – at least for analytical purposes – since they can be seen to represent private rather than public interests, and in some cases may work against environmental protection (Betsill and Corell, 2008; Jordan and van Tuijl, 2000; Pasha and Blaney, 1998). IEP scholars are particularly interested in the ways INGOs engage in processes of global environmental governance as well as whether and how they are reshaping state–society relations and world politics more broadly.

MNCs are distinguished from INGOs by their profit-seeking motives. Many IEP scholars, especially those who view capitalism as the driving logic behind global environmental politics (Paterson, this book), research the role of individual MNCs (for example, British Petroleum or Wal-mart) or MNCs within a particular economic sector (for example, forests or mining) in global environmental politics (Clapp, 1998a, this book; Clapp and Fuchs, 2009; Dauvergne and Lister, 2012; Falkner, 2008; Garcia-Johnson, 2000; Levy and Newell, 2005). IEP scholars analyze the ways MNCs engage in global decision-making processes as well as the effects of global environmental politics on corporate behavior. Over the years, understanding of MNC involvement in global environmental politics has become much more nuanced; scholars no longer assume *a priori* that environmental protection is counter to MNC interests. For example, several studies show that international norms can change the way MNCs view their own social and environmental practices and/or give rise to shareholder advocacy initiatives, which put pressure on MNCs to change these practices (Bernstein and Cashore,

2007; McAteer and Pulver, 2009). Critical political economists often highlight the role of MNCs in generating global environmental degradation and thus see MNC activity as a threat to the earth's long-term sustainability (Clapp, this book; Dauvergne, 2008).

A growing number of IEP scholars study transnational networks that connect different types of actors across borders and link local and global concerns. Examples include transnational advocacy networks (Keck and Sikkink, 1998); epistemic communities (Haas, 1992); transnational municipal networks (Betsill and Bulkeley, 2004); public–private partnerships (Andonova, 2010; Bäckstrand, 2008); transnational business coalitions (Meckling, 2011); and social movements (O'Brien et al., 2000). These networks provide a means for aggregating interests and pooling resources while enabling political action across multiple scales. The transnational networks approach acknowledges the diversity of actors engaged in global environmental politics, does not force researchers to draw fine distinctions between them, and more fully recognizes the complexity of international life today. Unlike INGOs and MNCs, transnational networks can include government actors and thus blur the boundaries between public and private/state and non-state actors in international relations (Bulkeley, 2005; Pattberg and Stripple, 2008). Many IEP scholars differentiate between public, private, and hybrid networks based on the types of actors involved and are interested in understanding how these differences shape the resources and sources of authority such networks bring to bear in global environmental politics.

This chapter begins by situating IEP research on transnational actors within the broader international relations discipline. Current discussions are part of ongoing debates about the centrality of states in world politics and international relations theories. The second section presents research on how transnational actors engage in global environmental politics, as this has been a central focus of IEP scholarship in this area. Following Andonova and Mitchell (2010), this section differentiates between the *intergovernmental political sphere*, where transnational actors interact with governments and intergovernmental organizations in multilateral decision-making processes, and the *transnational political sphere*, which has become an alternative site for global environmental politics and governance. The third section reviews debates about the impacts and effects of transnational actors on global environmental politics. I then explore some of these issues and debates through a brief case study of transnational municipal networks in global climate governance before concluding with some thoughts about future research on transnational actors.

Theoretical roots

The study of transnational actors in IEP came of age in the early 1990s, motivated in large part by the 1992 Rio Conference where tens of thousands of citizens from around the world gathered to interact with government delegates and to participate in their own parallel conference. *Agenda 21*, one of the major outcomes of the conference, recognized non-state actors as partners in the global struggle for sustainable development and raised awareness that governments alone cannot manage global environmental threats. This motivated a new research agenda focused on understanding what non-state actors do, how they engage in global environmental politics, and with what effects.

Current discussions of transnational actors in IEP are part of broader debates about the state-centrism of international relations theories. Although Risse (2002) argues that these debates can be traced back to discussions of integration theory in the 1950s which focused on supranational organizations, such as the European Community (for example, Deutsch, 1957; Haas, 1958), most IEP scholars point to the importance of Keohane and Nye's (1977; see also Nye and Keohane, 1971) work on transnational relations and complex interdependence, which highlighted the increasing importance of MNCs in world affairs. For various reasons, interest in transnational relations waned in the 1980s as international relations scholars shifted their attention to the study of intergovernmental regimes. The study of transnational relations was revived in the 1990s, with Risse-Kappen's (1995) edited volume, *Bringing Transnational Relations Back In*, playing a particularly important role in providing a set of analytical tools to guide future inquiry. This renewed interest in transnational actors coincided with the 'constructivist turn' in international relations, which informed much of the IEP (and broader IR) scholarship of this period (Haas, 1992; Keck and Sikkink, 1998; Princen and Finger, 1994; O'Brien et al., 2000; Wapner, 1996).

In the 1990s and early 2000s, interest in 'global governance' created additional space for considering the role of transnational actors in world politics. Scholars working in this area linked the prominence of transnational actors to processes of globalization, which simultaneously challenge the ability of nation-states to respond to complex, multi-scalar problems, such as climate change and biodiversity loss, and empower new actors to participate in decision-making (Matthews, 1997). Importantly, scholars like Rosenau (1995) came to view non-state actors as 'new spheres of authority' and differentiated between a 'state-centric' world dominated by interactions between nation-states and the 'multi-centric' world of 'sovereignty-free' actors (see also Ruggie, 2004; Wapner, 1996). As discussed below, this idea of a transnational political sphere generated debates in IR about the demise of the nation-state (Strange, 1996) and the privatization of authority (Clapp, 1998b; Cutler et al., 1999; Hall and Biersteker, 2002). Once again, IEP scholars have been at the forefront of these wider debates.

Today, IEP scholars apply multiple theoretical lenses to analyze transnational actors (see Paterson, this book). For example, many liberal institutionalists are interested in how transnational actors facilitate cooperation between states by altering the costs and benefits of action as well as in the role of transnational actors as partners in monitoring and implementing multilateral agreements. Critical political economists often view transnational actors in the context of the global capitalist order and argue that such actors, especially MNCs, exercise structural power and reinforce neoliberal values and norms. Constructivist thinkers emphasize the role of transnational actors in producing new ideas and norms, which in turn alter how states and other actors define their interests in global environmental politics.

Transnational actors in global environmental politics

Although IEP research on transnational actors is a relatively recent phenomenon, transnational actors have engaged in global environmental politics for more than a century. The first environmental INGOs appeared in the late 19th century, with early examples including the International Union of Forestry Research Organizations (1891) and the International Friends of Nature (1895) (Frank et al., 1999). Frank et al. (1999) explain the rise of INGOs as a response to a rationalized discourse around nature, which created demand for 'expert bodies' to help states and societies protect (and exploit) nature for human needs. In the mid-20th century, changes in the institutional architecture of global governance (for example, the formation of the United Nations) provided a forum through which transnational actors could actively engage in international politics, although some scholars argue that these new political spaces merely reflect the needs of states and international governance regimes (Raustiala, 1997). In contrast, Keck and Sikkink (1998) contend that transnational networks emerge as political space is closed at the national level and 'domestic NGOs bypass their state and directly search out international allies to try to bring pressure on their states from outside' (Keck and Sikkink, 1998: 12). Many critical scholars view the rise of transnational actors as part of a broader neoliberalization of global environmental politics, involving a rollback of state involvement and greater reliance on markets in environmental protection (Levy and Newell, 2005; Holmes, 2011).

The vast majority of IEP research on transnational actors focuses on how they engage in global environmental politics and with what effect. Most studies rely on qualitative case studies whereby researchers draw upon data obtained through interviews, archival research, participant observation, and in some cases questionnaires. IEP scholarship on transnational actors can be differentiated in terms of where political activity occurs. In the *intergovernmental political sphere*, scholars explore the ways transnational actors interact with governments and intergovernmental organizations in multilateral decision-making processes. Given that transnational actors often outnumber government representatives, particularly in multilateral environmental treaty negotiations, these processes can no longer be analyzed solely in terms of inter-state diplomacy. Analysts must take into consideration the myriad actors involved and the diversity of interests represented (Betsill and Corell, 2008; Schroeder and Lovell, 2012). IEP scholars also focus on the *transnational political sphere*, recognizing that transnational interactions constitute a new 'global public domain' that exists outside of the intergovernmental political sphere (Ruggie, 2004; see also Rosenau, 1995). Scholars working in this area argue that transnational actors exercise new forms of agency, where agency is understood as the capacity to prescribe behavior and affect outcomes (Biermann et al., 2010; Dellas et al., 2011).

Intergovernmental political sphere

A great deal of IEP scholarship on transnational actors, particularly in the 1990s, revolved around their participation in the major UN-sponsored global environmental conferences: the 1972 Stockholm Conference, the 1992 Rio Conference, and the 2002 World Summit on Sustainable Development (Johannesburg Conference). Stockholm is seen to be a watershed in terms of NGO participation in global governance, marking the beginning of a 'slow yet steady liberalization of the NGO system occurring over the following two decades' (Willets, 1996: 57). The Rio Conference recognized the role transnational actors play as partners with states in the global struggle to promote sustainable development (Chatterjee and Finger, 1994; Dodds, 2001). Transnational actors were central to the creation of 'partnerships' for sustainable development at the 2002 Johannesburg Summit (Andonova, 2010; Szulecki et al., 2011; Wapner, 2003).

IEP scholars also analyze transnational actors (typically environmental INGOs and business associations) in multilateral environmental treaty negotiations (Betsill and Corell, 2008; Gareau, 2012; Gulbrandsen and Andresen, 2004; Vormedal, 2008). In addition, IEP scholars have examined transnational campaigns to reform the lending practices of multilateral development banks such as the World Bank, as well as attempts to integrate environmental protection within international trade institutions such as the World Trade Organization (Mason, 2007; O'Brien et al., 2000; Park, 2005; Williams and Ford, 1999; Wright, 2000). In directing their attention to the international economic institutions, transnational actors recognize that global economic forces are often drivers of environmental degradation (see also Clapp, this book).

Many scholars working on transnational actors in the intergovernmental political arena try to understand and explain the types of strategies used (Newell, 2008; Vormedal, 2008). Much of this literature examines how transnational actors strategically use information (Keck and Sikkink, 1998). Some scholars differentiate between direct strategies, such as providing technical information and policy advice or serving on national delegations, which are targeted at state decision-makers, and indirect strategies, such as public protests or holding parallel events, which work through secondary channels such as the media and the public (Wright, 2000; Young, 1999). While environmentalists employ both strategies regularly, MNCs and business/industry INGOs tend to rely more heavily (though not exclusively) on direct strategies, which may reflect the close relationship between business and states in a neoliberal economic order (Burgiel, 2008).

Networking is another common strategy employed in the intergovernmental political arena (Orsini, 2011; Vormedal, 2008).[1] Much of the literature on environmentalists focuses on networks operating across scales, creating links between INGOs and grassroots organizations (Doherty and Doyle, 2012; Princen and Finger, 1994). Transnational advocacy coalitions often use these cross-scale linkages to invoke the 'boomerang' strategy; by working at both the international and domestic levels, they seek to pressure states from above and below (Hochstetler, 2002; Keck and Sikkink, 1998). Other transnational networks create connections between groups with similar substantive interests, with the expectation that such networks result in greater efficiency, facilitate the exchange of knowledge, and allow groups to develop a shared voice (Arts, 2001; Botetzagias et al., 2010). A number of transnational networks link transnational actors with different substantive interests and/or transnational actors with states in order to build strong political coalitions (Meckling, 2011; Newell, 2000; O'Brien et al., 2000).

IEP scholars find that transnational actors select strategies according to their particular resources and sources of leverage as well as the political opportunity structure within a given decision-making process (Betsill and Corell, 2008). Business/industry actors, it is argued, have 'tacit power' over states based on 'the role they play in the creation of economic growth and the production of energy' (Newell, 2000: 159). According to this line of reasoning, it is not their economic resources per se that make business/industry actors powerful but their central position in national economies and the global political economy (Levy and Newell, 2005; Vormedal, 2008). Alternatively, Chatterjee and Finger (1994) argue that business/industry has a privileged position in international environmental policymaking simply because 'money talks'. From this perspective, the possession of economic resources is seen to be an important source of leverage for MNCs and business coalitions. IEP scholars typically do not emphasize economic resources as a source of leverage for environmental INGOs, although some organizations have considerable economic resources that can be deployed in their efforts to shape decision-making processes (Skodvin and Andresen, 2003). More commonly, researchers argue that the leverage of INGOs in intergovernmental decision-making processes derives from their specialized knowledge and expertise and/or their moral claims to speak on behalf of under-represented groups in civil society or nature itself (this is especially true for some environmental, human rights, and development INGOs) (Bernstein, 2011; Betsill and Corell, 2008).

How transnational actors engage in intergovernmental environmental politics also is shaped by the rules of access, which differ between policy arenas. States routinely invoke their sovereign privilege to restrict transnational actors' participation in such processes by limiting the number of organizations that can physically attend meetings, deliberating behind closed doors, or withholding official documents (Friedman et al., 2005; Raustiala, 1997). In particular, IEP scholars observe that access is much more restrictive in the international economic institutions than in multilateral environmental treaty negotiations (Mason, 2007). In some cases, states may reach out to transnational actors and actively facilitate their involvement by creating formal institutional spaces and recognizing them as partners in achieving environmental goals, which creates opportunities for transnational actors to engage in direct strategies (Corell, 2008; Dodds, 2001). More frequently, states place restrictions on access; however, many transnational actors have become adept at working around these restrictions (Betsill and Corell, 2008).

Transnational political sphere

Over the past decade, many IEP scholars have shifted their attention to interactions between transnational actors outside of the intergovernmental political sphere. This work engages broader debates about state–society relations and has links to work on the concept of civil society in the fields of political theory and comparative politics, where civil society is generally conceptualized as a domain separate from the state in which individuals engage in voluntary association (Hall, 1995; Seligman, 1992). Recognizing that such voluntary association increasingly takes place across national boundaries, Wapner (1997: 66) defines global civil society as 'the domain that exists above the individual and below the state but also across state boundaries, where people voluntarily organize themselves to pursue various aims' (see also Lipschutz, 1992; Lipschutz with Mayer, 1996). Global governance scholars argue that these interactions constitute a 'new public domain' (Ruggie, 2004) or 'sphere of authority' (Rosenau, 1995), which challenges governments' monopoly on global decision-making. Today, many IEP scholars use the term 'transnational political sphere' to refer to the space in which transnational actors engage in political activity and develop new modes of global environmental governance (Andonova et al., 2009; Biermann, this book; Bulkeley et al., 2012; Pattberg and Stripple, 2008).

The transnational political sphere provides an alternative site for environmental advocacy and activism. For example, environmental INGOs frequently target MNCs and other private actors (such as foundations) directly using consumer boycotts, public awareness campaigns, and shareholder activism, bypassing states, which they believe are either unable or unwilling to regulate corporate environmental practices (Falkner, 2003, 2008; MacLeod and Park, 2011; McAteer and Pulver, 2009). Newell (2008) contends that this reflects the changing nature of power in a globalized world where INGOs recognize it may be more important to change the behavior of a few large companies or donor organizations than to persuade a large number of small states to sign another multilateral treaty.

While interactions between INGOs and MNCs are often adversarial, increasingly it is common for these actors to work cooperatively, creating dense networks of communication among themselves, which in turn facilitates mutual learning and opens up new opportunities for governing the global environment (Auld and Gulbrandsen, 2010; Falkner,

2008; Pattberg, 2005; Wapner, 1996). Even when transnational actors participate in intergovernmental conferences and treaty negotiations, some authors claim the significance of these activities lies not in their impact on governments but in the networks transnational actors create among themselves (Chatterjee and Finger, 1994; Friedman et al., 2005; Schroeder and Lovell, 2012).

In recent years, scholars have analyzed the internal structures and dynamics of transnational networks in order to understand why these actors pursue network connections and some of the resulting mobilization challenges (Betsill and Bulkeley, 2004; Botetzagias et al., 2010; Holmes, 2011; Orsini, 2011). These studies frequently reveal very real differences in the sources of leverage, access to resources, and motivations of different types of actors involved in a network. Some transnational networks have developed highly sophisticated institutional structures, which enable network members to coordinate activities, share information, identify priorities, and specialize in particular aspects of an issue or policy process, while others are more informal. The bonds that hold networks together vary considerably. For example, transnational advocacy networks are linked by shared principled beliefs (Keck and Sikkink, 1998) while epistemic communities are held together by shared causal beliefs as well (Haas, 1992). Keck and Sikkink (1998) argue that transnational business coalitions and transnational municipal networks are held together by shared instrumental goals.[2] Social movements, also held together by shared principled beliefs, are distinguished by the fact that they tend to mobilize their constituencies through protest or disruptive action and provide the opportunity for mass participation (Ford, 2003). IEP scholars frequently observe tensions within transnational networks (for example, along North–South dimensions), reminding us that transnational actors, like states, are themselves political entities with their own power relations (Alcock, 2008; Dodds, 2001; Duwe, 2001; Jordan and Van Tuijl, 2000).

For many global governance, constructivist, and critical scholars, the transnational political sphere is seen as an important site for the creation of 'transnational' or 'private' forms of environmental governance (Andonova et al., 2009; Biermann, this book; Bulkeley et al., 2012; Clapp, 1998b; Hoffmann, 2011). Transnational actors have been instrumental in establishing a broad range of governance arrangements, including certification schemes such as the Forest Stewardship Council (Auld and Gulbrandsen, 2010; Bernstein and Cashore, 2007; Pattberg, 2005); transnational municipal networks such as ICLEI Cities for Climate Protection (Betsill and Bulkeley, 2004; Toly, 2008); investor-driven governance networks such as the Coalition for Environmentally Responsible Economies (CERES) (MacLeod and Park, 2011); and public–private partnerships such as those established as an outcome of the 2002 Johannesburg Conference (Andonova, 2010; Bäckstrand, 2008; Szulecki et al., 2011). Many of these arrangements link actors operating from the local to the global level in both the public and private spheres.

Whereas transnational actors seek to influence the decisions and behaviors of states or IGOs in the intergovernmental political sphere, in this sphere they are seen as authoritative actors – or 'global governors' – in their own right, which has opened up a line of research focused on how transnational actors come to be seen as authoritative (Andonova et al., 2009; Betsill et al., 2011; Pattberg, 2005). Bernstein (2011: 20) argues, 'Legitimacy is also the glue that links authority and power. By justifying authority in the eyes of the governed, legitimacy empowers authorities and increases the likelihood their commands will be obeyed.' In contrast to state actors whose authority to govern often is taken for granted, transnational actors must establish their legitimacy to serve as global governors through appeals to moral arguments, expertise, relationship to key stakeholders, or problem-solving abilities (Avant et al., 2010; Hall and Bierstecker, 2002). Legitimacy claims may be accepted explicitly through delegation or implicitly through recognition and must be routinely reiterated as they are often contested.

A number of studies examine transnational actors' claims as to why they should have the right to govern and whether these claims are accepted by affected publics (Bouteligier, 2011; Dellas et al., 2011). Preliminary research on the legitimation process suggests that the types of claims and their acceptance are a function of resources available to the particular transnational actors as well as the social context in which legitimacy claims are made (Bernstein, 2011). In the realm of transnational climate governance, legitimacy claims are most frequently based on appeals to expertise within a broader discourse of efficiency and market governance, reflecting the dominance of 'liberal environmentalist' ideology within the climate governance system (Bernstein, 2002; Bulkeley et al., 2012).

Impacts and effects of transnational actors

In addition to studying how transnational actors engage in global environmental politics, IEP scholars also seek to evaluate the impacts and effects of this activity. This section reviews this area of IEP research and notes that researchers emphasize many different types of impacts and effects depending on their theoretical orientation.

For many researchers working within a pluralist or neoliberal institutionalist framework, transnational actors are seen as pressure groups or lobbyists advocating for particular policies and practices. In recent years, IEP scholars have made significant methodological advances in assessing the influence of transnational actors on the specific outputs of decision-making processes in the intergovernmental political arena using process tracing and counterfactual analysis to link transnational actor participation to observed effects on decision-making processes and outcomes and to rule out alternative explanations for these effects (Arts, 2001; Betsill and Corell, 2008; Newell, 2000). This research shows that transnational actors help states overcome collective action problems throughout the policy process. Transnational actors identify problems and put pressure on states (through both domestic and international channels) to take action during the agenda-setting phase (for example, Burgiel, 2008; Meckling, 2012). In the policy formulation phase, transnational actors often influence debates about particular proposals, shape the positions of key states, and/or convince states to adopt particular text or policy approaches in treaty documents (for example, Betsill, 2008; Humphreys, 2008; Vormedal, 2008). During the implementation phase, transnational actors help states monitor compliance with international agreements and/or carry out specific projects, often in the form of public–private partnerships (for example, Andonova, 2010; Grundbrandsen and Andresen, 2004). Of course it is important to recognize that some transnational actors (for example, those representing corporate interests) may impede implementation through non-cooperative strategies (Newell, 2000).

Transnational actors often fall short in their efforts to influence decision-making processes and outcomes, and IEP scholars seek to explain this variation (Gareau, 2012). Some studies emphasize the specific attributes and behaviors of the actors themselves (for example, professional skill, material resources, or networking), while others highlight the importance of context (for example, the institutional setting or prevailing norms). Many scholars combine elements of both, recognizing that they are closely related. In a meta-analysis of several case studies, Betsill and Corell (2008) found that the rules of access, stage of the negotiations, political stakes, institutional overlap, alliances with key states, and the extent to which issues are embedded in entrenched economic interests helped explain variation in NGO influence in international environmental negotiations. These conditions may affect environmental and business-oriented groups differently.

There has been less effort to systematically evaluate and explain the impact of transnational actors on policies and practices in the transnational political sphere (Dellas et al., 2011). IEP scholars have documented many instances in which transnational actors have been successful in reshaping corporate practices (for example, the Starfish Tuna Dolphin campaign) or establishing new forms of global environmental governance (Andonova et al., 2009; Pattberg and Stripple, 2008; Wright, 2000). However, questions remain about what attributes of transnational actors or the policy context contribute to these successes as well as the conditions under which transnational actors are effective in establishing new modes of governance.

Other IEP scholars are more interested in the broader impacts and effects of transnational actors on global environmental politics. For example, many constructivist thinkers explore the ways transnational actors shape norms and ideas about the environment. Borrowing a concept from sociology, some scholars argue that this occurs as transnational actors frame[3] (or re-frame) environmental problems (Humphreys, 2004; Toly, 2008), while others emphasize the role of transnational actors in broader socialization processes (MacLeod and Park, 2011; Park, 2005). These thinkers argue that transnational actors help create the social context within which actors make decisions about which policies to adopt and/or about their own practices, which in turn shapes how these actors define their interests. Many of these scholars rely on discourse analysis, which assumes that transnational actors' influence occurs through discursive acts of contestation and persuasion (Hajer and Versteeg, 2005).

Many critical scholars are interested in whether transnational actors contribute to a broader remapping of world politics. In particular, what is the impact of transnational activity on the power and authority of states in global environmental politics? Some scholars see the rise of transnational actors as a force usurping the authority of states and argue that the development of transnational forms of environmental governance indicates declining state power when it comes to managing the global environment (Lipschutz, 1992; Mathews, 1997). Other IEP scholars associate the rise of transnational actors with a reorientation of the state's role in global environmental politics and redefined notions of sovereignty (Bulkeley and Schroeder, 2012; Eckersley, 2004; Pattberg and Stripple, 2008; Wapner, 1998). Alternatively, a number of critical scholars use neo-Gramscian theory and/or the notion of governmentality to argue that transnational actors legitimize state power and agendas (at least for strong states) by reinforcing the hegemonic neoliberal economic order (Duffy, 2006; Ford, 2003; Gareau, 2012; Holmes, 2011).

There is ongoing debate about whether transnational actors contribute to the democratization of global environmental governance. As they become more involved in decision-making, transnational actors are seen to enhance participatory democracy by giving voice to a broad range of stakeholders whose interests may be otherwise underrepresented (Bäckstrand, 2006; Dombrowski, 2010). This participation, along with monitoring and reporting activities, also serves as an important accountability mechanism (Bäckstrand, 2008; Mason, 2007; Newell, 2008; Park, 2005). In addition, transnational actors contribute to deliberative democracy by creating new public spaces for reasoned argument and persuasion (Dryzek, 2000). However, these claims raise questions about the democratic legitimacy of transnational actors themselves. Whose interests do they represent and on what authority (Biermann and Pattberg, 2008)? Critics ask how transnational actors can contribute to democratization when they 'reflect as much inequality as they are trying to undo' (Jordan and Van Tuijil, 2000: 2061) and note the absence of strong mechanisms for holding transnational actors to account (Pasha and Blaney, 1998; Skodvin and Andresen, 2003).

A relatively small group of IEP scholars explores whether the involvement of transnational actors contributes to 'better' environmental outcomes. One line of reasoning revolves around the importance of knowledge and information in addressing the complexity of global environmental problems. Transnational actors play an important role by providing specialized knowledge and expertise as well as multiple types of knowledge (for example, local or traditional knowledge), which enhance society's resilience in the face of global change (Böhmelt and Betzold, 2013; Corell, 2008; Folke et al., 2005; Lövbrand, this book). Another argument suggests that the involvement of transnational actors confers greater legitimacy on intergovernmental and transnational decision-making processes, thereby enhancing the likelihood that policies will be implemented and rules will be followed (Bernstein and Cashore, 2007). Empirical studies suggest that transnational actors provide knowledge and information in a highly political context where knowledge is often seen to threaten certain interests leading to debates about how to interpret knowledge claims and whose knowledge counts as 'legitimate' (Bäckstrand, 2003; Bernstein, 2011). In addition, Hochstetler (2002) argues that domestic implementation of global environmental rules and regulations depends not only on what happened during the rule-making process but also on whether the implementing actor has accepted the relevant rules, along with its own capacities. Ultimately, linking transnational actors to environmental outcomes raises many of the same methodological challenges faced by scholars of international regimes and global governance (Young, this book).

Transnational municipal networks and global climate governance

In this section, I explore many of the issues and debates discussed above as they apply to the study of transnational municipal networks (TMNs) and global climate governance. TMNs are voluntary associations premised on the assumption that local governments play an important role in addressing global climate change. While the contribution of any single local authority may be small, collectively local action to reduce greenhouse gas emissions can be significant given that half the world's population currently lives in urban areas. Cities are key sites in the production of emissions through energy use, transportation, waste production, and consumption patterns. Many local authorities have direct jurisdiction over emissions-producing activities as well as considerable experience addressing environmental impacts in the energy, transportation, and planning sectors. Local climate governance also can stimulate social and technological innovation and generate political support for addressing climate change in the face of a stalled multilateral treaty-making process (Bulkeley, 2010, 2013; Hoffmann, 2011).

TMNs have been a prominent feature in global climate change politics since the early 1990s, but were overlooked by IEP scholars for many years because they did not fit neatly in the (central) state/non-state actor divide that has dominated international relations thinking (Betsill and Bulkeley, 2004; Pattberg and Stripple, 2008). Such networks engage in activities similar to other non-state actors such as INGOs – they lobby national governments, for example – but their members are public entities and they draw on public forms of authority to carry out many of their initiatives. Today, IEP scholars recognize that TMNs reflect and are constitutive of a complex system of global climate governance involving hybrid forms of authority and new modes of governance. Research focuses on understanding their role in both the intergovernmental and transnational political arenas.

Examples of climate-related TMNs include ICLEI-Local Governments for Sustainability's Cities for Climate Protection (CCP) program; the Climate Alliance; Energie-cités; the C40 Cities Climate Leadership Group; and the International Solar Cities Initiative. The first TMNs (CCP, Climate Alliance and Energie-cités) were established in the early 1990s and tended to engage small and medium municipalities in the global North, especially in Europe and North America. In the late 1990s, these networks expanded to reach out to cities in the global South. More recent TMNs such as C40 have focused on capital cities as well as large 'global' cities, many of which are located in low- and middle-income countries. The number of cities enrolled in TMNs has grown tenfold over the last decade, to include several thousand local authorities, with some joining multiple networks (Kern and Bulkeley, 2009; Bulkeley, 2010). TMNs have emphasized climate mitigation with much less attention on adaptation, although that is slowly changing (Bulkeley and Betsill, 2013).

The CCP program has received the most attention from IEP scholars (for example, Betsill and Bulkeley, 2004, 2006; Bulkeley and Betsill, 2003; Lindseth, 2004; Toly, 2008). CCP, which was launched by ICLEI in 1993, was organized around a series of five milestones designed to help local authorities design and implement climate protection measures: (1) conduct an emissions inventory, (2) set a reduction target, (3) develop a climate action plan, (4) implement measures from the plan, and (5) monitor and report on progress. At its peak, the CCP network included more than 1,000 members worldwide, representing more than 15 percent of global greenhouse gas emissions (Bulkeley, 2010). Today, the CCP program has been integrated within ICLEI's broader sustainability agenda, but its governance model focused on measurable targets and timetables has been replicated in other climate-related TMNs (Hoffmann, 2011).

TMNs operate in both the intergovernmental and transnational political arenas. ICLEI was originally established to represent local environmental concerns in intergovernmental organizations and has had observer status to the UN climate negotiations since 1995. Today ICLEI coordinates the 'local governments and municipal authorities' constituent group in the negotiation process. Inside the conference halls, TMN representatives lobby national governments to adopt more progressive climate policies and especially to enhance recognition that cities are key actors in the global response to climate change and to mobilize resources for local climate action (Bulkeley, 2010). TMNs also organize side events designed to raise awareness about local climate action as part of a general strategy to 'lead by example' and to provide networking opportunities for local authorities (Hoffmann, 2011; Kern and Bulkeley, 2009).

IEP scholarship has placed greater emphasis on TMNs in the transnational political arena, where they are seen as new spheres of authority or sites of transnational climate governance (Betsill and Bulkeley, 2006; Bulkeley, 2005; Hoffmann, 2011; Pattberg and Stripple, 2008). This research demonstrates that TMNs have been instrumental in engaging local authorities with climate change by providing key resources, such as knowledge, policy tools, and political support, as well as mechanisms through which best practices are identified and promoted. The majority of evidence for why municipalities join transnational networks comes from studies of 'pioneer' cities. Betsill and Bulkeley (2004) found that municipalities were drawn to the CCP when entrepreneurial individuals perceived that the network could help them reframe existing policy agendas and gain recognition for work they were already doing (see also Bulkeley and Betsill, 2003). This same study found that the level of participation among member municipalities varied and argued that creating thick and dense webs of interaction was key to maintaining the network. The creation and exchange of material and non-material resources is the glue that holds networks together and maintains the authority of TMNs rather than an end in itself. Members with individual political champions and the ability to capitalize on these resources appear more likely to establish these open connections, leading Kern and Bulkeley (2009: 311) to conclude that TMNs are largely 'networks of pioneers for pioneers'.

IEP scholars argue that TMNs have contributed to an urban climate policy agenda as well as a broader rescaling of global climate governance where municipal governments are seen as legitimate governors (Andonova and Mitchell, 2010; Bulkeley, 2005). There is strong evidence that 'the rhetoric of climate protection has entered policy rationale in the sectors of land-use planning, transport, and energy management in the built environment' (Betsill and Bulkeley, 2004: 487). The early TMNs facilitated a wave of 'municipal voluntarism' in which pioneer cities developed an evidence-based approach to climate planning and policy, while later networks have promoted 'strategic urbanism' whereby climate protection is integrated with wider goals and a broader range of private interests are engaged in urban climate governance (Bulkeley and Betsill, 2013). At the same time, there is a notable gap between rhetoric and reality and considerable variation in the extent to which member cities are able to implement policies and practices promoted through these networks. The impact of TMNs on local climate policies is mediated by the position of key individuals in local policy networks as well as institutional variables related to funding and regulatory authority (Bulkeley and Betsill, 2003; Kern and Bulkeley, 2009).

Conclusions

The subject of transnational actors has generated considerable attention in IEP scholarship since the early 1990s. This literature provides a rich picture of the types of transnational actors engaged in global environmental politics and their various forms of political activity. Thus far, IEP scholarship has tended to focus on specific types of transnational actors – INGOs, MNCs, and transnational networks – often without consideration of their diversity. All transnational actors are not alike; they differ in terms of their interests, resources, and strategies. Moreover, they participate in many different political arenas, each likely to involve different dynamics that in turn shape the ways that transnational actors participate, the goals they pursue, the strategies they use, and the likelihood that they will achieve those goals. It is also important to acknowledge the diversity within each category of transnational actor. Just as many MNCs and business sectors have different orientations to global environmental issues, environmental INGOs often have disagreements about the drivers of environmental degradation and appropriate solutions as well as distinct positions on issues of global equity and markets (Okereke and Charlesworth, this book).

There is also a need for IEP scholars to develop more sophisticated approaches to analyzing transnational actors in the transnational political sphere. Given the rescaling of global environmental politics (Andonova and Mitchell, 2010), it is reasonable to anticipate that transnational actors will continue to reorient themselves away from the intergovernmental political sphere where national governments appear either unable or unwilling to agree on ambitious strategies for dealing with global environmental degradation. IEP scholars should continue to ask questions about how transnational actors come to be seen as authoritative in this political arena and how they exercise that authority. In addition, future research should build on recent methodological advances in the study of transnational actor influence in the intergovernmental political arena to assess when and under what conditions transnational actors are successful in shaping transnational political processes.

Future research might also consider how transnational interactions affect the individual actors that engage in such relations. For example, do individual organizations enhance their autonomy and ability to act 'at home' by participating in transnational networks or do such networks actually take away from their autonomy and control? As noted above, IEP scholars disagree on how transnational engagement affects state sovereignty and authority, with some arguing that such interactions can actually enhance the ability of states to confront global environmental issues and/or legitimize state power and agendas by reinforcing a neoliberal economic order. Others contend that sovereignty is compromised when states must abide by rules generated through transnational processes. Similar questions can be raised for non-state actors. It may be that different network structures are related to different types of effects on individual participants.

Finally, scholars of transnational actors should look beyond the environmental issue area and explore how findings about transnational actors in IEP research compare to similar research in other areas of international relations. Are there insights from other areas of transnational relations research (for example, in the field of human rights) that would enhance our study of transnational environmental politics?

Notes

1. For many transnational actors, the significance of networking goes beyond its utility as a strategy to influence intergovernmental decision-making processes. I return to this point in the next section.
2. For a critique of this distinction, see Sell and Prakesh (2004).
3. A frame is 'an interpretive schemata that simplifies and condenses the "world out there" by selectively punctuating and encoding objects, situations, events, experiences, and sequences of action within one's present or past environment' (Snow and Benford, 1992: 137).

Annotated bibliography

Betsill, M. M. and E. Corell (eds) (2008) *NGO Diplomacy: The Influence of Non-governmental Organizations in International Environmental Negotiations* (Cambridge, MA: The MIT Press).

Betsill and Corell's edited volume has contributed to methodological advances in the analysis of non-state actors in global environmental politics. The editors introduce a framework for analyzing NGO influence in multilateral environmental treaty negotiations, which is then applied in a series of case studies covering the issues of climate change, desertification, biodiversity, deforestation, and whaling. The volume concludes with a meta-analysis of the factors that condition NGO influence in these decision-making processes and identifies a number of hypotheses for future research.

Keck, M. E and K. Sikkink (1998) *Activists Beyond Borders: Advocacy Networks in International Politics* (Ithaca, NY: Cornell University Press).

Keck and Sikkink's book carefully examines the work of activist networks in the areas of human rights, environment, and violence against women and has prompted a great deal of work in the area of transnational networks to the field of IEP. Moreover, their idea of the 'boomerang' pattern in which national NGOs reach out to international parnters in order to exert external pressure on their own governments has been important in highlighting the linkages between domestic and international politics.

Levy, D. L. and P. Newell (eds) (2005) *Business in International Environmental Governance: A Political Economy Approach* (Cambridge, MA: The MIT Press).

Levy and Newell's edited volume brings together several important works as well as some new material analyzing the role of business interests in international environmental governance. Using a political economy approach, the contributions demonstrate how business and industry have both shaped and been shaped by international environmental policies and policymaking processes. Case studies examine environmental politics in the areas of climate change, ozone depletion, tropical logging, and the development of environmental management standards.

Wapner, P. (1996) *Environmental Activism and World Civic Politics* (Albany, NY: State University of New York Press).

Wapner's detailed study of three environmental organizations (the World Wildlife Fund, Greenpeace, and Friends of the Earth) offers excellent insight into how these groups operate internally. In addition, this book was important in moving the scholarly discussion of NGOs in international environmental politics beyond questions about how NGOs influence states. Wapner argues that NGOs shape world politics in the realm of 'global civil society' by promoting an environmental sensibility.

Works cited

Alcock, F. (2008) 'Conflicts and Coalitions within and Across the ENGO Community', *Global Environmental Politics* 8, 4, 66–91.
Andonova, L. B. (2010) 'Public-Private Partnerships for the Earth: Politics and Patterns of Hybrid Authority in the Multilateral System', *Global Environmental Politics* 10, 2, 25–54.
Andonova, L. B., M. M. Betsill and H. Bulkeley (2009) 'Transnational Climate Governance', *Global Environmental Politics* 9, 2, 52–73.
Andonova, L. B. and Ronald B. Mitchell (2010) 'The Rescaling of Global Environmental Politics', *Annual Review of Environment and Resources* 18, 1, 255–282.
Arts, B. (2001) 'Impact of ENGOs on International Conventions', in Bas, A., M. Noortmann and B. Reinalda (eds) *Non-State Actors in International Relations* (Aldershot, UK: Ashgate Publishing Ltd), 195–210.
Auld, G. and L. H. Gulbrandsen (2010) 'Transparency in Nonstate Certification: Consequences for Accountability and Legitimacy', *Global Environmental Politics* 10, 3, 97–120.

Avant, D. D., M. Finnemore, and S. K. Sell (eds) (2010) *Who Governs the Globe?* (Cambridge: Cambridge University Press).
Bäckstrand, K. (2003) 'Civic Science for Sustainability: Reframing the Role of Experts, Policy-Makers and Citizens in Environmental Governance', *Global Environmental Politics* 3, 4, 24–41.
Bäckstrand, K. (2006) 'Democratizing Global Environmental Governance? Stakeholder Democracy after the World Summit on Sustainable Development', *European Journal of International Relations* 12, 4, 467–498.
Bäckstrand, K. (2008) 'Accountability of Networked Climate Governance: The Rise of Transnational Climate Partnerships', *Global Environmental Politics* 8, 3, 74–102.
Bernstein, S. (2002) 'Liberal Environmentalism and Global Environmental Governance', *Global Environmental Politics* 2, 3, 1–16.
Bernstein, S. (2011) 'Legitimacy in Intergovernmental and Non-state Global Governance', *Review of International Political Economy* 18, 1, 17–51.
Bernstein, S. and B. Cashore (2007) 'Can Non-state Global Governance be Legitimate? An Analytical Framework', *Regulation & Governance* 1, 4, 347–371.
Betsill, M. M. (2008) 'Environmental NGOs and the Kyoto Protocol Negotiations: 1995–1997', in M. M. Betsill and E. Corell (eds) *NGO Diplomacy: The Influence of Non-governmental Organizations in International Environmental Negotiations*, (Cambridge, MA: The MIT Press), 43–66.
Betsill, M. M. and H. Bulkeley (2004) 'Transnational Networks and Global Environmental Governance: The Cities for Climate Protection Program', *International Studies Quarterly* 48, 471–493.
Betsill, M. M. and H. Bulkeley (2006) 'Cities and the Multilevel Governance of Global Climate Change', *Global Governance* 12, 2, 141–159.
Betsill, M. M. and E. Corell, (eds) (2008) *NGO Diplomacy: The Influence of Non-governmental Organizations in International Environmental Negotiations* (Cambridge, MA: The MIT Press).
Biermann, F., M. M. Bestill, J. Gupta, N. Kanie, L. Level, D. Liverman, H. Schroeder, B. Siebenhüner and R. Zondervan (2010) 'Earth System Governance: A Research Agenda', *International Environmental Agreements* 10, 277–298.
Biermann, F. and P. Pattberg (2008) 'Global Environmental Governance: Taking Stock, Moving Forward', *Annual Review of Environment and Resources* 33, 1, 277–294.
Böhmelt, T. and C. Betzold (2013) 'The Impact of Environmental Interest Groups in International Negotiations: Do ENGOs Induce Stronger Environmental Commitments?', *International Environmental Agreements: Politics, Law and Economics* 13, 2, 127–151.
Botetzagias, I., P. Robinson and L. Venizelos (2010) 'Accounting for Difficulties Faced in Materializing a Transnational ENGO Conservation Network: A Case-Study', *Global Environmental Politics* 10, 1, 115–151.
Bouteligier, S. (2011) 'Exploring the Agency of Global Environmental Consultancy Firms in Earth System Governance', *International Environmental Agreements: Politics, Law and Economics* 11, 1, 43–61.
Bulkeley, H. (2005) 'Reconfiguring Environmental Governance: Towards a Politics of Scales and Networks', *Political Geography* 24, 8, 875–902.
Bulkeley, H. (2010) 'Cities and the Governing of Climate Change', *Annual Review of Environment and Resources* 35, 1, 229–253.

Bulkeley, H. (2013) *Cities and Climate Change*. Routledge Critical Introductions to Urbanism and the City (London and New York: Routledge).

Bulkeley, H., L. Andonova, K. Bäckstrand, M. Betsill, D. Compagnon, R. Duffy, A. Kolk, M. Hoffmann, D. Levy, P. Newell, T. Milledge, M. Paterson, P. Pattberg and S. VanDeveer (2012) 'Governing Climate Change Transnationally: Assessing the Evidence from a Database of Sixty Initiatives', *Environment and Planning C: Government and Policy* 30, 4, 591–612.

Bulkeley, H. and M. M. Betsill (2003) *Cities and Climate Change: Urban Sustainability and Global Environmental Governance* (London: Routledge).

Bulkeley, H. and M. M. Betsill (2013) 'Revisiting the Urban Politics of Climate Change', *Environmental Politics* 22, 1, 136–154.

Bulkeley, H. and H. Schroeder (2012) 'Beyond State/Non-state Divides: Global Cities and the Governing of Climate Change', *European Journal of International Relations* 18, 4, 743–766.

Burgiel, S. (2008) 'Non-state Actors and the Cartegena Protocol on Biosafey', in M. Betsill and E. Corell (eds) *NGO Diplomacy: The Influence of Non-governmental Organizations in International Environmental Negotiations* (Cambridge, MA: The MIT Press), 67–100.

Chatterjee, P. and M. Finger (1994) *The Earth Brokers: Power, Politics and World Development* (London: Routledge).

Clapp, J. (1998a) 'Foreign Direct Investment in Hazardous Industries in Developing Countries: Rethinking the Debate', *Environmental Politics* 7, 4, 92–113.

Clapp, J. (1998b) 'The Privatization of Global Environmental Governance: ISO 14000 and the Developing World', *Global Governance* 4, 3, 295–316.

Clapp, J. (2003) 'Transnational Corporate Interests and Global Environmental Governance: Negotiating Rules for Agricultural Biotechnology and Chemicals', *Environmental Politics*, 12, 4, 1–23.

Clapp, J. A. and D. A. Fuchs (eds) (2009) *Corporate Power in Global Agrifood Governance* (Cambridge, MA: The MIT Press).

Corell, E. (2008) 'NGO Influence in the Negotiations of the Desertification Convention', in M. Betsill and E. Corell (eds) *NGO Diplomacy: The Influence of Non-governmental Organizations in International Environmental Negotiations* (Cambridge, MA: The MIT Press), 101–118.

Cutler, A. C., V. Haufler and T. Porter (eds) (1999) *Private Authority and International Affairs* (Albany, NY: SUNY Press).

Dauvergne, P. (2008) *The Shadows of Consumption: Consequences for the Global Environment* (Cambridge, MA: The MIT Press).

Dauvergne, P. and J. Lister (2012) 'Big Brand Sustainability: Governance Prospects and Environmental Limits', *Global Environmental Change* 22, 1, 36–45.

Dellas, E., P. Pattberg and M. Betsill (2011) 'Agency in Earth System Governance: Refining a Research Agenda', *International Environmental Agreements: Politics, Law and Economics* 11, 1, 85–98.

Deutsch, K. W. (1957) *Political Community in the North Atlantic Area: International Organization in Light of Historical Experience* (Princeton: Princeton University Press).

Dodds, F. (2001) 'From the Corridors of Power to the Global Negotiating Table: The NGO Steering Committee of the Commission on Sustainable Development', in M. Edwards and J. Gaventa (eds) *Global Citizen Action* (Boulder, CO: Lynne Rienner Publishers), 203–213.

Doherty, B. and T. Doyle (2012) 'Beyond Borders?: Transnational Politics, Social Movements and Modern Environmentalisms', *Environmental Politics* September, 37–41.
Dombrowski, K. (2010) 'Filling the Gap? An Analysis of Non-governmental Organizations Responses to Participation and Representation Deficits in Global Climate Governance', *International Environmental Agreements: Politics, Law and Economics* 10, 4, 397–416.
Dryzek, J. (2000) *Deliberative Democracy and Beyond: Liberals, Critics, Contestations* (Oxford: Oxford University Press).
Duffy, R. (2006) 'Non-governmental Organisations and Governance States: The Impact of Transnational Environmental Management Networks in Madagascar', *Environmental Politics* 15, 5, 37–41.
Duwe, M. (2001) 'The Climate Action Network: Global Civil Society at Work?' *Reciel* 10, 2, 1–14.
Eckersley, R. (2004). *The Green State: Rethinking Democracy and Sovereignty* (Cambridge, MA: The MIT Press).
Falkner, R. (2003) 'Private Environmental Governance and International Relations: Exploring the Links', *Global Environmental Politics* 3, 2, 72–87.
Falkner, R. (2008) *Business Power and Conflict in International Environmental Politics* (Basingstoke: Palgrave Macmillan).
Folke, C., T. Hahn, P. Olsson and J. Norberg (2005) 'Adaptive Governance of Social-Ecological Systems', *Annual Review of Environment and Resources* 30, 441–473.
Ford, L. H. (2003) 'Challenging Global Environmental Governance: Social Movement Agency and Global Civil Society', *Global Environmental Politics* 3, 2, 120–134.
Frank, D. J., A. Hironaka, J. W. Meyer, E. Schofer and N. B. Tuma (1999) 'The Rationalization and Organization of Nature in World Culture', in J. Boli and G. Thomas (eds) *Constructing World Culture: International Nongovernmental Organizations since 1875* (Stanford: Stanford University Press), 81–99.
Friedman, E. J., K. Hochstetler and A. M. Clark (2005) *Sovereignty, Democracy, and Global Civil Society: State-Society Relations at UN World Conferences* (Albany, NY: SUNY Press).
Garcia-Johnson, R. (2000) *Exporting Environmentalism: U.S. Multinational Chemical Corporations in Mexico and Brazil* (Cambridge: The MIT Press).
Gareau, B. J. (2012) 'The Limited Influence of Global Civil Society: International Environmental Non-governmental Organisations and the Methyl Bromide Controversy in the Montreal Protocol', *Environmental Politics* 21, 1, 37–41.
Gulbrandsen, L. H. and S. Andresen (2004) 'NGO Influence in the Implementation of the Kyoto Protocol: Compliance, Flexibility Mechanisms, and Sinks', *Global Environmental Politics* 4, 4, 54–75.
Haas, E. B. (1958) *The Uniting of Europe: Political, Social and Economic Forces 1950–1957* (Stanford: Stanford University Press).
Haas, P. M. (1992) 'Introduction: Epistemic Communities and International Policy Coordination', *International Organization* 46, 1, 1–35.
Hajer, M. and W. Versteeg (2005) 'A Decade of Discourse Analysis of Environmental Politics: Achievements, Challenges, Perspectives', *Journal of Environmental Policy & Planning* 7, 3, 175–184.

Hall, J. A. (ed.) (1995) *Civil Society: Theory, History, Comparison* (Cambridge, MA: Blackwell and Cambridge, UK: Polity).
Hall, R. B. and T. J. Biersteker (eds) (2002) *The Emergence of Private Authority in Global Governance* (Cambridge: Cambridge University Press).
Hochstetler, K. (2002) 'After the Boomerang: Environmental Movements and Politics in the La Plata River Basin', *Global Environmental Politics* 2, 4, 35–57.
Hoffmann, M. J. (2011) *Climate Governance at the Crossroads: Experimenting with a Global Response after Kyoto* (Oxford: Oxford University Press).
Holmes, G. (2011) 'Conservation's Friends in High Places: Neoliberalism, Networks, and the Transnational Conservation Elite', *Global Environmental Politics* 11, 4, 1–21.
Humphreys, D. (2004) 'Redefining the Issues: NGO Influence on International Forestry Negotiations', *Global Environmental Politics* 4, 2, 51–74.
Humprheys, D. (2008) 'NGO Influence on International Policy on Forest Conservation and the Trade in Forest Products', in M. Betsill and E. Corell (eds) *NGO Diplomacy: The Influence of Non-governmental Organizations in International Environmental Negotiations* (Cambridge, MA: The MIT Press), 119–148.
Jordan, L. and P. Van Tuijl (2000) 'Political Responsibility in Transnational NGO Advocacy', *World Development* 28, 2051–2065.
Keck, M. E. and K. Sikkink (1998) *Activists Beyond Borders: Advocacy Networks in International Politics* (Ithaca, NY: Cornell University Press).
Keohane, R. O. and J. S. Nye, Jr. (1977) *Power and Interdependence* (Boston: Little, Brown).
Kern, K. and H. Bulkeley (2009) 'Cities, Europeanization and Multi-level Governance: Governing Climate Change through Transnational Municipal Networks', *Journal of Common Market Studies* 47, 2, 309–332.
Krasner, S. (ed.) (1983) *International Regimes* (Ithaca, NY: Cornell University Press).
Levy, D. L. and P. Newell (eds) (2005) *Business in International Environmental Governance: A Political Economy Approach* (Cambridge, MA: The MIT Press).
Lindseth, G. (2004) 'The Cities for Climate Protection Campaign (CCPC) and the Framing of Local Climate Policy', *Local Environment* 9, 4, 325–336.
Lipschutz, R. (1992) 'Reconstructing World Politics: The Emergence of Global Civil Society', *Millennium* 21, 389–420.
Lipschutz, R. D. and J. Mayer (1996) *Global Civil Society and Global Environmental Governance* (Albany: State University of New York Press).
Macleod, M. and J. Park (2011) 'Financial Activism and Global Climate Change: The Rise of Investor-Driven Governance Networks', *Global Environmental Politics* 11, 2, 54–75.
Mason, M. (2007) 'Representing Transnational Environmental Interests?: New Opportunities for Non-Governmental Organisation Access Within the World Trade Organisation?' *Environmental Politics* 13, 3, 37–41.
Mathews, J. T. (1997) 'Power Shift', *Foreign Affairs* January, 51–66.
McAteer, E. and S. Pulver (2009) 'The Corporate Boomerang: Shareholder Transnational Advocacy Networks Targeting Oil Companies in the Ecuadorian Amazon', *Global Environmental Politics* 9, 1, 26–29.
Meckling, J. (2011) 'The Globalization of Carbon Trading: Transnational Business Coalitions in Climate Politics', *Global Environmental Politics* 11, 2, 26–51.
Newell, P. (2000) *Climate for Change: Non-state Actors and the Global Politics of the Greenhouse* (Cambridge: Cambridge University Press).

Newell, P. (2008) 'Civil Society, Corporate Accountability and the Politics of Climate Change', *Global Environmental Politics* 8, 3, 122–154.

Newell, P., P. Pattberg and H. Schroeder (2012) 'Multiactor Governance and the Environment', *Annual Review of Environment and Resources* 37, 1, 365–387.

Nye, J. S. and R. O. Keohane (1971) 'Transnational Relations and World Politics: An Introduction', *International Organization* 25, 3, 329–349.

O'Brien, R., A. M. Goetz, J. A. Scholte and M. Williams (2000) *Contesting Global Governance: Multilateral Economic Institutions and Global Social Movements* (Cambridge: Cambridge University Press).

Orsini, A. (2011) 'Thinking Transnationally, Acting Individually: Business Lobby Coalitions in International Environmental Negotiations', *Global Society* 25, 3, 311–329.

Park, S. (2005) 'How Transnational Environmental Advocacy Networks Socialize International Financial Institutions?: A Case Study of the International Finance Corporation', *Global Environmental Politics* 5, 4, 95–119.

Pasha, M. K. and D. L. Blaney (1998) 'Elusive Paradise: The Promise and Peril of Global Civil Society', *Alternatives* 23, 417–450.

Pattberg, P. (2005) 'The Institutionalization of Private Governance: How Business and Nonprofit Organizations Agree on Transnational Rules', *Governance* 18, 4, 589–610.

Pattberg, P. and J. Stripple (2008) 'Beyond the Public and Private Divide: Remapping Transnational Climate Governance in the 21st Century', *International Environmental Agreements: Politics, Law and Economics* 8, 4, 367–388.

Princen, T. and M. Finger (eds) (1994) *Environmental NGOs in World Politics: Linking the Local and the Global* (London: Routledge).

Raustiala, K. (1997) 'States, NGOs and International Environmental Institutions', *International Studies Quarterly* 41, 719–740.

Risse, T. (2002) 'Transnational Actors and World Politics', in W. Carlsnaes, T. Risse and B. A. Simmons (eds) *Handbook of International Relations* (Thousand Oaks, CA: SAGE Publications), 255–274.

Risse-Kappen, T. (1995) 'Bringing Transnational Relations Back In: Introduction', in T. Risse-Kappen (ed.) *Bringing Transnational Relations Back In: Non-State Actors, Domestic Structures and International Institutions* (Cambridge: Cambridge University Press), 3–36.

Rosenau, J. N. (1995) 'Governance in the Twenty-first Century', *Global Governance* 13, 13–43.

Ruggie, J. G. (2004) 'Reconstituting the Global Public Domain: Issues, Actors and Practices', *European Journal of International Relations*, 10, 4, 499–531.

Schroeder, H. and H. Lovell (2012) 'The Role of Non-nation-state Actors and Side Events in the International Climate Negotiations', *Climate Policy* 12, 1, 23–37.

Seligman, A. B. (1992) *The Idea of Civil Society* (New York: The Free Press).

Sell, S. K. and A. Prakash (2004) 'Using Ideas Strategically: The Contest Between Business and NGO Networks in Intellectual Property Rights', *International Studies Quarterly* 48, 43–175.

Skodvin, T. and S. Andresen (2003) 'Nonstate Influence in the International Whaling Commission, 1970–1990', *Global Environmental Politics* 3, 4, 61–86.

Snow, D. A. and R. D. Benford (1992) 'Master Frames and Cycles of Protest', in A. D. Morris and C. M. Mueller (eds) *Frontiers in Social Movement Theory* (New Haven, CT: Yale University Press), 133–155.

Strange, S. (1996) *The Retreat of the State: The Diffusion of Power in the World Economy* (Cambridge: Cambridge University Press).
Szulecki, K., P. Pattberg and F. Biermann (2011) 'Explaining Variation in the Effectiveness of Transnational Energy Partnerships', *Governance* 24, 4, 713–736.
Toly, N. J. (2008) 'Transnational Municipal Networks in Climate Politics: From Global Governance to Global Politics', *Globalizations* 5, 3, 341–356.
Vormedal, I. (2008) 'The Influence of Business and Industry NGOs in the Negotiation of the Kyoto Mechanisms?: the Case of Carbon Capture and Storage in the CDM', *Global Environmental Politics* 8, 4, 96–97.
Wapner, P. (1996) *Environmental Activism and World Civic Politics* (Albany, NY: State University of New York Press).
Wapner, P. (1997) 'Governance in Global Civil Society', in O. R. Young (ed.) *Global Governance: Drawing Insights from the Environmental Experience* (Cambridge, MA: MIT Press), 66–84.
Wapner, P. (1998) 'Reorienting State Sovereignty: Rights and Responsibilities in the Environmental Age', in K. Litfin (ed.) *The Greening of Sovereignty in World Politics* (Cambridge, MA: The MIT Press), 275–297.
Wapner, P. (2003) 'World Summit on Sustainable Development: Toward a Post-Jo'burg Environmentalism', *Global Environmental Politics* 3, 1, 1–10.
Willetts, P. (1996) 'From Stockholm to Rio and Beyond: The Impact of the Environmental Movement on the United Nations Consultative Arrangements for NGOs', *Review of International Studies* 22, 57–80.
Williams, M. and L. Ford (1999) 'The World Trade Organisation, Social Movements and Global Environmental Management', *Environmental Politics* 8, 268–289.
Wright, B. G. (2000) 'Environmental NGOs and the Dolphin-Tuna Case', *Environmental Politics* 9, 4, 82–103.
Young, O. R. (1989) *International Cooperation: Building Regimes for Natural Resources* (Ithaca, NY: Cornell University Press).
Young, Z. (1999) 'NGOs and the Global Environmental Facility: Friendly Foes?' *Environmental Politics* 8, 243–267.

9
Environmental Security

Larry A. Swatuk

Introduction

The chapter presents a critical genealogy of environmental security studies, with a particular emphasis on trends in scholarship over the past 25 years. At the heart of environmental security studies is a profound contradiction: while 'human transformation of the environment is a global-scale problem', and while humans are connected by 'pervasive flows of people, goods, money, ideas, images and technology across borders', we remain 'fragmented by the political division into sovereign states' (Conca and Dabelko, 2010: 1–2). This state-centered fragmentation further divides us by values, perspectives, and approaches to 'security', be it national, regional, or individual. As questions regarding climate change, planetary boundaries (Rockstrom et al., 2009), mitigation and adaptation measures, and appropriate governance structures have moved to global center stage in environmental politics debates, the conversation around the environment and security has, in my view, never been more divisive (Bernauer et al., 2012; Floyd and Matthew, 2013; Matthew et al., 2010).

Proponents of environmental security argue that if environmental change is a potential source of social conflict, and if societies face dangers from environmental change, then security policies – indeed, the very concept itself – must be redefined to account for these threats (Conca and Dabelko, 1998, 2010). While new thinking about security had been ongoing throughout the 1980s (Mathews, 1989; Ullman, 1983; World Commission on Environment and Development, 1987), the end of the Cold War provided real intellectual space for considering how the environment might be accounted for in various approaches to security (Dalby, 2002b). Following the terrorist attacks of 9/11, however, environmental security once again was shunted to the sidelines by mainstream security concerns made manifest in the war on terror. As climate change has moved up the international policy agenda, many of the academic debates regarding the meaning and practice of 'security' have been resurrected (Barnett and Adger, 2007; Burke et al., 2009; Gartzke, 2012).

According to Floyd and Matthew (2013), 'environmental security has not evolved over the last two decades as a homogeneous field of analysis, but rather as a polysemous category encompassing a wide range of analytical and normative meanings and positions'. The 1990s saw two interrelated discussions: one involving the redefinition of security (Baldwin, 1997; Booth, 1991; Buzan, 1991; Buzan et al., 1998; Krause and Williams, 1997; Lipschutz, 1995; Walt, 1991); the other involving questions about how environmental change threatens (individual, state, global) security (Deudney and Matthew, 1999; Myers, 1989, 1993; Ohlsson, 1999; Renner, 1989). As this chapter will show, more than a decade into the new millennium, there continues to be little consensus on either of these issues. An added variation that has marked environmental security studies throughout the 21st century has been the focus on climate change. Whereas biosphere-level processes initially were regarded as too amorphous and varied to concentrate popular and professional concerns over security (Homer-Dixon, 1991, 1999), there have been concentrated efforts across all scholarly camps to bring climate change to the forefront of environmental security analysis (Barnett and Adger, 2007; Dalby, 2009; Floyd, 2008; Moran, 2011). Yet, because these debates are ongoing at the policy level, the practical application of various ideas regarding environmental security by a wide array of actors embodies all of the contradictions and controversies extant in the debates. This will be demonstrated in our case study below.

This chapter proceeds as follows. In the next section I outline the traditional, realist, approach to 'security'. Following that I provide a number of critiques of this approach: institutionalist, (neo)realist, structuralist, and poststructural/postmodernist. I argue that the first three work within a problem-solving framework in an effort to 'rethink' security, whereas the last presents a 'critical' challenge to both the analysis and practice of security. I then turn to an extended discussion of the environment and security. I frame this section in terms of scholars working within the ambit of the dominant paradigm – what I call *Environmental Security Studies* – and those who challenge this approach – *Critical Environmental Security Studies*. What I show there is the way climate change as a security issue has driven these two positions further apart, with a good deal of the problem-solving literature (that is, Realist, Liberal Institutionalist) focusing even more so on either resource capture or/and planning for extreme events through containment at source, meaning, primarily, in the 'Third World' (Dinar, 2011; Moran, 2011).

At the same time, there are faint echoes of a middle position that argues that the nature and understanding and therefore the actions around security have become more nuanced and flexible. So, whereas securitization argues that extraordinary measures are unleashed through speech acts (Waever, 1997; Williams, 2003), there are scholars arguing that 'securitization' moves have actually helped decision-makers think in a nuanced way about suitable (state-level, multilateral) responses to different sorts of security threats (Trombetta, 2009). In other words, not every acknowledged threat requires the sword; some do indeed require ploughshares (Dinar, 2011; Dinar et al., 2013). Lastly, I show how the analysis and practice of 'environmental security' lead in very contradictory directions in a short case study from South Africa.

Rethinking security

The traditional approach to security studies, known as Realism, focuses on the causes of war and the conditions of peace between and among states. Given the systemic condition of anarchy, wherein sovereign states pursue policies of self-help, states by their very existence threaten each other.[1] 'Security', then, may be defined as the protection of a state (and, by logical extension, its citizens) from the threat of other self-regarding states acting in their own interests. This protection is facilitated by military preparedness. Diplomacy is considered the practice of 'war by other means'. In this rendering, the 'state' is both means and end (or 'referent object') of 'security'. State security is achieved through the exercise of 'power', itself also both means and ends, to wit: a powerful state is made secure by the exercise of power. Power involves multiple factors, including centrally the threat and use of military force (Morgenthau, 1978).

Since the acquisition and projection of power is the main currency of the inter-state system, self-regarding states usually find themselves caught up in global and regional security dilemmas. The end result may be an overall decrease in regional or global security, depending on the nature of the arms produced. For Realists, stability of the inter-state system depends on achieving a 'balance of power'.

To many, great power 'security' threatens the lives of everybody. Critiques of the Realist approach to security proliferated during the 1970s and 1980s. There are many different ways to categorize these challenges to Realist analysis. One is to group them according to general theoretical position. Liberal institutionalists, for example, work within the accepted framework of inter-state relations but challenge the logic of states pursuing unilateral security policies (Hurrell, 1995; Ruggie, 1998). Liberal institutionalists argue in support of both a broadened understanding of 'security' (defined in military, economic, and ecological terms) and a multilateral framework for addressing multidimensional challenges to global security (Commission on Global Governance, 1995).

Many scholars working within the ambit of 'national security' have modified their position in acknowledgement of the multidimensional nature of threats to the state (Buzan, 1991), or threats to individuals (Matthew et al., 2010), or to the international system itself, from weak, fragile, and failing states (Rotberg, 2003). So while security means (individual or collective) freedom from these threats, the framework for analysis remains fundamentally state-centric.

Structuralist analyses may or may not work within a statist ontology. For example, dependency and world systems theorists generally retain a statist language while focusing attention on broad, structural processes like global capitalism. So, states may be divided into core, semi-peripheral, and peripheral (Wallerstein, 1974). In the context of late 20th-century and early 21st-century globalization, whole regions are seen to merit core, intermediate, or peripheral zone status (Hettne, 1997). In terms of security,

> [r]egions in the core zone, North America, Europe and East Asia centred on Japan, are thus economically more advanced and normally growing, and they have stable – if not always democratic – regimes which manage to avoid interstate as well as intra-state conflicts... Regions in the intermediate zone... are under 'core guidance'... Regions in the peripheral zone, in contrast, are politically turbulent and economically stagnant. War, domestic unrest, and underdevelopment constitute a vicious circle which make them sink to the bottom of the system, creating a zone of war and starvation.
> (Hettne, 2001: 95–96)

Thus, for many structuralists, insecurity is a consequence of a state's or region's history of incorporation into the global capitalist system (Swatuk, 2001: 278–283).

Focusing attention on other structural processes and conditions – for example, race, class, and gender – helps reveal a panoply of threats and insecurities individuals and groups suffer despite (or possibly because of) the existence of a 'strong', militarily-secure state (Sutton et al., 2008; Tickner, 1995). Herein lies a more serious challenge to traditional renderings of security: the challenge is not merely to redefine the content of security (from military to military-plus) but to reconsider the very subject itself. If traditional approaches to security ignore or exacerbate those global socio-political-economic inequalities that constitute the basis for most forms of insecurity, whose interest does the traditional approach serve? When combined with the specter of nuclear Armageddon, the answer would seem to be 'no one' (Booth, 2007). Yet, Steven Walt (1991: 222), reflecting upon the 1980s 'renaissance' in security studies, admonishes those who would ask such difficult questions:

> Security studies seeks cumulative knowledge about the role of military force. To obtain it, the field must follow the standard canons of scientific research: careful and consistent use of terms, unbiased measurement of criticial concepts, and public documentation of theoretical and empirical claims.

But is security studies the dispassionate search for objective truth? Poststructuralist and post-modernist perspectives suggest that it is nothing of the sort. To the contrary, they suggest that, as a discourse, it acts as the handmaiden of the powerful (Walker, 1997: 62). Traditional security studies have long been tied to the strategic interests of the world's most powerful states. So, answers given to questions regarding the security of (powerful) states lead directly to policies that serve the few and imperil the many. The post-modern/structural critique constitutes a broad but barely unified church, including, inter alia, feminism, environmental politics, political ecology, development studies, and critical strands of international political economy (IPE). In Waever's (1997) terms, they are part of the 'fourth debate' in international relations (IR). Most of these challengers may be thought of as constitutive of a nascent, open-ended 'critical theoretical' enterprise (George, 1994). Critical theory is to be contrasted with 'problem-solving theory' which, in Cox's terms, takes 'prevailing social and power relationships and the institutions into which they are organised... as the given framework for action' (Cox, 1986: 206). For several key reasons, the environment stands at the epicenter of this fourth debate.

Environmental degradation, resource depletion, species loss – for critical theorists these conditions are emblematic of the crisis of contemporary society (Booth, 2007). Along with weapons of mass destruction and the Holocaust (Baumann, 1989; George, 1994), they are logical consequences of modernity whose overriding idea – progress – has for several hundred years been dependent upon a belief in the rightness of 'harnessing' nature, of rendering useful to 'rational man' all that is 'natural'. Given that the modern age arose out of a white, Western European Enlightenment, too often and for too long, this constructed category has included not just the physical environment but animals, women, inferior European men, and all people of color (Pettman, 1996).

There are two points of importance for the topic at hand. First, the national security state remains the primary institutional structure of the modern era. To paraphrase Eckersley (2004), it is not going anywhere anytime soon. 'Security studies' as the handmaiden of the most powerful of these states is founded on key binaries: the possibilities of the political (inside the state where order is possible); the location of threats (outside the state where anarchy reigns); the means of discerning threats and proper responses (through the identification of objective facts by a rational 'knower'); the construction of a theory of security through practice so that one eventually arrives at an accurate representation of truth (that is, knowledge claims independent of subjectivity); the masculine exercise of 'power over' the 'irrational other' as proper form of response (in defense of the weak and feminine, but also, given the condition of anarchy, any other response would itself be weak and feminine).

For critical theorists, the practice of this form of 'social science' produces simultaneously, wealth for the few, poverty for the many; creative invention and possibly irreversible environmental destruction; and increased leisure and choice for some, increased toil and a lack of options for most (Schnurr and Swatuk, 2012). Any 'celebration of the age of rational science and modern technological society', therefore, 'cannot simply be disconnected from' its myriad, historically specific negative consequences. They are, in fact, integrally related (George, 1994: 141; Peterson, 2003: 22–29). In terms of 'the environment', the very forms and practices giving rise to its degradation cannot, therefore, be its salvation (Barnett, 2001; Dalby, 2009). As will be seen below, 'the state is not only unnecessary from a Green point of view, it is positively undesirable' (Paterson, 1995: 238; cf. Eckersley, 2004; Barry and Eckersley, 2005).

The second, and related, point is that for critical theorists, these outcomes are not the accidental result of policies based on poorly understood science. Rather, they are the direct outcome of a form of knowledge production dependent upon a positivist/empiricist methodology, a discipline whose ontological frameworks regarding both 'human nature' and social organization asks the wrong questions, or frames the right questions the wrong way, and whose epistemological claims 'to know' 'obstruct a relational, multi-dimensional understanding of the social' (Peterson, 2003: 39; also, Booth, 2007).

While constructivist methodologies take us some way toward a more nuanced understanding of the ways in which human language, thought, and action create social reality – so, in my view, effectively contesting positivist claims of passive subjects studying objective reality – it is the poststructural, postmodernist perspective which lends most insight into this process (for example, Peterson, 2002).

In the face of dramatic change – for example, climate change, deforestation, the collapse of fisheries, the collapse of, first, the Soviet empire and, a decade or so later, the 'twin towers', leading to the rise of the 'war on terror' – we desire explanations that will ensure stability or bring order. In other words, we seek to contain the 'threat'. But explanations regarding causality diverge markedly: for Realists, 'nature' remains somehow 'out there', as does a collapsed Soviet Union, failed and failing states, and terrorist cells – an untamed other beyond the purview of ordered civilization within a (democratic) state. Security requires containment. For a majority of critical theorists, however, these events and conditions are linked. They are emblematic of the multidimensionality of and multiple paradoxes within the late modern era, wherein the state system, industrial capitalism, technological innovation (particularly its military dimension), individual liberty manifesting as consumerism, and the tyranny of bureaucracy are equally complicit (Hall et al., 1992; Berman, 2006; Dalby, 2009). 'Stability' in this case requires a fundamental re-ordering of late modern society (Peterson, 2003). This is an ongoing academic debate whose impact on people, places, and things is, as we will see below, very real. Having mapped the terrain of and debates within security studies, let us now turn to environmental security.

Environment and security

Positing negative outcomes of environmental decay for humanity and/or the planet, particularly from an over-consumption perspective, has a proud lineage stretching back to the beginnings of the Industrial Age (Rousseau, 1953; Leopold, 1949; Mumford, 1934). It has been particularly strong since the 1960s, with seminal texts being provided by Rachel Carson (1962) and the Sprouts (1965) (see, also, Ehrlich, 1968). As with realism, a variety of historical examples – for example, the 'lessons of Easter Island' – provide a foundational narrative of sorts. Since the end of the Cold War, however, the security focus has been dominated by state-centric approaches interrogating the implications of environmental degradation *In the Global South* for security of states *In the Global North*, an argument described by supporters and critics alike as 'Malthusian' or 'neo-Malthusian' (Dalby, 2002a; Gray, 2002: 180–182). Discussions related to climate change and its impact on security generally serve to reinforce this perspective (Bernauer et al., 2012).

The end of the Cold War meant that, for a time, Western statemakers were at a loss regarding a revised role for their militaries. If the West had arrived at 'the end of history', as hypothesized by Fukuyama (1992), from where would future threats emanate? Given 20 years of high-profile summitry regarding the human impact on the global environment, state-makers logically looked to the 'threat potential' of the environment.

Throughout the 1990s several extensive research programs were undertaken regarding the hypothesized link between environmental degradation, resource scarcity, and violent conflict, commonly called 'environmental security'. There have been a number of attempts to review and synthesize this vast literature (Barnett, 2001; O'Brien and Barnett, 2013; Gleditsch, 1998, 2012), with the annual report of the Woodrow Wilson Center's Environmental Change and Security Project (ECSP) acting as a valuable 'rough guide' to the field (see http://wwics.si.edu). Floyd and Matthew (2013), Matthew (2002), and Matthew et al. (2010) are useful and important guides which review the research in terms of studies explicitly concerned with the state and environmentally induced acute conflict probability and those that are more holistic, historical, and critical in their approaches.

In essence, the first group are engaged in an intra-paradigm debate regarding problem-solving theory. They are primarily concerned with methodology, data collection, appropriate scope, and definitions, in the hope that their research programs will lead to both the accumulation of knowledge (Homer-Dixon, 2003: 90) and policy relevance.

The second group, in contrast, are centrally concerned with critical theory, though they come at it in a wide variety of ways (Dinar, 2011; Schnurr and Swatuk, 2012; O'Brien and Barnett, 2013). While no less concerned with scope and method, they are equally concerned with engaging those working within the problem-solving camp regarding ontological and epistemological questions. Their central concern is that problem-solving theory begins, rather like traditional security studies, from a false premise: that global society is primarily a world of self-regarding states, some of which have more (political, military, technological) power and are more economically developed than are others. A state's place in the system is largely the result of its own efforts, and political order – indeed, 'political life' – is only possible within the boundaries of the sovereign state (Walker, 1997: 62).

Many scholars, mostly of the neo-institutionalist or political ecology variety, fall somewhere in between the two – at once holding poststructuralist positions regarding, for example, an ontology centered on individuals or ecosystems, and making policy advice for states and global institutions (Watts, 2013). In the next section, I detail the findings of these two different groups. Following Krause and Williams (1997), I label these 'environmental security studies' and 'critical environmental security studies'.

Environmental security studies

Gleditsch (1998) summarized historical concerns with the link between environmental degradation, resource scarcity, and violence in terms of fights over territory, raw materials, continental shelves and islands, energy, and food. In traditional Realist analysis, state power depends fundamentally on the natural resources contained within its territorially delimited space (Morgenthau, 1978). It also depends on the ability to access key resources not contained within the state. In the late-modern industrial era, national power requires the ability of a state to transform these renewable and non-renewable resources into tradable consumables. Natural resources may be enhanced, depleted, or transformed over time. According to Klare (2002), competition and control over critical natural resources will be the guiding principle behind the use of military force in the 21st century. The popular press throughout the world continues to play on these fears. The physical location of these events in the global South and its link to warfare and general human misery in the popular imagination have been facilitated by magazine and newspaper reporting focusing on the more bizarre and fantastical aspects of West African conflicts, with Robert Kaplan's 1994 article 'The Coming Anarchy' for the *Atlantic Monthly* being influential in policy circles (see Ellis, 1999, for a trenchant critique; also de Soysa, 2013, and Jones Luong and Weinthal, 2010, for contrasting perspectives on 'the resource curse'). In the 1990s, a number of systematic studies emerged to examine the hypothesized relationship between the environment and conflict.

Renewable resource degradation and violent conflict

The Homer-Dixon-directed projects on 'Population, Environment and Security' and on 'Environmental Change and Security' are undoubtedly the most well known in North America of these systematic programs of research (Homer-Dixon, 1991, 1994, 1999; Homer-Dixon and Blitt, 1998). Research focused on the causal link between violent conflict and the depletion of renewable resources, in particular agricultural land, water, forests, and fisheries, arguing that the social effects of large-scale environmental changes such as global warming and ozone depletion 'will not be seen until well into the [21st] century' (in Conca and Dabelko, 1998: 289). 'Researchers sought to answer two questions. First, does environmental scarcity contribute to violence in developing countries? Second, if it does, how does it contribute?' (Homer-Dixon, 1998: 279).

Three hypotheses were used to link conflict with environmental change:

- decreasing supplies of physically controllable resources (for example, water and agricultural land) would provoke 'simple-scarcity' conflicts;
- 'group identity' conflicts would result from large population movements caused by environmental stress;
- severe environmental scarcity would simultaneously increase economic deprivation and disrupt key social institutions (most importantly, the state) and cause 'deprivation' conflicts.

Key to their analysis is the definition of scarcity. Scarcity can come about in one of three ways: (i) as a result of increased demand (demand-induced), for example, through population growth or increased per capita consumption; (ii) as a result of decreased supply (supply-induced), for example, the erosion of cropland; and/or (iii) as a result of unequal access to and distribution of a resource (structural) 'that concentrates it in the hands of relatively few people while the remaining population suffers from serious shortages' (Homer-Dixon, 1998: 280). These push-and-pull factors are said to often coexist and interact (Homer-Dixon and Blitt, 1998: 6). In the event of scarcity, two processes are set under way: resource capture by those with the means to do so and ecological marginalization of those without.

So, what is important is not merely the absolute supply of a resource. 'What we should investigate, rather, is the resource's supply *Relative To*, first, demand on the resource, and, second, the social distribution of the resource' (Schwarz et al., 2000: 79). According to Bernauer et al. (2012), this body of research represents 'the baseline model for neo-Malthusian environment-conflict arguments' (see also Dalby, 2002).

A key finding of the Homer-Dixon project is that 'environmental scarcity does not inevitably or deterministically lead to social disruption and violent conflict'. The causal pathway from environmental degradation to social violence is neither direct nor unilinear. On the one hand, 'environmental scarcity... increases society's demands on the state while decreasing its ability to meet those demands' (Homer-Dixon, 1998: 281). But, on the other hand, 'environmental scarcity produces its effects within extremely complex ecological-political systems' (Homer-Dixon, 1999: 178). Therefore, a key finding is that 'environmental scarcity is not sufficient, by itself, to cause violence; when it does contribute to violence, research shows, it always interacts with other political, economic, and social factors. Environmental scarcity's causal role can never be separated from these contextual factors, which are often unique to the society in question' (Homer-Dixon, 1999: 178). More recent research seems to support this conclusion (Gleditsch, 2012). In an extensive review of the literature, Bernauer et al. (2012) conclude that 'the effect of environmental changes on violent conflict appears to be contingent on a set of intervening economic and political factors that determine adaptive capacity'.

'Maldevelopment' as contextual factor

The international research team under the direction of Günther Baechler examined these contextual factors, particularly the perceived links between maldevelopment, environmental transformation, and conflict. While increasing population pressure on renewable resources remains important, population increase is but one of a number of consequences of maldevelopment in developing and transitional societies: '*Development And Security Dilemmas* are connected to a syndrome of problems which produces environmental conflicts of varying intensity and nature' (1998: 24). Maldevelopment, in Baechler's terms, is most often the result of the ways in which developing and transitional societies have experienced modernization, the outcome typically being a weak state form imposed on a multi-ethnic society, but dominated by one ethnic group dependant on one or a few primary commodities for revenue. This, they argue, is the most likely setting in which conflict will occur (Bannon and Collier, 2003; Collier and Hoeffler, 2004; Le Billon, 2001, 2012).

In contrast to Homer-Dixon, Baechler suggests a somewhat longer causal chain, one that begins from the conditions that gave rise to environmental degradation. The important finding here is that states in both the North and South help constitute instability in the South.

With regard to war-torn Sierra Leone, Richards (1996: xvii) states, ' "we" and "they" have made this bungled world of Atlantic-edge rainforest-cloaked violence together'. Maldevelopment is a consequence of historical processes and contemporary global links (also, Duffield, 2001).

Based on 40 case studies, Baechler (1998: 37–38) draws several conclusions: (i) where causality is concerned, the political context matters more than the environmental; (ii) environmental conflicts often catalyze cooperation where 'political compromises are seen as desirable and technical solutions are feasible'; (iii) there is 'little empirical support for the first hypothesis that environmental scarcity causes violent conflicts or wars between states'; (iv) 'environmental scarcity causes large population movements, which in turn cause conflicts'; and (v) environmental scarcity can simultaneously increase economic deprivation and disrupt key social institutions (these are Homer-Dixon's terms); however, this is no guarantee that the problem will result in violent conflict.

Nevertheless, most scholars working within a state-centric framework continue to assert the logical sensibility that environmental change at whatever physical or temporal scale does increase the likelihood of conflict, be it interpersonal, communal, intra-state, or inter-state, primarily in the global South (Hsiang et al., 2013; Kahl, 2006).

Back to the future

With the ascendance of climate change onto the international policy agenda, mainstream security analysts have basically taken Homer-Dixon's 1999 findings and stood them on their head, that is, climate change will create myriad uncertainties surrounding renewable resources, thereby leading to violent conflict. And while the U.S. Department of State's *2014 Climate Action Report* demonstrates nuance in understanding the threats and reducing vulnerabilities, it is still ineluctably cast within the framework of 'preparing the United States for the impacts of climate change' (US Dept of State, 2014: 7). In other words, the state must be secured from the environment.

In a recent survey of the literature, Scheffran et al. (2012) begin by pointing to the 'multiple pathways and feedbacks between the climate system, natural resources, human security and societal stability'. The authors describe 'vulnerability' as having three characteristics: (i) exposure to climate change; (ii) sensitivity to climate change; and (iii) adaptive capacity. Like the earlier studies regarding renewable resources and acute conflict probability, the authors point to the weak and fragile states of the global South as being least resilient and most seriously affected by the anticipated climate change impacts. The authors conclude that 'there is reason to believe that [increasing global mean temperature] might overwhelm adaptive capacities and response mechanisms of both social and natural systems and thus lead to "tipping points" toward societal instability and an increased likelihood of violent conflict'.

In a detailed analysis of 60 quantitative studies whose findings 'reliably infer causal association', Hsiang et al. (2013) provide compelling evidence that 'climatological variables have a large effect on the risk of violence or instability in the modern world'. While acknowledging 'widespread contradictory evidence' (for example, Theisen, 2012; Slettebak, 2012; Gartzke, 2012; Nordås and Gleditsch, 2007; Buhaug et al., 2010; Buhaug, 2010), the authors nevertheless conclude that 'climatic anomalies of all temporal durations, from the anomalous hour to the anomalous millennium have been implicated in some form of human conflict'. By 'conflict', the authors mean (i) personal violence and crime; (ii) group-level violence and political instability; and (iii) institutional breakdown. The authors present seven plausible hypotheses regarding climate change–conflict mechanisms (Hsiang et al., 2012: 11):

- 'when climate events cause economic productivity to decline, the value of engaging in conflict is likely to rise relative to the value of participating in normal economic activities';
- 'declines in economic productivity reduce the strength of governmental institutions... curtailing their ability to suppress crime and rebellion';
- 'when climate events increase actual (or perceived) social and economic inequalities in a society, this could increase conflict by motivating attempts to redistribute assets';
- 'if climatic events cause large population displacements or rapid urbanization, this might lead to conflicts over geographically stationary resources that are unrelated to the climate but become relatively scarce where populations concentrate';
- 'changes in climate might also affect the logistics of human conflict, for example, by altering the physical environment (for example, road quality) in which disputes or violence might occur';
- 'climate anomalies might result in conflict because they can make cognition and attribution more difficult or error-prone, or they may affect aggression through some physiological mechanism.'

As with the earlier studies, the authors 'do not conclude that climate is the sole, or even primary, driving force in conflict', but their findings do show there to be 'substantial effects' (Hsiang et al., 2012: 12). The obvious questions remain unanswered, however: if changes to mean global temperature are anthropogenic in origin, (i) how do we put a stop to these changes; and (ii) in the absence of successfully addressing (i), how do we avoid the spiral toward violence?

Governance, violent conflict, and peace-building

Homer-Dixon's research suggests certain states are prone to environmentally induced conflict. According to him, 'The character of the state is particularly important: a representative state will receive [demands from civil society] and react quite differently to a non-representative state such as apartheid South Africa' (Homer-Dixon, 1998: 281). Moreover, economically poor states lacking both financial and human capital, and being ethnically diverse, seem less able to adapt to severe environmental challenges. They are said to lack adaptive capacity (Homer-Dixon and Blitt, 1998: 9). A cursory look at their case studies gives some idea as to the sorts of states Homer-Dixon and his colleagues consider short on adaptive capacity: Mexico, the Philippines, South Africa, Pakistan, and Rwanda. There are many studies conducted at different scales whose findings support Homer-Dixon (for example, Kahl, 2006; Gartzke, 2012). It is these states that, today, feature in so much research and speculation about the impacts of climate change (Moran et al., 2011).

A clear trend across these studies is the distance they have traveled away from a fascination with conflict potential toward deliberately pursuing peace-building opportunities through cooperation on environmental issues (Dinar, 2011; Dinar et al., 2013). For Lietzmann and Vest (1999: 40), 'violence is by no means the automatic outcome of conflict. Countless issues of conflict, particularly at the local or regional level are resolved cooperatively; only a limited number of conflicts reach a higher conflict intensity'. Environmental stress plays different roles along the 'conflict dynamic': as a structural source, a catalyst, or a trigger (Lietzmann and Vest, 1999: 41). To ensure that environmental stress does not reach levels at which violent conflict becomes likely, 'the development of early warning indicator systems, data bases and decision support systems are feasible and warranted'. De Soysa (2013), for example, argues 'that much of the problem of violence can be solved by policies in the realm of global governance, for it is the developmental and governance traps and the lack of "institutional capital" that leads to persistent poverty, human insecurity, and continued dependence on natural resources'.

Environmental 'peacemaking' is a decade-old trend in the environmental security studies literature (Conca and Dabelko, 2002; Conca and Wallace, 2009; Daoudy, 2010; Maas et al., 2013). Conca and Dabelko (2002), for example, suggest two specific ways in which cooperation on environmental problems can help build peace: by altering the 'strategic climate' within which states operate; and by helping construct post-Westphalian forms of governance that may ultimately tie states into cooperative agreements and practices to facilitate 'learning'. Environmental peacemaking may be understood as one way of bringing parties in conflict together 'to work on environmental issues in ways that build confidence and reduce political tensions'. Certain resources may be more conducive to cooperative behavior than others. For example, shared water resources seem to offer pathways to peaceful cooperation, whereas forests and minerals – due to the particular nature of the resource – appear more prone to induce conflict (Adelphi Research and Woodrow Wilson Centre, 2004: 2, 7–10; De Jong et al., 2007; Dinar, 2011; Dinar et al., 2013; Swatuk, 1996).

Critical environmental security studies

Almost as soon as 'the environment' appeared on the policy map of state security apparatuses, dissenting and critical voices could be heard questioning the appropriateness of linking environmental issues to (national) security practices. Dalby (1997, 1998, 2002b, 2009), Peluso and Watts (2001), Barnett (2001, 2007), Gleditsch (1997, 1998, 2012), Matthew et al. (2010), and Levy (1995) are among those who have made important critical interventions in the environment and security debate. Yet, it was Deudney's 1990 article which remains, to my mind, the most compelling argument against this link, with all others taking their cue in one way or another from him. His discussion centers on three basic points. First, the structures developed to ensure *National* security are of little help as far as environmental problems – be they local or global – are concerned. National security is 'safeguarded' through a system of organized violence highly dependent on secrecy and technological expertise, whereas solutions to environmental problems require transnational cooperation, openness, and creativity.

Second, given that national security discourses render all those people, places, and things outside the state as a potential 'threat' – as an unknowable, untrustable 'other' – 'it seems doubtful that the environment can be wrapped in national flags without undercutting the "whole earth" sensibility at the core of environmental awareness' (in Conca and Dabelko, 1998: 309).

Third, while 'few ideas seem more intuitively sound than the notion that states will begin fighting each other as the world runs out of usable natural resources', global systems of trade, the substitutability of many raw materials, and 'the very multitude of interdependency in the contemporary world, particularly among the industrialised countries, makes it unlikely that intense cleavages of environmental harm will match interstate borders...Resolving such conflicts will be a complex and messy affair, but the conflicts are unlikely to lead to war' (in Conca and Dabelko, 1998: 310, 312).

Each of these three claims has been subject to, in Matthew's terms, 'a decade of environmental security research, debate and policy experimentation' (Matthew, 2002: 109). Military organizations have gone to great lengths to argue the value of their institutions and methods to environmental preservation. Others have revealed an abiding tendency, however, to 'securitize the environment' rather than 'green the military' (Van de Veer and Dabelko, 1998; more generally, see Williams, 2003). Most importantly, in my estimation, however, is Deudney's claim that environmentalism calls into question the very idea of a world of self-regarding states seeking security through violent practices.

The problem of 'the state'

A common theme running through the critical environmental security studies literature is what to do about the state – as a constellation of power and practice and as an analytical concept. Many of those who first argued for an expanded definition of security, including environmental issues, did so from the vantage point of disciplines not wedded to the state – environmental studies; human geography; ecology; philosophy; anthropology; biology; feminist theory; and gender studies – and/or from physical locations outside traditional networks of power and privilege. While many people working in these fields bring traditional methodologies to bear on their research, their starting points are generally conceptions of space and time at variance with state-based analyses. The world they 'see', therefore, is one quite different than those working within IR's dominant paradigm, be they (neo)Realist or (neo)Institutionalist. And what they see, most often, is the negative consequences of modern industrial life dominated by a hierarchy of states.

What this disparate group shares, therefore, is a general desire to problematize practices and institutions that are considered 'natural' and/or immutable by, for example, most political scientists, security studies 'specialists' (be they experts or practitioners), or policymaking elites. For example, to many ecologists environmental security is about securing environmental health (within specific ecosystems; or at the level of the planetary biosphere) and, by extension, human well-being for humans is part of the biosphere, not separate from it (Stoett, 2012). To ensure this 'security' requires a holistic understanding of the ways in which humans interact with 'nature'. It requires at minimum 'global environmental governance', not divisive national security policies (Peluso and Watts, 2003).

At the same time, however, they recognize the very real power wielded by state-makers and maintainers (Pettman, 1991). If 'anarchy is what you make of it', as Wendt suggests, so too is order. States, as powerful actors operating within a largely Realist logic, make a rather brutal reality. For this reason, most people concerned with the negative side of the modern era, in particular those activities undertaken in the name of 'order' and 'stability', are forced to work with states so that they might at minimum modify their practices. Yet the vituperative debate between climate deniers and their political and economic patrons, on the one side, and almost the entirety of the academic and scientific world concerned with the human-induced effects of climate change, on the other, shows how difficult it is to find common purpose (*Cf* Mulligan, 2012 on US energy policy). Even where 'enlightened' state makers find themselves seated with critical environmental security scholars, the tendency is to move toward a lowest common denominator: that is, extreme climate events will negatively impact Africa and South Asia and we must help them to adapt (see the numerous postings on the Wilson Center's *New Security Beat* for specific examples).

In terms of dealing with the state as a constellation of power and practice, most academics, if called upon, are only too willing to give testimony in the halls of state power. The environment, being outside the traditional purview of security practitioners, requires a different kind of expertise, so opening up the heavy (and heavily fortified) doors of the state to, among others, ecologists, geographers, and historians (Matthew, 2002: 118–119). In terms of dealing with the state as an analytical category, once again preferences differ. Barnett (2001), Lonergan (2000), Matthew (2002), and Schnurr and Swatuk (2012) are among those who, as far as the environment goes, seek to decenter the state. The state is but one form of social organization that changes through time. More appropriate referents of security are the biosphere and the individual, together linked by the concept of 'human security' (Barnett and Adger, 2007; Matthew et al., 2010). Eckersley (2004), in contrast, argues that we must begin from the fact that the state is going nowhere soon, and work toward 'greening' it.

A new language for new understanding?

A focus on human security located within a discourse of modernity takes us only part way toward a more inciteful understanding of the forces at work in creating what McKibbon (1998) labels 'Earth II'. What many critical scholars feel needed is a language of dissent – one which challenges consistently the truth claims made by the purveyors of global (political) order (Peluso and Watts, 2001; Schnurr and Swatuk, 2012). For George (1994: 166), this

> politics of dissent... is *post*modern in the sense that it seeks to confront, at every level, those aspects of modernity that undermine any potential people might have to produce, in their everyday lives, resistances to power relations that silence, demean and oppress them.

Language is key to this enterprise, for it is not neutral (Peterson, 2003: 41). Because 'development', 'security', and the 'environment' are dominated by 'expert' groups vested with specific technical/managerial knowledge, challenging their way of knowing and ordering the world requires, almost unavoidably, the use of their language. When a term like 'sustainable development' or 'security' is being used, its multiple meanings ensure (i) that different groups continue to talk past each other; and (ii) reaching no consensus, or a consensus so vague as to allow participants in the discourse to walk away equally satisfied, the dominant group's approach remains ascendant.

Magnusson (1994) articulates world politics 'as a problem of urban politics'. Gadgil and Guha (1995) encourage us to think of resource use at global level in terms of biosphere and ecosystem people. McNeill et al. (1991) describe the 'ecological shadow' of a country as 'the environmental resources it draws from other countries and the global commons'. Wachernagel and Rees (1996; also, Rees and Wachernagel, 1994) use the term 'ecological footprint' to help us better understand resource flows beyond 'the state' (see Dalby, 2002a, for details). These approaches all deploy a new and different language to that of state-centric security and environmental security studies.

Holistic, historically rooted and conscious approaches are also key to the critical environmental security studies enterprise. Traditional approaches to environmental security present, at best, a linear and 'bifurcated' understanding of history, with 'developed' and 'developing' states acting as simplified markers in this 'historical' process. States, societies, and empires rise and fall through time and do so as a result of decisions taken within their spatial boundaries and/or as they impact upon one another through practices such as trade and war. Successful states persist due to their own 'ingenuity' (to use Homer-Dixon's term), while failed and fragile states are cursed by their resource abundance (De Soysa, 2013) and/or demographic 'youth bulge' (Roche, 2010).

For (human) geographers, (social) anthropologists, sociologists, environmental historians, and ecologists, among others, this is only one part of a more complex story. Because political science and international relations focus primarily on the modern state, their frameworks are partial. As such, their understandings and answers to important questions are equally partial. Dalby, for example, articulates a history of environmental change that draws our attention to the way processes of modernization – of (industrial) imperial conquest leading to profound social and ecological transformation throughout the world – implicate 'developed' states in perceived 'resource scarcities' in 'developing' states.

He argues in support of the importance of understanding nature in the long term (Dalby, 2002a: 72). This, he suggests, could lead toward thinking of political space in terms of ecopolitics, rather than through the conventional markers of states. Most recently, critical environmental security scholars have taken to using the term 'the Anthropocene' to denote the current, post-Holocene era where earth systems are directly affected by human activity (Crutzen, 2002). The Anthropocene, characterized by instability and rapid change, is thought to have been initiated with the advent of the Industrial Revolution. Some scholars, however, suggest that its effects can be traced to the beginnings of settled agriculture and the emergence of complex civilizations (Rockstrom et al., 2009). Whatever its origins, the primary point, however, is clear: the Anthropocene characterizes an 'age of uncertainty', in contrast to the Holocene era's long period of earth systems stability.

What the critical school makes clear is that, among other things, 'environmental security has been written by the rich omnivores in their comfortable offices and libraries' (Dalby, 2002a: 184). This must change, for it suggests that traditional approaches to environmental security are being folded into dominant security routines designed in the main to ensure the stability of the state system, and US primacy therein. 'Stability' versus 'transformation' ensures that 'environmental security' will remain an essentially contested concept for the foreseeable future, particularly where challenges to 'carboniferous capitalism' aim to pin the source of global insecurity on the West. That the discourse surrounding climate change privileges adaptation over mitigation suggests that the world's most powerful actors are in no way prepared for whole-earth approaches to security.

Climate change and environmental security

Recently, the world has passed through the 400 ppm/atmospheric CO_2 threshold. What scientists tell us is that we are to expect a wide variety of climate-induced weather extremes and earth systems changes as the planet experiences a warming of some two-degrees Celsius within a generation. Put succinctly, these changes constitute threats to human communities from the environment in the form of direct (for example, super storms, sea level rise, polar ice melt, crop failure, new disease vectors) and indirect (for example, border disputes, resource capture, forced migrations – all stressing existing social systems) hazards (United Nations Security Council, 2007). Moreover, climate change is likely to induce a long period of uncertainty regarding patterns and processes in the earth's systems and the ensuing impacts on people, places, and things. Paradoxically, this shared threat has only served to further divide problem-solving and critical theory perspectives.

State-centric narratives have moved toward adaptation and maintain the discourse within the framework of self-regarding sovereign states acting on their own and/or multilaterally. This involves a number of moves that may or may not be considered as security practice: (i) containment of radical approaches through drawn-out and divisive multilateral dialog; (ii) scenario planning within state-houses and state-associated think tanks; (iii) preparation for adapting to change through resource capture (for example, Asian state land grabs throughout Africa) or barrier refinement (for example, amendments to Canadian immigration policy; refashioning of the eastern seaboard of the United States after superstorm Sandy). There is also a good deal of 'head in the sand' practice, as, for example, the United States aims for 'energy security' through continent-wide fracking for oil and natural gas (Mulligan, 2012). Whereas Trombetta (2009) paints a hopeful picture regarding the ways in which speech acts regarding 'climate security' may be leading to new – non-extraordinary, non-military – useful, and important security practices, the evidence seems to suggest that states everywhere are preparing for a brave new world through their standard routines.

Critical security specialists continue to be exasperated by such shortsighted and unenlightened approaches to climate change, arguing that while adaptation measures are important, we must not lose sight of opportunities for mitigation, in particular moving away from fossil-fuel dependence (Dalby, 2012). In the context of a changing climate, critical environmental security points to those most seriously affected (Schnurr and Swatuk, 2012) and argues for a shift toward environmental justice (Stoett, 2012). This position recognizes that part of the difficulty in dealing effectively with threats from a changing climate are structural (the nature of the state; the nature of production and consumption) and locational (the geography of extreme event impact is not uniform). Put simply, those with the greatest capacity to act in the interests of 'planetary security' are therefore best placed to act in their own interests. Thus, for the world's most vulnerable – both states and peoples – it seems that things will only get worse. One wonders what sort of catastrophe will be necessary to move decision-makers toward meaningful, inclusive action. Until then, former British Foreign Secretary Margaret Beckett's words regarding climate change's effect on 'our collective security in a fragile and increasingly interdependent world' will continue to ring hollow (quoted in Detraz and Betsill, 2009).

Environmental security in practice: State security, human insecurity?

The environmental security landscape is complicated even where one approaches the issues with a clear (problem-solving or critical) framework. It is so complex that a positive intervention might be made just about anywhere. But what adds up? What sort of approach will lead to the kind of environmental security that critical environmental security scholars are interested in – a stable Anthropocene, a just and sustainable distribution of the world's resources for people and nature – and avoid the type of environmental conflict that state-centric approaches so fear, leading empowered actors toward sub-optimal, zero-sum tragedy of the commons types of approaches? While the answer to this is far from clear, different perspectives counsel different approaches. To illustrate their strengths, weaknesses, and contradictions, I present two different interpretations of 'environmental security' in relation to one geographical setting: the Pongola River Basin in KwaZulu Natal (KZN), South Africa.

State security and environmental change

The Rio Earth Summit marked the last moment of South Africa's isolation from the UN system. The country's foreign policy quickly moved from bilateral isolationism toward engaged multilateralism (Vale and

Swatuk, 2002: 180). South Africa ratified the UNFCCC in 1997 and since that time it has tried to behave like a 'responsible global citizen' (Government of South Africa, n.d.: 8). Recently it released the National Climate Change Response White Paper, wherein the government describes not only a series of integrated national programs and policies but the importance of multilateral approaches (for example, through the African Union and Group of 77 countries) and developed country supports (financial, technical) if the country is to meet its targets of 34 percent and 42 percent below 'business as usual' emission levels by 2020 and 2025 (Government of South Africa, n.d.: 25).

The South African government describes climate change as real, present, and a major threat to state stability. It acknowledges South Africa's role as a significant emitter of greenhouse gases (GHGs) (Government of South Africa, n.d.: 26), with energy use emissions, mostly in the form of coal-fired electricity generation, accounting for 80 percent of total emissions. The government commits itself to moving toward a less emissions-intensive energy mix, noting 'in the short term, the emergence of bio-fuels...are also important' (Government of South Africa, n.d.: 26).

Host to COP 17 at Durban, South African government policy addresses climate security from many different directions:

- Multilateralism with a narrative focused on mitigation (on behalf of the most vulnerable, such as fellow Southern African Development Community (SADC) member states and neighbors Seychelles and Mauritius).
- National policy toward sustainable energy, including regional power pool development (combining hydro from the North with coal in the South), a turn toward nuclear energy, and carbon offsets through biofuel production (among other things).
- Local level adaptive strategies, such as the Working for Water program, which aims at the removal of water-hogging alien floral species.

Numerous planning and strategy documents across national and provincial government ministries and departments are in place.

These new approaches derive in part from two different but related perceptions of 'threat'. The first regards South African industrial practice as a threat to overall state stability. High levels of GHG emissions drive climate change which will have a disproportionate impact on developing states like South Africa, in particular the poor majority of citizens. The second regards emerging global norms in favor of cleaner production as a potential barrier to South African economic growth. South Africa's dependence on high ash/high sulfur coal for thermal power would threaten its ability to trade through strictures regarding cradle-to-cradle production (Vale and Swatuk, 2002). Carbon offsets are currently being sought through bio-fuel production (Department of Minerals and Energy, 2007).

South Africa is a major producer of sugar cane. Commercial agriculture in general accounts for 12 percent of GDP and roughly 30 percent of employment. Sugar cane accounts for approximately 0.7 percent of GDP, and directly employs 136,000 workers. Indirectly it is estimated to involve almost one million workers along the commodity value chain (see http://kzntopbusiness.co.za/site/agriculture). In KZN province, 430,000 hectares (ha) are under sugar cane, with only 31,000 ha belonging to black farmers. Of the 1,741 commercial farmers in the province, approximately 200 are from the black majority. There are an estimated 47,334 so-called developing farmers growing sugar cane (see www.nda.agric.za/docs/FactSheet/Sugar06.pdf). According to Green and Mhlongo (n.d.), there is widespread support among smallholder farmers for biodiesel production in KZN. Canola, sunflower, and soybeans are primary crops for biodiesel, whereas sugar cane, sugar beets, and sorghum are primary crops for ethanol. Commercial farmers in the province are particularly interested in growing sugar cane for bioethanol. According to the South African government, its climate security approach must adopt a pro-poor adaptation agenda (Government of South Africa, n.d.: 1). It would seem that biofuels offer a win-win solution: state security for human security through a deliberate and concerted shift away from dirty coal.

The people and environmental insecurity

Whereas state-centric analysis accepts the state as the primary unit of analysis in questions of environmental change and security, critical environmental security begins from the position of those most vulnerable, at grassroots level (Stoett, 2012), and may or may not regard the state as a security threat. In the Pongola River Basin, official unemployment figures are 41.1 percent (Census, 2011: 26). In Jozini, the main town located near to the Pongolapoort Dam wall, more than 16,000 households have no access to piped water. With a household average of 5 people per dwelling, this adds up to nearly 50 percent of Jozini's 186,502 people. With regard to sanitation, nearly 9,000 households have no formal toilet system of any kind. In contrast, the region around the dam is populated with privately owned game lodges and upstream from the impoundment there are extensive commercial sugar cane plantations, all drawing water for irrigation from the dam.

From a critical security perspective, the threat of environmental change pales in comparison to current environmental insecurities, almost all of which derive from historical processes of resource capture and ecological marginalization (Schnurr, 2012). Aside from the questionable practice of turning a complex biosphere into a cane field, and using vast quantities of water in an otherwise dry region to grow only sugar, promoting biofuels as a means to managing climate change privileges those already in place to take advantage of newly minted policies, plans, and government subsidies resulting from the multilateral negotiation processes. Left out are the most vulnerable – the landless poor who originally hoped to benefit from the shift to majority rule in South Africa in 1994. Today, still, 20 years after the end of apartheid, 94 percent of productive agricultural water is in the hands of the few thousand white commercial farmers who have remained on the land, many of whom have significantly expanded their holdings (personal communication, Department of Water Affairs, KZN, August 2013). As goes water, so goes land (see also Schnurr, 2012).

Thus, the economic benefits accruing to the central South African state as a result of inter-governmental negotiations on emissions limits reinforces a system that privileges the few at the expense of the many. Critical environmental security seeks environmental justice (Stoett, 2012). In KZN, this means security through, among other things, land redistribution, economic water allocation, and secure tenure on these lands. Global climate negotiations around emissions limits blinds us to these ongoing vulnerabilities.

What should be done? For those at the frontline of environmental change, improving livelihoods and alleviating poverty are the appropriate frameworks for dealing with complex vulnerabilities, including environmental insecurity (Deligiannis, 2012). A first step, therefore, would be to ask the question where are the poor and why are they where they are? What are their resource vulnerabilities? What sorts of environmental insecurities do they face? And to challenge policy from this point of view.

Future research

In thinking about lines of future research, I highlight nine items below: four statements where there is broad (but not universal) agreement in the research; five where there are countless differences of opinion.

Drawing on the evidence provided in this chapter, it seems to me that what we know and can generally agree about 'environmental security' is the following: (i) climatic anomalies – that is, significant deviations from the mean – precipitate violence at a variety of levels, from the inter-personal to the inter-state; (ii) populations in the global South face significant interrelated challenges – population increase, environmental change, governance, societal – that may lead to violence but can also lead toward cooperation and peace; (iii) middle-income countries, particularly those in transition from totalitarian/authoritarian systems to some semblance of democracy, are highly vulnerable to intra-state violence; (iv) environmental change is not considered the primary cause of violence but is an important element among several in a network of 'conjunctural causation'.

It seems to me that there is a good deal of disagreement regarding the following: (i) the role of 'carboniferous capitalism' – that is, as source of the threat – in precipitating violence and instability outside of the small band of prosperous OECD countries; (ii) the role of poor governance in precipitating the same; (iii) the appropriate responses of the 'most secure' states to the 'least secure', particularly in light of (i) and (ii); (iv) the place, if any, for the military in responding to threats beyond national borders (for example, through containment or pre-emptive action); and (v) the appropriate theoretical framework through which to understand these and related dynamics.

As suggested throughout this chapter, it is doubtful that consensus regarding causality, conceptualization of the primary referent object, primary sources of threat and vulnerability, and appropriate actions will ever be reached. As Mao once said, 'Let a hundred flowers blossom and a hundred schools of thought contend'. Two decades of research have yielded important insights and many debates. In my view, I would like to see research continue along the lines of the nine items I have highlighted above: deepen and challenge our understanding of those things that we agree upon; probe further into those issues where we fundamentally disagree. Each item raises numerous research questions, all of which will contribute meaningfully to the next two decades of work regarding the meaning and practice of environmental security.

Note

1. See, also, Paterson's discussion in chapter 2 regarding those studies taking international anarchy as their points of departure.

Annotated bibliography

Dalby, S. (2009) *Security and Environmental Change* (Cambridge: Polity Press).

Following on from his seminal 2002 book, *Environmental Security*, Dalby seeks to reorient questions of 'security' away from the sovereign state toward people ('human security') and the biosphere ('planetary security'). On the one hand, he is keen to show how state-centric approaches to security, founded as they are on principles of sovereignty, self-help, and order among the great powers, compound the threats emanating from environmental change. Power emanates from a state's ability to successfully engage in 'carboniferous capitalism'; preserving order, therefore, depends on a continuing and often expanding ability to burn fossil fuels. Given that the people most vulnerable to climate change impacts live within the Tropics in weak or middle-income states, the actions taken to maintain the current international system of states systematically weakens those least able to help themselves. Thus, on the other hand, Dalby aims to show how emerging 'environmental threats' are self-imposed. Put differently, social and natural systems are part of a mutually constituted and dialectical relationship. In the Anthropocene, security can only be achieved through a planetary politics that regards the most vulnerable as the primary referent of security.

Kahl, C. (2006) *States, Scarcity, and Civil Strife in the Developing World* (Princeton, NJ: Princeton University Press).

Kahl's erudite study is the logical follow-up to Thomas Homer-Dixon's 1999 study *The Environment, Scarcity and Violence*. Kahl provides a detailed critical review of the literature on environmental change and the probability of violence, arguing that the neo-Malthusian position is the most persuasive. He presents a theoretical framework wherein demographic change (for example, population growth) and environmental stress (for example, degraded resource base) can lead either to inter-group violence and state failure/state exploitation or to adaptation and state regeneration. The path followed depends upon two factors: 'groupness' (that is, the degree to which a state is united or divided across a number of variables such as race and class) and 'institutional inclusivity' (that is, another term for degree of democracy, with the more democratic the state the less likely violence will directly ensue). Kahl applies this theory to the cases of the Philippines and Kenya, but drawing in other cases as he sees fit. What he shows is the possibility for a 'perfect storm': population increase, environmental stress, divided societies, and a weak state are a guaranteed recipe for the emergence of inter-group violence. The Central African Republic might offer empirical support for his argument today. Interestingly, Kahl also argues that the 'resource curse' theory does not contradict his own framework. South Sudan seems to offer testimony in this regard today.

Gleditsch, N. P. (2012) 'Special Issue on *Climate Change and Conflict*', *Journal of Peace Research*, 49, 1.

Gleditsch has consistently and importantly contributed to the literature on resources and conflict. This special issue consists of 17 essays covering a wide range of topics, so serving as a good introduction to the subject of climate change and conflict. Gleditsch provides a useful introduction to the issue of climate change and conflict. This paper is followed by an analysis of global public opinion on the importance of climate change as a security issue. In the rest of the text, there is a heavy emphasis on Africa – East Africa in general, Kenya in particular, the Sahelian region, and sub-Saharan Africa. While their findings show a correlation between weather variability and conflict, most also show that institutions matter. Where land rights are embedded and relatively equitable, or where local government maintains good relations with civil society, cooperation is more likely than conflict in the face of climate change. This is also the general finding with regard to the several global-level studies (focusing on climate variation and economic growth; natural disasters and conflict) and those that focus particularly on water in a transboundary setting (in Africa, in Central Asia, in the Middle East).

Floyd, R. and R. A. Matthew (eds) (2013) *Environmental Security: Approaches and issues* (London and New York: Routledge).

For those interested in environmental security, this state-of-the-art collection should be your first stop. It covers in a detailed chapter every issue discussed in cursory fashion in the main body of this chapter. Following an introduction by the editors, there are 15 substantive chapters divided into two sections. The 'approaches' section focuses on theoretical frameworks with chapters covering topics such as the resource curse, political ecology and ecological security, environmental cooperation for peace-building, human security, and gender. Part two focuses on key issues such as water, conservation, population, sustainable development, food, energy and climate change. Rita Floyd provides an afterword entitled 'Whither environmental security studies?' In her estimation, the text illustrates the 'many different competing approaches to environmental security': from those wishing to maintain a narrow, traditional focus where environmental change raises challenges to national security amenable to military involvement, to those who wish to see a reimagining of security that strips the military of its relevance altogether. Is there a way to bring coherence to such diverse perspectives, or is it futile to even try? Floyd suggests that since we are all talking about 'security', then we must all be talking about some sort of threat. A starting point, therefore, would be to ask the question 'Is the threat objectively present?' The answers given will reveal much and no doubt lead to many further questions. But it seems to me to be a sensible place to start.

Works cited

Adelphi Research and Woodrow Wilson Center (Environmental Change and Security Project) (2004) *Environment, Development and Sustainable Peace: Finding Paths to Environmental Peacemaking*. Report of a conference held at Wiston House, Wilton Park, UK, 16–19 September.

Alker, H. and P. M. Haas (1993) 'The Rise of Global Ecopolitics', in N. Choucri (ed.) *Global Accord: Environmental Challenges and International Responses* (Cambridge: MIT Press), 133–171.

Baechler, G. (1998) 'Why Environmental Transformation Causes Violence: A Synthesis', *Environmental Change and Security Project Report* 4, Spring, 24–44.

Baechler, G. (1999) *Violence Through Environmental Discrimination: Causes, Rwanda Arena, and Conflict Model* (Dordrecht: Kluwer).

Baldwin, D. (1997) 'The Concept of Security', *Review of International Studies* 23, 1, 5–26.
Bannon, I. and P. Collier (eds) (2003) *Natural Resources and Violent Conflict: Options and Actions* (Washington, D.C.: World Bank).
Barnett, J. (2001) *The Meaning of Environmental Security: Ecological Politics and Policy in the New Security Era* (London: Zed Books).
Barnett, J. (2007) 'Environmental Security and Peace', *Journal of Human Security* 3, 4–16.
Barnett, J. and W. N. Adger (2007) 'Climate Change, Human Security and Violent Conflict', *Political Geography* 26, 6, 639–655.
Barry, J. L. and R. Eckersley (eds) (2005) *The State and the Global Ecological Crisis* (Cambridge, MA: MIT Press).
Baumann, Z. (1989) *Modernity and the Holocaust* (Cambridge: Polity).
Berman, B. (2006) 'The Ordeal of Modernity in an Age of Terror', *African Studies Review* 49, 1, 1–14.
Bernauer, T., T. Böhmelt and V. Koubi (2012) 'Environmental Changes and Violent Conflict', *Environmental Research Letters* 7, 1, 015601 (8pp).
Booth, K. (1991) 'Security and Emancipation', *Review of International Studies* 17, 4, 313–326.
Booth, K. (2007) *Theory of World Security* (Cambridge: Cambridge University Press).
Buhaug, H. (2010) 'Climate Not to Blame for African Civil Wars', *PNAS* 107, 38, 16477–16482.
Buhaug, H., N. P. Gleditsch and O. M. Theisen (2010) 'Implications of Climate Change for Armed Conflict', in R. Means and A. Norton (eds) *The Social Dimension of Climate Change: Equity and Vulnerability in a Warming World* (Washington, DC: World Bank), 75–101.
Buzan, B. (1991) *People, States and Fear*, 2nd edition (Boulder: Lynne Rienner).
Buzan, B., O. Waever and J. de Wilde (1998) *Security: A New Framework for Analysis* (Boulder: Lynne Rienner).
Callaghy, T., R. Kassimir and R. Latham (eds) (2001) *Intervention and Transnationalism in Africa* (Cambridge: Cambridge University Press).
Chengeta, Z., J. Jamare and N. Chishakwe (2003) *Assessment of the Status of Transboundary Natural Resources Management Activities in Botswana* (Gaborone: Printing and Publishing Company Botswana).
Collier, P. and A. Hoeffler (2004) 'Greed and Grievance in Civil War', *Oxford Economic Papers* 56, 4, 563–595.
Commission on Global Governance (1995) *Our Global Neighborhood* (Oxford: Oxford University Press).
Conca, K. and G. D. Dabelko (eds) (1998) *Green Planet Blues: Environmental Politics from Stockholm to Kyoto*, 2nd edition (Boulder: Westview).
Conca, K. and G. D. Dabelko (eds) (2002) *Environmental Peacemaking* (Washington, DC: Johns Hopkins University Press).
Conca, K. and G. D. Dabelko (eds) (2010) *Green Planet Blues: Four Decades of Environmental Politics*, 4th edition (Boulder: Westview).
Cox, R. W. (1986) *Production, Power and World Order* (New York: Columbia University Press).
Crutzen, P. J. (2002) 'Geology of Mankind: The Anthropocene', *Nature* 415, 23.

Dalby, S. (1997) 'Contesting an Essential Concept: Reading the Dilemmas in Contemporary Security Discourse', in K. Krause and M. C. Williams (eds) *Critical Security Studies Concepts and Cases* (Minneapolis: University of Minnesota Press), 3–32.

Dalby, S. (1998) 'Ecological Metaphors of Security: World Politics in the Biosphere', *Alternatives* 25, 3, 291–320.

Dalby, S. (2002a) *Environmental Security* (Minneapolis: University of Minnesota Press).

Dalby, S. (2002b) 'Security and Ecology in the Age of Globalization', *Environmental Change and Security Project Report*, 8, Summer, 95–108.

Dalby, S. (2009) *Security and Environmental Change* (Cambridge: Polity Press).

Dalby, S. (2012) 'Afterword: Ecoviolence, Security, Geopolitics', in M. A. Schnurr and L. A. Swatuk (eds) *Environmental Change, Natural Resources and Social Conflict* (Palgrave: Macmillan), 224–230.

Deligiannis, T. (2012) 'The Evolution of Environment-Conflict Research: Toward a Livelihood Framework', *Global Environmental Politics* 12, 1, 78–100.

De Jong, W., D Donovan and K. Abe (2007) *Extreme Conflicts and Tropical Forests* (Dordrecht, The Netherlands: Springer).

De Soysa, I. (2013) 'Environmental Security and the Resource Curse', in R. Floyd and R. A. Matthew (eds) *Environmental Security: Approaches and Issues* (London: Routledge).

Detraz, N. and M. M. Betsill (2009) 'Climate Change and Environmental Security: For Whom the Discourse Shifts', *International Studies Perspectives* 10, 303–320.

Deudney, D. (1990) 'The Case against Linking Environmental Degradation and National Security', *Millennium* 19, 3, 461–476.

Deudney, D. and R. A. Matthew (eds) (1999) *Contested Grounds: Security and Politics in the New Environmental Politics* (New York: SUNY Press).

De Villiers, B. (1999) *Peace Parks: The Way Ahead* (Pretoria: HSRC).

Dinar, S. (ed.) (2011) *Beyond Resource Wars: Scarcity, Environmental Degradation and International Cooperation* (Cambridge, MA: MIT Press).

Dinar, A., S. Dinar, S. McCaffrey and D. McKinsey (2013) *Bridges Over Water: Understanding Transboundary Water Conflict, Negotiation and Cooperation*, 2nd edition (Singapore: World Scientific Publishing).

Duffield, M. (2001) *Global Governance and the New Wars* (London: Zed Books).

Eckersley, R. (2004) *The Green State: Rethinking Democracy and Sovereignty* (Cambridge, MA: MIT Press).

Ellis, S. (1999) *The Mask of Anarchy* (New York: New York University Press).

Erhlich, P. (1968) *The Population Bomb* (New York: Ballantine).

Escobar, A. (1996) 'Constructing Nature: Elements for a Poststructural Political Ecology', in R. Peet and M. Watts (2002) (eds) *Liberation Ecologies: Environment, Development and Social Movements* (London: Routledge).

Esty, D. C., J.A. Goldstone, T.R. Gurr, B. Harff, M. Levy, G.D. Dabelko, P.T. Surko, and A.N. Unger (1999) 'State Failure Task Force Report: Phase II Findings', *Environmental Change and Security Project Report*, 5, Summer, 49–72.

Ferguson, J. (1990) *The Anti-Politics Machine: 'Development', Depoliticization and Bureaucratic Power in Lesotho* (Cambridge: Cambridge University Press).

Floyd, R. (2008) 'The Environmental Security Debate and its Significance for Climate Change', *The International Spectator* 43, 51–65.

Floyd, R. and R. A. Matthew (eds) (2013) *Environmental Security: Approaches and Issues* (London: Routledge).
Fukuyama, F. (1992) *The End of History and the Last Man* (New York: Free Press).
Gadgil, M. and R. Guha (1995) *Ecology and Equity: The Use and Abuse of Nature in Contemporary India* (London: Routledge).
Gartzke, E. (2012) 'Could Climate Change Precipitate Peace?', *Journal of Peace Research* 49, 1, 177–192.
George, J. (1994) *Discourses of Global Politics: A Critical (Re)introduction to International Relations* (Boulder: Lynne Rienner).
Gleditsch, N. P. (1997) *Conflict and the Environment* (Dordrecht: Kluwer).
Gleditsch, N. P. (1998) 'Armed Conflict and the Environment: A Critique of the Literature', *Journal of Peace Research* 35, 3, 381–400.
Gleditsch, N. P. (2012) 'Whither the Weather? Climate Change and Conflict', *Journal of Peace Research* 49, 1, 3–9.
Government of South Africa (2011) *National Climate Change Response White Paper* (Pretoria: Department of Environmental Affairs).
Grove, R. (1997) *Ecology, Climate and Empire: Colonialism and Global Environmental History, 1400–1940* (Cambridge: White Horse).
Hall, S., D. Held and T. McGrew (eds) (1992) *Modernity and Its Futures* (Cambridge: Polity Press).
Hanlon, J. (1986) *Beggar Your Neighbours: Apartheid Power in Southern Africa* (London: James Currey).
Hettne, B. (1997) 'Development, Security and World Order: A Regionalist Approach', *The European Journal of Development Research* 9, 1, 83–106.
Hettne, B. (2001) 'Regional Cooperation for Security and Development in Africa', in P. Vale, L. A. Swatuk and B. Oden (eds) *Theory, Change and Southern Africa's Future* (Basingstoke: Palgrave).
Homer-Dixon, T. (1991) 'On the Threshold: Environmental Changes as Causes of Acute Conflict', *International Security* 16, 2, 76–116.
Homer-Dixon, T. (1994) 'Environmental Scarcities and Violent Conflict: Evidence from Cases', *International Security* 19, 1, 5–40.
Homer-Dixon, T. (1998) 'Environmental Scarcities and Violent Conflict: Evidence from Cases' (abridged version), in K. Conca and G. D. Dabelko (eds) *Green Planet Blues* (Boulder, CO: Lynne Reinner Press), 287–297.
Homer-Dixon, T. (1999) *The Environment, Scarcity and Violence* (Princeton: Princeton University Press).
Homer-Dixon, T. (2003) 'Debating Violent Environments', *Environmental Change and Security Project Report*, 9, 89–96.
Homer-Dixon, T. and J. Blitt (eds) (1998) *Ecoviolence: Links Among Environment, Population and Security* (Lanham, MD: Rowman and Littlefield).
Hsiang, S. M., M. Burke and E. Miguel (2013) 'Quantifying the Influence of Climate on Human Conflict', *Science*, 341, 1235367.
Hurrell, A. (1995) 'International Political Theory and the Global Environment', in K. Booth and S. Smith (eds) *International Relations Theory Today* (Oxford: Clarendon Press).
Jeffery, R. and B. Vira (eds) (2001) *Analytical Issues in Participatory Natural Resources Management* (Basingstoke: Palgrave).
Jones Luong, P. and E. Weinthal (2010) *Oil is Not a Curse: Ownership Structure and Institutions in Soviet Successor States* (Cambridge: Cambridge University Press).

Kahl, C. (2006) *States, Scarcity, and Civil Strife in the Developing World* (Princeton, NJ: Princeton University Press).

Kaplan, R. (1994) 'The Coming Anarchy', *The Atlantic Monthly*, February.

Klare, M. (2001) *Resource Wars: The New Landscape of Global Conflict* (New York: Henry Holt and Company).

Krause, K. and M. C. Williams (eds) (1997) *Critical Security Studies* (Minneapolis: University of Minnesota Press).

Le Billon, P. (2001) 'The Political Ecology of War: Natural Resources and Armed Conflicts', *Political Geography* 20, 561–584.

Le Billon, P. (2012) *Wars of Plunder: Conflicts, Profits and the Politics of Resources* (London: C. Hurst & Co).

Leopold, A. (1949) *A Sand County Almanac: And Sketches Here and There* (Oxford: Oxford University Press).

Levy, M. A. (1995) 'Is the Environment a National Security Issue?' *International Security* 20, 2, Fall, 35–62.

Lietzmann, K. M. and G. Vest (1999) 'Environment and Security in an International Context: Executive Summary Report, NATO/Committee on the Challenges of Modern Society Pilot Study', *Environmental Change and Security Project Report* 5, Summer, 34–48.

Lipschutz, R. (1995) *On Security* (New York: Columbia University Press).

Lonergan, S. (2000) 'Human Security, Environmental Security and Sustainable Development', in M. R. Lowi and B. R. Shaw (eds) *Environment and Security: Discourses and Practices* (London: Palgrave), 66–83.

Lowi, M. R. and B. R. Shaw (2000) *Environment and Security: Discourses and Practices* (London: Palgrave).

Maas, A., A. Carius and A. Wittich (2013) 'From Conflict to Cooperation? Environmental Cooperation as a Tool for Peace-Building', in R. Floyd and R. A. Matthew (eds) *Environmental Security: Approaches and Issues* (Oxford: Routledge).

Magnusson, W. (1994) 'Social Movements and the Global City', *Millennium* 23, 3, 621–645.

Mathews, J. T. (1989) 'Redefining Security', *Foreign Affairs* 68, 2, 162–177.

Matthew, R. A. (2000) 'Environment and Security in an International Context: Critiquing a Pilot Study from NATO's Committee on the Challenges of Modern Society', *Environmental Change and Security Project Report* 6, Summer, 95–98.

Matthew, R. A. (2002) 'In Defense of Environment and Security Research', *Environmental Change and Security Project Report* 8, Summer, 109–124.

Matthew, R. A., J. Barnett, B. McDonald and K. L. O'Brien (eds) (2010) *Global Environmental Change and Human Security* (Cambridge, MA: MIT Press).

McKibbon, B. (1998) 'A Special Moment in History', *The Atlantic Monthly*, May.

Midlarsky, M. (1998) 'Democracy and the Environment: An Empirical Assessment', *Journal of Peace Research* 35, 3, 341–362.

Moran, D. (ed.) (2011) *Climate Change and National Security: A Country-Level Analysis* (Washington, DC: Georgetown University Press).

Morgenthau, H. (1978) *Politics Among Nations*, 5th edition (New York: Knopf).

Myers, N. (1989) 'Environment and Security', *Foreign Policy* 47, 23–41.

Myers, N. (1993) *Ultimate Security: The Environmental Basis of Political Stability* (New York: W.W. Norton).

Mulligan, S. (2012) 'Official Secrets and Popular Delusions: Security at the End of the Fossil Fuel Age?', in M. A. Schnurr and L. A. Swatuk (eds) *Environmental Change, Natural Resources and Social Conflict* (Palgrave: Macmillan), 141–158.

Mumford, L. (1934) *Technics and Civilization* (New York: Harcourt Brace and World).

Nordås, R. and N. P. Gleditsch (2007) 'Climate Change and Conflict', *Political Geography* 26, 6, 627–638.

O'Brien, K. and J. Barnett (2013) 'Global Environmental Change and Human Security', *Annual Review of Environmental Resources* 38, 373–391.

Ohlsson, L. (1999) *Environment, Scarcity and Conflict: A Study of Malthusian Concerns* (Goteborg: PADRIGU).

Paterson, M. (1995) 'Green Politics', in S. Burchill and A. Linklater with R. Devetak, M. Paterson and J. True (eds) *Theories of International Relations* (New York: St. Martins Press).

Peet, R. and M. Watts (eds) (1996) *Liberation Ecologies: Environment, Development, Social Movements* (London and New York: Routledge).

Peluso, N. and M. Watts (eds) (2001) *Violent Environments* (Cornell: Cornell University Press).

Peluso, N. and M. Watts (2003) 'Violent Environments: Responses', *Environmental Change and Security Project Report*, 9, 89–96.

Peterson, V. S. (2003) *A Critical Rewriting of Global Political Economy: Integrating Reproductive, Productive and Virtual Economies* (London and New York: Routledge).

Pettman, J. J. (1996) *Worlding Women: A Feminist International Politics* (London and New York: Routledge).

Pettman, R. (1991) *International Politics* (Boulder: Lynne Rienner).

Rees, W. E. and M. Wackernagel (1994) 'Ecological Footprints and Appropriated Carrying Capacity: Measuring the Natural Capital Requirements of the Human Ecology', in A. Jannson, M. Hammer, C. Folke and R. Constanza (eds) *Investing in Natural Capital* (Washington, DC: Island Press).

Renner, M. (1989) *National Security: The Economic and Environmental Dimensions*. Worldwatch Paper No. 89 (Washington, DC: Worldwatch Institute).

Richards, P. (1996) *Fighting for the Rainforest: War, Youth and Resources in Sierra Leone* (Oxford: James Currey).

Roche, S. (2010) 'From Youth Bulge to Conflict: The Case of Tajikistan', *Central Asia Survey* 29, 4, 405–419.

Rockstrom, J. et al. (2009) 'Planetary Boundaries: Exploring the Safe Operating Space for Humanity', *Ecology and Society* 14, 2, article 32.

Rotberg, R. (2003) *When States Fail: Causes and Consequences* (Princeton, NJ: Princeton University Press).

Ruggie, J. G. (1998) *Constructing the World Polity: Essays on International Institutionalism* (London: Routledge).

Scheffran, J., M. Brzoska, J. Kominek, P. M. Link and J. Schilling (2012) 'Climate Change and Violent Conflict', *Science* 336, 6083, 869–871.

Schnurr, M. A. (2012) 'Inventing Makhathini: Creating a Prototype for the Dissemination of Genetically Modified Crops into Africa', *Geoforum* 43, 784–792.

Schnurr, M. A. and L. A. Swatuk (eds) (2012) *Environmental Change, Natural Resources and Social Conflict: Towards Critical Environmental Security* (Palgrave: Macmillan).

Schwarz, D. M., T. Deligiannis and T. Homer-Dixon (2000) 'The Environment and Violent Conflict: A Response to Gleditsch's Critique and Some Suggestions for Future Research', *Environmental Change and Security Project Report* 6, Summer, 77–94.

Slettebak, R. T. (2012) 'Don't Blame the Weather! Climate-related Natural Disasters and Civil Conflict', *Journal of Peace Research* 49, 1, 163–176.

Sprout, Harold and Margaret Sprout (1965) *The Ecological Perspective on Human Affairs with Special Reference to International Politics* (Princeton: Princeton University Press).

Statistics South Africa (2012) *Census 2011 Municipal Report – KwaZulu-Natal* (Pretoria: Statistics South Africa).

Stoett, P. (2000) *Human and Global Security: An Exploration of Terms* (Toronto: University of Toronto Press).

Stoett, P. (2012) 'What are we Really Looking for? From Eco-violence to Environmental Injustice', in M. A. Schnurr and L. A. Swatuk (eds) *Environmental Change, Natural Resources and Social Conflict* (Palgrave: Macmillan), 15–32.

Swatuk, L. A. (1996) *Power and Water: The Coming Order in Southern Africa* (Bellville: Centre for Southern African Studies).

Swatuk, L. A. and P. Vale (1999) 'Why Democracy is Not Enough: Southern Africa and Human Security in the Twenty-first Century', *Alternatives* 24, 3, 361–389.

Theisen, O. M. (2012) 'Climate Clashes? Weather Variability, Land Pressure, and Organized Violence in Kenya, 1989–2004', *Journal of Peace Research* 49, 1, 81–96.

Tickner, J. A. (1995) 'Re-visioning Security', in K. Booth and S. Smith (eds) *International Relations Theory Today* (Cambridge: Policy Press).

Tir, J. and P. F. Diehl (1998) 'Demographic Pressure and Interstate Conflict: Linking Population Growth and Density to Militarized Disputes and Wars, 1930–1989', *Journal of Peace Research* 35, 3, 319–339.

Trombetta, M. J. (2009) 'Environmental Security and Climate Change: Analysing the Discourse', *Cambridge Review of International Affairs* 21, 4, 585–602.

Ullman, R. (1983) 'Redefining Security', *International Security* 8, 129–153.

United Nations Security Council (2007) 'Security Council Holds First-Ever Debate on Impact of Climate Change on Peace, Security, Hearing over 50 Speakers', *Department of Public Information: News and Media Division*, 17 April.

United States Department of State (2014) *United States Climate Action Report 2014* (Washington, DC: U.S. Dept. of State).

USAID/Regional Center for Southern Africa (1997) *Regional Integration Through Partnership and Participation: RCSA Strategic Plan 1997–2003* (Gaborone: USAID/RCSA).

Vale, P. (2003) *Security and Politics in South Africa: The Regional Dimension* (Boulder: Lynne Rienner).

Vale, P. and L. A. Swatuk (2002) 'Sunset or Sunrise? Regime Change and Institutional Adjustment in South Africa – A Critical Analysis', in H. Hveen and K. Nordhaug (eds) *Public Policy in the Age of Globalization: Responses to Environmental and Economic Crises* (Basingstoke: Palgrave).

Van der Linde, H. et al. (2000) *Beyond Boundaries: A Framework for Transboundary Natural Resource Management in Sub-Saharan Africa* (Washington, DC: Biodiversity Support Program).

Van Deveer, S. and G. D. Dabelko (1998) 'Redefining Security Around the Baltic: Environmental Issues in Regional Context', *Global Governance* 5, 2, 221–249.

Wackernagel, M. and W. Rees (1996) *Our Ecological Footprint: Reducing Human Impact on the Warth* (Philadelphia: New Society).
Waever, O. (1997) *Concepts of Security* (Copenhagen: Institute of Political Science).
Walker, R. B. J. (1997) 'The Subject of Security', in K. Krause and M. C. Williams (eds) *Critical Security Studies: Concepts and Cases* (Minneapolis: University of Minnesota Press), 61–82.
Wallerstein, I. (1974) *The Capitalist World Economy* (Cambridge: Cambridge University Press).
Walt, S. M. (1991) 'The Renaissance of Security Studies', *International Security Studies Quarterly* 35, 211–239.
Watts, M. J. (2013) 'A Political Ecology of Environmental Security', in R. Floyd and R. A. Matthew (eds) *Environmental Security: Approaches and Issues* (London: Routledge).
Williams, M. C. (2003) 'Words, Images, Enemies: Securitization and International Politics', *International Studies Quarterly* 47, 4, 511–531.
World Commission on Environment and Development (1987) *Our Common Future* (New York: Oxford University Press).
Worldwatch Institute (2005) *State of the World 2005: Global Security* (London: Earthscan).

10
Global Governance and the Environment

Frank Biermann

'Global governance' has become a key term in academic and policy debates alike. While an Internet search conducted in 1997 revealed only 3,418 references to 'global governance' and in January 2004 less than 90,000, in March 2014 the World Wide Web contained over 800,000 pages that mentioned the term. Global governance has become a rallying call for policy advocates who hail it as panacea for the evils of globalization; a global menace for opponents who fear it as the universal hegemony of the many by the powerful few; and an analytical concept that has given rise to much discussion among scholars of international relations (IR), including the launch of the journal *Global Governance* in 1995.

Global governance is also part and parcel of many of the chapters in this book on international environmental politics (IEP). Global governance is largely a response to economic, social, and ecological globalization (Clapp, this book). It is characterized by the increased participation of non-state actors in world politics (Betsill, this book), including the growing relevance of knowledge and science for global decision-making (Lövbrand, this book). The notion of global governance is also inseparable from questions of sustainability (Happerts and Bruyninckx, this book), institutional effectiveness (Young, this book), and environmental justice (Okereke and Charlesworth, this book).

This chapter explores the notion of 'global environmental governance' in some detail. I proceed in three steps. I first sketch the historic trajectory and different uses of the term 'global environmental governance'. I then highlight three key characteristics of global environmental governance that make it different from traditional international environmental relations. The following section offers, as case studies, examples from recent policy debates on the reform of the existing

What is global environmental governance?

As is common with many chapters of this book, most of what is conceptualized today as 'global environmental governance' is not without predecessors. Instead, the concept builds on a substantial pedigree of studies in international environmental cooperation. Especially the 1972 Stockholm Conference on the Human Environment had led to a first wave of academic studies on intergovernmental environmental cooperation and organization (for example, Kennan, 1970; Johnson, 1972; Caldwell, 1984 with further references; see also Stevis, this book). The most relevant precursor to global environmental governance studies is the debate on international environmental regimes of the 1980s (see Young, this book), including discussions on the creation of environmental regimes, their maintenance and effectiveness (reviewed in Young et al., 2008; Andresen and Hey, 2005). Important earlier research also addressed intergovernmental environmental organizations (reviewed in Biermann and Siebenhüner, 2009) and non-state environmental organizations (Wapner, 1996), both of which have received new attention in the global environmental governance discourse.

Yet despite the prominence of the notion of global environmental governance, and the long pedigree of this field, the term itself has remained vaguely defined. At the root is the notion of 'governance' as it emerged in domestic politics (reviewed in Van Kersbergen and Van Waarden, 2004). Governance is often described as a new form of regulation that differs from traditional hierarchical state activity ('government'). Generally, 'governance' implies notions of self-regulation by societal actors, of private–public cooperation in the solving of societal problems and of new forms of multilevel policy. The more recent notion of 'global' governance builds on these earlier debates among political scientists working on domestic issues, and tries to capture similar developments at the international level. Clear definitions of global governance, however, have not yet been agreed upon: global (environmental) governance still means different things to different authors (reviewed in detail in Jordan, 2008; Adger and Jordan, 2009; Biermann and Pattberg, 2012). Broadly speaking, there are two types of meaning for global environmental governance.

One group of authors sees global environmental governance primarily as a new phenomenon in world politics that can be described and analyzed as such. A part of this literature employs a rather narrow definition, with some earlier writers seeking to restrict the new term to more traditional forms of world politics. Lawrence S. Finkelstein, for instance, defined the concept of global governance – not limited to environmental issues – in his widely cited essay *What is Global Governance?* as 'doing internationally what governments do at home' and as 'governing, without sovereign authority, relationships that transcend national frontiers' (Finkelstein, 1995: 369). The problem with such narrow phenomenological understandings is the need to distinguish the term from traditional international relations. Often it is not clear what is gained by using the term 'global governance' instead of 'international relations'.

Other writers try to address this problem by broadening the term to encompass a much larger range of social and political interactions. James Rosenau's definition is possibly the broadest, stating that global governance is 'the sum of the world's formal and informal rules systems at all levels of community [that] amount to what can properly be called global governance' (Rosenau, 2002: 4). In an earlier paper, Rosenau had defined global governance equally broadly as 'systems of rules at all levels of human activity – from the family to the international organization – in which the pursuit of goals through the exercise of control has transnational repercussions' (Rosenau, 1995: 13). Also the UN Commission on Global Governance (1995: 2–3) described its subject domain broadly as

> the sum of the many ways individuals and institutions, public and private, manage their common affairs. It is a continuing process through which conflicting or diverse interests may be accommodated and cooperative action taken. It includes formal institutions and regimes empowered to enforce compliance, as well as informal arrangements that people and institutions either have agreed to or perceive to be in their interest.

Such all-encompassing definitions, however, hardly leave room for anything that is not global governance. Given the increasing international interdependence at all levels, few political rules will have no repercussions beyond the borders of the nation-state.

A different strand of literature views global (environmental) governance as a political program, mainly in an affirmative sense that demands 'more global governance' as a counterweight to the negative consequences of economic and ecological globalization. Typically, this involves calls for the creation of new institutions, such as multilateral environmental treaties and conventions, of new and more effective international environmental organizations, and of new forms of financial mechanisms to account for the dependence of current international regimes on the goodwill of national governments. The UN Commission on Global Governance (1995) also used this understanding of the term and elaborated a plethora of more or less far-reaching reform proposals to deal with problems of modernization: global (environmental) governance is seen here as a solution, as a tool that politicians need to develop and employ to solve the problems that globalization has brought about. For environmental governance, various new or revised reform proposals have been submitted, especially in the run-up to the 2012 United Nations Conference on Sustainable Development ('Rio plus 20') (see, for example, Biermann et al., 2012; Biermann, 2012; Haas, 2012; Kanie et al., 2012).

Other authors followed this normative definition of global (environmental) governance but without its affirmative connotation. Some observers in the tradition of political realism distrust global governance as the attempt of the United Nations and others to limit the unilateral freedom of action of powerful states (see, for example, www.globalgovernancewatch.org). Writers in the tradition of neo-Marxism and international political economy place global governance in the context of larger transformations of the global capitalist system, seeing it as a purposeful strategy by hegemonic forces to manage various economic and political crises (Overbeek et al., 2010). A third group of writers view global governance through the lens of North–South power conflicts. The Geneva-based South Centre, for example, had cautioned already that in 'an international community ridden with inequalities and injustice, institutionalizing "global governance" without paying careful attention to the question of who wields power, and without adequate safeguards, is tantamount to sanctioning governance of the many weak by the powerful few' (South Centre, 1996: 32).

Which definition or conceptualization is preferable? All definitions offered in the current debate have pros and cons depending on the specific context in which they are used. Given the increasing complexity and interdependence of world society in the face of economic and ecological globalization, more effective global regimes and organizations are needed, and there is nothing wrong to call this political reform program 'global governance'. Also, today's international relations differ from the 1950s and 1960s in many respects, and it seems appropriate to denote these new forms of international regulation as 'global governance'.

The notion of 'global environmental governance' in itself remains dynamic. In parts of the literature the term is today increasingly replaced by a new concern for the potential transformation of entire planetary systems and the resulting need to adopt a more holistic and integrated perspective. As Nobel laureate Paul Crutzen and Veerabhadran Ramanathan warn, 'humankind will remain a major geological force for many millennia, maybe millions of years, to come. To develop a worldwide accepted strategy leading to sustainability of ecosystems against human-induced stresses will be one of the great tasks of human societies' (Crutzen and Ramanathan, 2004: 286). In 2001, major global change research programs urged that an

> ethical framework for global stewardship and strategies for Earth System management are urgently needed. The accelerating human transformation of the Earth's environment is not sustainable. Therefore, the business-as-usual way of dealing with the Earth System is not an option. It has to be replaced – as soon as possible – by deliberate strategies of good management that sustain the Earth's environment while meeting social and economic development objectives.
> (Amsterdam Declaration, 2001)

This new concern is sometimes referred to as 'earth system' governance, a term that has found its entry into more than 350,000 websites at present. Earth system governance is usually defined as the sum of the formal and informal rule systems and actor-networks at all levels of human society that are set up to steer societies toward preventing, mitigating, and adapting to environmental change and earth system transformation (Biermann, Betsill et al., 2009). The normative context of earth system governance is sustainable development, that is, a development that meets the needs of present generations without compromising the ability of future generations to meet their own needs (see Happerts and Bruyninckx, this book; Vieira, 2012).

This notion of earth system governance adds to the concept of global environmental governance a connotation that links governance research to the eventual core concern of environmental politics: the ongoing transformation of the entire earth system, from global warming, large-scale changes in biogeochemical cycles to unprecedented rates of species loss. Earth system governance bridges levels from global to local as well as academic communities from natural-science-oriented modeling and scenario building to political science and philosophy. The core concern of the study of earth system governance is the long-term prevention of a transformation of the entire earth system through human influences.

Key trends in global environmental governance

I now elaborate on key characteristics and trends of global environmental governance that make it different from traditional international relations. The first characteristic is that global environmental governance describes political processes that are no longer confined to nation-states but characterized by increased participation of actors that have so far been largely active at the subnational level. This increased participation has given rise, secondly, to new forms of institutions in addition to the traditional system of legally binding documents negotiated by states. Thirdly, the emerging global governance system is characterized by an increasing segmentation of different layers and clusters of rule-making and rule-implementing, fragmented both vertically between supranational, international, national, and sub-national layers of authority and horizontally between different parallel rule-making systems maintained by different groups of actors. I elaborate on these three new trends in more detail below.

None of these trends is entirely new. Some non-state actors such as the Catholic Church have been influential and engaged in treaty-making with governments for centuries. Politics among nations has always been a multilevel process, with governmental delegations being forced to seek support from domestic constituencies. Also, not all areas of politics follow the new paradigm of global governance, and the term may not aptly describe conflicts, for instance, in the area of war and peace. On the other hand, global governance is there. It is more frequent, and it is on the rise. It is a reaction to the complexities of modern societies and to the increasing economic, cultural, social, and ecological globalization (Clapp, Betsill, this book). Whereas globalization denotes the harmonization and mutual dependence of once separate, territorially defined spheres of human activity and authority, global governance catches the political reaction to these processes. New degrees of global interdependence beget the increasing institutionalization of decision-making beyond the confines of the nation-state, with a resulting transformation of the ways and means of global politics. Quantity – the increasing number of functional areas that require global regulation and of international regulatory regimes – creates shifts in quality: new types of actors have entered the stage; new types of institutions have emerged; with new types of problems of regulatory fragmentation and diffuse sites of authority as a result.

Trade integration, for example, required international regulation of more and more 'trade-related' issue areas beyond the key concerns of custom liberalization (Eckersley, 2004; Zelli, 2007). The impacts of this drive for institutionalization then brought the world trade regime on the radar screen of a variety of new actors beyond the traditional world of interstate politics: unions, business associations, or environmentalists pay close attention to the emergence of the world trade regime and become actors of global governance in their own right. The globalization of environmental problems, from global climate change to the loss of biodiversity, creates new interdependencies between nation-states that require new regulatory institutions at the global level. These institutions, however, do not remain isolated from the continuing debates within nation-states, a situation which results in governance systems that stretch from local environmental politics to global negotiations and back.

Increased participation: Diversity through inclusion

Global environmental governance differs, first, from international politics through the high participation of different actors that were earlier confined to the national sphere (Dellas et al., 2011; Betsill, this book). The traditional system of international politics was characterized as politics among states. Non-state actors were either non-existent, or lacked sufficient power beyond territorial borders. There have been exceptions – such as the Catholic Church with its highly centralized system of authority or the transnational antislavery movement in the 19th century – yet those remained rare and confined to specific historic circumstances. The notion of global governance departs from traditional state-centered politics in accepting a host of non-state entities as new influential actors in transnational relations. The field of environmental policy provides ample illustrations for this evolution of a 'multi-actor governance system'.

For one, the role of non-state activist groups and political organizations has increased substantially (see Stevis, this book; Betsill, this book; Gupta and Mason, this book), leading to what Paul Wapner recognized 18 years ago as a 'world civic society' (Wapner, 1996, see also Betsill and Corell, 2008; Gupta, 2003; Newell et al., 2012; Tamiotti, 2001). Carefully orchestrated campaigns of environmentalists have changed the foreign policy of powerful states, initiated new global rules, or influenced global norm-setting processes (Alcock, 2008; Skodvin and Andresen, 2003; Wapner, 2002). Schroeder et al. (2012) and Schroeder and Lovell (2012) have studied, for instance, the participation of non-state actors in climate negotiations and found a dramatic increase in participation by non-state representatives, both inside and outside national delegations.

Also networks of scientists have assumed a new role in providing complex technical information that is indispensable for policymaking on issues marked by both analytic and normative uncertainty. While the new role of experts in world politics is evident in many policy areas, it is particularly prevalent in the field of environmental policy (Gupta et al., 2012; Lövbrand, this book). New international networks of scientists and experts have emerged, in a mix of self-organization and state-sponsorship, to provide scientific information on the kind of environmental problem at stake and the options for decision-makers to cope with it. Such scientific advice for political decision-making is not new in world politics; negotiations on fishing quotas, for example, have long been assisted by the International Council for the Exploration of the Sea. These early examples, however, have been significantly increased in both number and impact, which is mirrored in the substantial academic interest in global scientific networks in recent years (Mitchell et al., 2006; Jasanoff and Long Martello, 2004; Lövbrand, this book).

Third, business actors have taken a more prominent direct role in international decision-making as well (Clapp, 2009, this book; Falkner, 2003; Tienhaara et al., 2012). Again, the influence of major companies on international affairs is not new, and in some social theories, business actors have been granted center stage in global affairs. However, this 'old' influence by the corporate sector was mainly indirect through its influence on national governments. Today, many corporations take a more visible, direct role in international negotiations as immediate partners of governments, for example, in the framework of the United Nations and of the Global Compact that major corporations have concluded with the world organization (Tienhaara et al., 2012). Another example is the influence of consultancy firms, who have a major impact on the design and implementation of novel types of global governance such as the Clean Development Mechanism and emissions trading (Bouteligier, 2011).

Fourth, global governance is marked by a growing influence of intergovernmental organizations (Biermann and Siebenhüner, 2009). In the field of environmental policy, more than 200 international organizations have been set-up in the form of secretariats to the many international environmental treaties concluded in the last two decades. Whether the creation of a new 'world environment organization' would help or harm global environmental governance has been debated for more than 40 years, with no conclusive answer (see the related discussion below).

Increased privatization: Negotiation through partnerships

In addition to the emergence of new actors, global environmental governance is also defined by new forms of cooperation that go beyond the traditional intergovernmental negotiation of international law. In many new institutions, private actors are becoming part of norm-setting and norm-implementing institutions and mechanisms in global governance, which denotes the shift from intergovernmental regimes to public–private and increasingly private–private cooperation and policymaking at the global level (see, in particular, Betsill, this book; Gupta and Mason, this book; as well as Andonova et al., 2009; Bulkeley et al., 2012; Bulkeley and Jordan, 2012; Hoffmann, 2011; Okereke et al., 2009; Pattberg and Stripple, 2008). Some private actors have negotiated their own standards, such as in the Forest Stewardship Council or the Marine Stewardship Council, two standard-setting bodies created by major corporations and environmental advocacy groups without direct involvement of governments (Pattberg, 2007; Gulbrandsen, 2010).

It sometimes seems that traditional intergovernmental policymaking through diplomatic conferences is being replaced by such networks, which some see as being more efficient and transparent. Yet the distribution of global public policy networks is often linked to the particular interests of private actors that have to respond to their particular constituencies, and serious questions of the legitimacy of private standard-setting remain. For example, the World Commission on Dams has been hailed as a new and effective mechanism that has quickly generated widely accepted standards, which had earlier been difficult to negotiate due to the persistent resistance of affected countries. Yet this success of private standard-setting has also given rise to other voices that point to the inherent problems of legitimacy, which are part of private policymaking that cannot relate back to democratic elections or other forms of direct representation (Dingwerth, 2007).

Public–private partnerships, in particular, have been proposed as complement to what is often seen as ineffective state action. Such partnerships received much attention at the 2002 Johannesburg World Summit on Sustainable Development and its focus on 'multi-stakeholder' partnerships of governments, non-governmental organizations and corporations – the so-called Partnerships for Sustainable Development (Bäckstrand et al., 2012; Glasbergen et al., 2007; Pattberg et al., 2012). These partnerships were formally supported in the implementation plan agreed in Johannesburg, with over 220 partnerships with 235 million US dollars committed before the summit. Yet, the effectiveness of these transnational public policy networks remains contested (see for more detail Mert, 2014). Some observers view the new emphasis on transnational public–private partnerships as problematic, since voluntary arrangements might privilege more powerful actors, in particular industrialized countries and major corporations, and consolidate the privatization of governance and dominant neo-liberal modes of globalization. In addition, some argue that transnational public policy networks lack accountability and legitimacy. Yet others see transnational public policy networks as an innovative form of governance that addresses deficits of inter-state politics by bringing together key actors of civil society, governments and business. In this perspective, transnational public policy networks are important new mechanisms to help resolve a variety of current governance deficits (see for a critical perspective Pattberg et al., 2012).

Increased segmentation: Complexity through fragmentation

In addition, global environmental governance is marked by a new segmentation of policymaking, both vertically (multilevel governance) and horizontally (multipolar governance). First, the increasing institutionalization of world politics at the global level does not occur and is indeed not conceivable, without continuing policymaking at national and sub-national levels (Hoffmann, 2011; Betsill and Bulkeley, 2006). Global standards need to be implemented and put into practice at the local level, and global norm-setting requires local decision-making to set the frames for global decisions. This results in the coexistence of policymaking at the sub-national, national, regional, and global levels in more and more issue areas, with the potential of both conflicts and synergies between different levels of regulatory activity.

Likewise, the increasing institutionalization of global environmental governance does not occur in a uniform manner that covers all parts of the international community to the same extent. In the case of the 1987 Montreal Protocol on Substances that Deplete the Ozone Layer, for example, various recent amendments have provided for new standards and timetables that are not accepted by all parties to the original agreement from 1987. This leads to a substantial multiplicity of sub-regimes within the overall normative framework. The most prominent example of such horizontal fragmentation of policies is climate governance. Here, we observe the emergence of parallel policy approaches that include equally important segments of international society and could even develop into divergent regulatory regimes (for example, Falkner et al., 2010).

Divergent policy approaches within a horizontally and vertically fragmented policy arena pose significant challenges (Biermann, Pattberg et al., 2009). Lack of uniform policies may jeopardize the success of fragmented approaches adopted by groups of countries or at different levels of decision-making. The possibly strong economic implications of a stringent climate policy adopted by one group of states may have severe ramifications for other policy arenas such as the world trade regime. On the other hand, a fragmented policy arena may also have advantages. Distinct policy arenas allow for the testing of innovative policy instruments in some nations or at some levels of decision-making, with subsequent diffusion to other regions or levels. Also, sensible international policies could mitigate the negative political consequences of a horizontally and vertically fragmented governance architecture, and innovative policies may assist in the step-by-step convergence of parallel approaches.

These challenges of interlinkages within a fragmented governance architecture are some of the most interesting research domains at present (Zelli et al., 2012). While some authors argue in favor of a diversity of approaches and a more positive evaluation of the fragmentation of the international governance (Abbott, 2012; Keohane and Victor, 2011; Young, this book), others are more critical and call for political measures to increase the consistency and coherence of the overall governance architecture (Biermann, Pattberg, et al., 2009). Important new research has also addressed the political responses to potential problems of regime conflicts (Zelli, 2010) and ways forward in 'interplay management' (Oberthür and Gehring, 2006; Oberthür and Stokke, 2011; Van Asselt, 2014).

Current reform debates

Global environmental governance is a political response to economic, cultural, social and ecological globalization. It is not initiated and developed by a centralized decision-making body, but by an amalgam of centers of authority at various levels. The efficacy of the current system of global environmental governance has been the subject of intense debate. It is not only a normative discussion on 'more global governance' but also a debate on 'better global governance'. As a case study for my analysis of global environmental governance, I will now sketch two reform debates in this field; each attends to a particular aspect of global environmental governance that has been highlighted above.

Participation and privatization: Institutionalizing civil society involvement

The first example of a reform debate deals with the increased participation of non-state actors in global environmental governance. Many observers have argued for a stronger involvement of civil society in global governance, among others to increase the accountability and legitimacy of the slow and often bureaucratic processes of the intergovernmental decision-making system. Yet this call for more participation of civil society and a generally larger role of private actors has not been without friction. Developing countries, in particular, often object to increases in the influence of non-governmental organizations in international forums because they view these groups as being more favorable to Northern agendas, perspectives, and interests. Developing countries argue that most associations are headquartered in industrialized countries, that most funds donated to their cause stem from Northern organizations, both public and private, and that this situation influences the agenda of these groups to be more accountable to Northern audiences (Dombrowski, 2013). However, many observers argue that these biases in the work of non-governmental actors should not lead to a decrease in the participation of civil society, but rather to the establishment of mechanisms that ensure a balance of opinions and perspectives.

One example is the recent institutionalization and formalization of the advice of scientists and other experts on climate change (see also Gupta et al., 2012; Lövbrand, this book). The key institution here is the Intergovernmental Panel on Climate Change (IPCC). The evolution of the IPCC is typical for the functioning of global governance: it has been initiated not by governments but by international organizations – the World Meteorological Organization (WMO) and the United Nations Environment Programme. It comprises of private actors – experts, scientists, and their professional organizations – which are nonetheless engaged in a constant dialog with representatives from governments. The summary conclusions of IPCC reports are drafted by scientists, but are submitted to line-by-line review by governmental delegates. The reports from the IPCC are partially commissioned by public institutions – the UN climate convention – but are structured and organized by the expert community itself.

Typical for global governance has also been the continuous struggle for influence in this body, especially between industrialized and developing countries. When IPCC was set up in 1988, only a few experts and scientists from developing countries were involved. This has led, as many observers from developing countries argued, to a notable lack of credibility, legitimacy, and saliency of these reports in the South. Continuous complaints from developing countries led to several reforms since 1989, which resulted in an increasing institutionalization of the involvement of private actors. For example, current IPCC rules of procedure now require each working group of scientists to be chaired by one developed and one developing country scientist. Each chapter of assessment reports must have at least one lead author from a developing country. Participation of developing country scientists in IPCC thus appears much more visible than previously, and the institutionalization of the involvement of scientists in IPCC has helped to increase the legitimacy of the panel in the South.

For other areas, different types of institutionalizing the involvement of civil society or citizens might be needed. One way to increase the accountability and legitimacy of intergovernmental institutions is, for example, to give a more effective voice to civil society through inviting their representatives to join national delegations in international negotiations (Schroeder et al., 2012). One example is the unique representation of labor unions and employers associations in the International Labor Organization (ILO). In the ILO, each state is represented with four votes, two of which are assigned to governments and one each to business associations and labor unions. The ILO procedure, if adopted for environmental institutions, would attend to the basic problem of private participation in environmental governance – namely that environmental groups can often not adequately compete with the financial power of business associations, and non-state organizations of developing countries lack standing vis-à-vis the financially well-endowed organizations of industrialized countries. However, an adoption of the ILO formula in global environmental governance will also be difficult given the higher degree of complexity compared to ILO's more clear-cut 'business versus labor' type of conflicts.

An alternative approach is the establishment, and empowerment, of separate decision-making or consultative organs in international institutions and organizations. One proposal has been advanced by the Commission on Global Governance, which argued for an international Forum of Civil Society within the United Nations. This forum would comprise of 300–600 'organs of global civil society' (Commission on Global Governance, 1995: 258) to be self-selected from civil society. Similar separate chambers for civil society representatives could be also established in international environmental institutions.

A radically different approach is to create a United Nations parliamentary assembly of representatives of national parliaments, convening on a regular basis at the seat of the United Nations, possibly in the form of a second chamber of the United Nations to complement the chamber of state representatives (Commission on Global Governance, 1995: 257; Falk and Strauss, 2001). A parliamentary assembly would not need to have decision-making power, but could be restricted to a mere consultative role. Even then, the double identity of the parliamentarians – as members of their national parliament and of the world parliament – could guarantee some feedback across scales of negotiations and decision-making.

Another approach is the establishment of deliberative global citizens' assemblies. This could be a separate assembly within the United Nations system, or numerous specialized assemblies under distinct institutions, for example, under an environmental treaty. A global deliberative assembly would not include representatives of particular countries or interests, and would not be based on election. Instead, members of the assembly would be identified based on random selection, hence merely representing their own ideas in a mutual process of deliberation among fellow citizens (Dryzek et al., 2011: 36–37; see also Bäckstrand et al., 2010; Dryzek and Stevenson, 2011). A related idea – advanced by Baber and Bartlett (2009 and 2013) – is to set up a global deliberative process to generate global environmental norms through deliberations of numerous citizen juries all over the world. If implemented properly, such a series of citizen juries could help ascertain globally acceptable norms in contested areas of global environmental governance. To become effective international law, however, these norms would eventually need to be affirmed and implemented by governments and other governance institutions. Even if the global network of citizen juries did not result in binding law, its normative force, if grounded in sufficient numbers of professionally run citizen juries, could be substantial.

In sum, there are numerous avenues for increasing the participation of citizens in global governance, from special chambers of civil society to a parliamentary assembly or a global deliberative citizens' assembly. All proposals could serve as a blueprint for either a general high-level body within the UN system, such as a second chamber or consultative chamber in addition to the UN General Assembly, or as issue-specific special assemblies within the framework of environmental regimes and agencies. At present, however, there is not much political support among governments for either proposal, and any new assembly would likely be restricted to a mere consultative role, possibly linked with rights to expect a particular treatment or recognition of the advice within the empowered space of intergovernmental decision-making.

Governance fragmentation: The debate on a world environment organization

Another current reform debate in the field of global environmental governance concerns the organizational and institutional fragmentation of global policies. Many observers have pointed to the paradoxical situation that strong and powerful international bodies oriented toward economic growth – such as the World Trade Organization, the World Bank or the International Monetary Fund – are hardly matched by UNEP, the modest UN program for environmental issues. The same imbalance is revealed when UNEP is compared to the plethora of influential UN specialized agencies in the fields of labor, shipping, agriculture, communication, or culture. As a mere program, UNEP has no right to adopt treaties or any regulations upon its own initiative, it cannot avail itself of any regular and predictable funding, and it is subordinated to the UN Economic and Social Council.

This situation has led to a variety of proposals to grant the environment what other policy areas long had: a strong international agency with a sizeable mandate, significant resources, and sufficient autonomy. Proposals to create an international agency on environmental protection have been debated for now over 40 years (overviews in Biermann and Bauer, 2005). These proposals use a variety of names for such a new world environment organization, such as 'Global Environment Organization', 'United Nations Environment Organization', or 'United Nations Environmental Protection Organization'. What all proposals have in common is the argument for setting up a new specialized intergovernmental organization within the system of the United Nations that would focus on environmental policies.

The first proposal for a world environment organization dates back to US foreign policy strategist George F. Kennan (1970), who argued for an International Environmental Agency encompassing 'a small group of advanced nations'. Several authors supported this idea at that time. As one outcome of this debate, the United Nations established in 1973 the United Nations Environment Programme (UNEP) (Ivanova, 2010). Yet the creation of a UN environment program was a more modest reform than the strong international environmental organization that some observers had called for at that time. UNEP is not an intergovernmental organization but a subsidiary body of the General Assembly reporting through the Economic and Social Council, lacking institutional clout vis-à-vis other organizations and multilateral environmental agreements. Most studies argue that UNEP is lacking efficiency and effectiveness (see for further references Andresen and Rosendal, 2009; Bauer, 2009; Ivanova, 2010).

Since then, various authors have published elaborate proposals arguing for the establishment of a world environment organization to replace, or upgrade, UNEP (overviews in Biermann and Bauer, 2005). More sceptical voices and critics of a new organization also came forward. Sebastian Oberthür and Thomas Gehring (2005), for example, argued based on institutional theory against establishing a new organization that would integrate existing regimes and institutions; in their view, this reform would be not only unfeasible but also have few benefits combined with high costs in terms of negotiations and reorganization. Konrad von Moltke (2005) proposed as a more feasible and likely alternative to a new agency decentralized institutional clusters that would deal with diverse sets of environmental issues rather than entrusting all problems to one central organization.

Virtually all these proposals for a world environment organization can be categorized in three ideal type models, which differ regarding the degree of change that is required. First, the least radical proposals advise upgrading UNEP to a specialized UN agency with full-fledged organizational status. Proponents of this approach have referred to the World Health Organization or the International Labor Organization as suitable models. The new agency in this model is expected to improve the facilitation of norm-building and norm-implementation processes. This strength would in particular derive from an enhanced mandate and better capabilities of the agency to build capacities in developing countries. This differs from UNEP's present 'catalytic' mandate that prevents the program from engaging in project implementation. Furthermore, additional legal and political powers could come with the status of a UN special agency. For example, its governing body could approve by qualified majority vote certain regulations that could be binding, under certain conditions, on all members (comparable to the International Maritime Organization), or could adopt drafts of treaties negotiated by sub-committees under its auspices (comparable to the International Labor Organization). Such powers would exceed those entrusted to UNEP, which cannot adopt legal instruments.

Second, some observers argue for a more fundamental reform to address the substantive and functional overlap between the many international institutions in global environmental governance. These advocates of a more centralized governance architecture call for the integration of several existing agencies and programs into one all-encompassing world environment organization. Such an integration of environmental regimes could loosely follow the model of the World Trade Organization, which has integrated diverse multilateral trade agreements. According to some scholars, this integration could even include established intergovernmental organizations, although historic evidence suggests that this goes beyond the politically conceivable.

The third and most far-reaching model is that of a hierarchical intergovernmental organization on environmental issues that would be equipped with majority decision-making as well as enforcement powers vis-à-vis states that fail to comply with international agreements on the protection of global commons. Support for such a powerful international agency remains very scarce and mainly restricted to a few nongovernmental activist organizations.

Most scholars have focused in recent years on reform proposals that are feasible in the current political context (Biermann et al., 2012). The discussion among reform proponents now more or less oscillates between proponents of a streamlined, centralizing umbrella organization and an upgraded version of UNEP with the full status of a UN special organization.

In sum, while a world environment organization is still not a reality after 40 years of debate, the idea of creating such a new agency is one of the most long-standing, and most vivid, reform debates in the field. At the 2012 UN Conference on Sustainable Development, a specialized agency status for UNEP was supported by the member states of the European Union as well as the African Union and a number of other developing countries. Resistance remained strong, however, from the United States, Japan, Russia, and even Brazil, which had earlier been a supporter of a world environment organization but now seemed afraid of an imbalance in favor of the environmental pillar of sustainable development. These countries argued that this question requires further debate and analysis (Vijge, 2013). Yet with numerous nations now behind a concrete proposal for a world environment organization, the establishment of a new agency, based on the existing UN Environment Programme, becomes more likely, at least in the medium term. Yet the extent to which this new agency would in fact advance the effectiveness of global environmental governance, is certain to remain a hotly debated issue.

Conclusion

Global environmental governance is not only a tremendous political challenge. It is also one of the most crucial areas of research in the social sciences. The current knowledge about global environmental governance shows that more theoretical debate as well as empirical research is needed. I will now, in concluding, emphasize five needs for further discussion (drawing on Biermann and Pattberg, 2012; Biermann, Betsill et al., 2009).

First, we need to gain a better understanding of processes of change in governance and more generally of the institutional dynamics that play a role in the emergence, evolution and eventual effectiveness of institutions (Young, 2010, this book). In broader terms, this is the challenge of analyzing at the global level the adaptiveness and resilience of social-ecological systems, as well as learning processes in governance. This line of research should pay attention also to the larger discursive struggles about what constitutes effective and legitimate global environmental governance.

Second, we need to know more about the overarching systems of principles, rules, and norms that go beyond single institutions – that is, the overall 'regime complexes' (Keohane and Victor, 2011) or 'governance architectures', which I understand as the entire interlocking web of widely shared principles, institutions and practices that shape decisions by stakeholders at all levels in a given global policy domain (see for more detail Biermann, Pattberg et al., 2009).

Third, there is a need for a better understanding of the accountability and legitimacy in global environmental governance (for example, Bernstein, 2005; Bernstein and Cashore, 2007; Mason, 2008; Newell, 2008). Understanding accountability and legitimacy is important in its own right, but also with a view to the effectiveness of institutions. A related issue for research are the promises and potential pitfalls of the growing calls for more transparency as an environmental governance mechanism (Gupta, 2010; Gupta and Mason, 2014, this book). A stronger focus on questions of accountability and legitimacy is especially important given the rising role of non-state actors in global governance (Betsill, this book). More research in this area is needed, including on policy-relevant dimensions (see for a more detailed discussion Biermann and Gupta, 2011).

Fourth, we need better methodologies and approaches to integrate social science knowledge on environmental governance and institutions into more formal approaches in earth system science, such as modeling and scenario development (Hochstetler and Laituri, this book). For the more quantitatively oriented modeling and scenarios research programs, it becomes increasingly important to include data on international regimes and organizations or transnational governance mechanisms. Numerous approaches exist in International Relations and global environmental governance research to understand and measure the effectiveness of institutions and organizations to steer societies toward sustainability. Yet what is lacking so far is a concerted effort to bring this social science knowledge into formal modeling and scenario approaches.

Fifth and finally, with the increasing relevance of global environmental governance, allocation mechanisms and criteria – and thus more broadly questions of equity and justice – will become central questions to be addressed by social scientists (Okereke and Charlesworth, this book). At stake are not only the costs of mitigating global problems. Given the large-scale and potentially disastrous consequences of global environmental change, questions of fairness in adaptation will gain prominence. Compensation and support through the global community of the most affected and most vulnerable regions, such as small island states, will not only be a moral responsibility, but also politically and economically prudent. This situation calls for allocation modes that all stakeholders in North and South perceive as fair. Questions of allocation among nations are especially contested in global environmental governance. In particular, the causes and consequences of different allocation mechanisms are still not sufficiently understood. Little systematic analysis has been devoted to studying allocation as independent variable and to analyzing allocation mechanisms in relation to variant effectiveness of international institutions. Hence, given the growing relevance of global environmental change, allocation is certain to become a major concern for researchers and practitioners of global environmental governance alike.

In sum, despite all efforts in the last 20 years, numerous core mechanisms, processes, and actors in global environmental governance are not yet sufficiently understood in the social sciences. To critically explore the emergence, effectiveness, equity and legitimacy of institutions and processes of global environmental governance remain therefore one of the fundamental research domains for students in the field of International Relations.

Annotated bibliography

Biermann, F., M. M. Betsill, J. Gupta, N. Kanie, L. Lebel, D. Liverman, H. Schroeder and B. Siebenhüner, with contributions from K. Conca, L. da Costa Ferreira, B. Desai, S. Tay and R. Zondervan (2009) *Earth System Governance: People, Places and the Planet*. 'Science and Implementation Plan of the Earth System Governance Project. ESG Report 1' (Bonn: The Earth System Governance Project). Available at www.earthsystemgovernance.org

This document is the science and implementation plan of a long-term research program in the area of global and local environmental governance, called the Earth System Governance Project. The document, which has been extensively peer-reviewed and discussed by numerous social scientists, lays down numerous research questions for the current decade, organized around five analytical themes: architectures of governance; agency in environmental governance, both of states and non-state actors; adaptiveness of governance systems; accountability and legitimacy; and allocation and access, including questions of equity and fairness.

Biermann, F. and P. Pattberg (eds) (2012) *Global Environmental Governance Reconsidered* (Cambridge, MA: MIT Press).

This book presents the core findings of the Global Governance Project, a ten-year network of scholars of global environmental governance, focussing on the new actors, new transnational steering mechanisms, and new contexts of institutional fragmentation and dispersed authority in environmental governance.

Park, J., K. Conca and M. Finger (eds) (2008) *The Crisis of Global Environmental Governance: Towards a New Political Economy of Sustainability* (Abingdon, Oxon: Routledge).

This book presents a critical perspective on global environmental governance that is rooted in political economy, regulation theory, and post-sovereign international relations, covering among others the tensions between (economic) globalization and sustainability, the ongoing current marketization of environmental governance, and vital questions of societal justice.

Young, O. R. (2002) *The Institutional Dimension of Environmental Change: Fit, Interplay and Scale* (Cambridge, MA: MIT Press).

This book represents the key outline of the research strategy of the Institutional Dimensions of Global Environmental Change (IDGEC) core project of the International Human Dimensions Programme on Global Environmental Change, written by IDGEC's initiator and long-time chair.

Young, O. R., L. A. King and H. Schroeder (eds) (2008) *Institutions and Environmental Change: Principal Findings, Applications and Research Frontiers* (Cambridge, MA: MIT Press).

This book represents the key research results of the Institutional Dimensions of Global Environmental Change (IDGEC) core project of the International Human Dimensions Programme on Global Environmental Change. The comprehensive chapters address, among other issues, questions of institutional diagnostics, causality and performance, along with questions of institutional fit, scale, and interplay.

Works cited

Abbott, K. W. (2012) 'The Transnational Regime Complex for Climate Change', *Environment and Planning C: Government and Policy* 30, 4, 571–590.

Adger, W. N. and A. J. Jordan (eds) (2009) *Governing Sustainability* (Cambridge: Cambridge University Press).

Alcock, F. (2008) 'Conflicts and Coalitions within and Across the ENGO Community', *Global Environmental Politics* 8, 4, 66–91.

Amsterdam Declaration (2001) *Challenges of a Changing Earth*. Declaration of the Global Change Open Science Conference Amsterdam, signed by the chairs of the International Geosphere-Biosphere Programme, the International Human Dimensions Programme on Global Environmental Change, the World Climate Research Programme, and the International Biodiversity Programme DIVERSITAS, Amsterdam, 13 July 2001. Available at http://www.essp.org/index.php?id=41&L=

Andonova, L. B., M. M. Betsill and H. Bulkeley (2009) 'Transnational Climate Governance', *Global Environmental Politics* 9, 2, 52–73.

Andresen, S. and E. Hey (2005) 'The Effectiveness and Legitimacy of International Environmental Institutions', *International Environmental Agreements: Politics, Law and Economics* 5, 3, 211–226.

Andresen, S. and K. Rosendal (2009) 'The Role of the United Nations Environment Programme in the Coordination of Multilateral Environmental Agreements', in F. Biermann, B. Siebenhüner and A. Schreyögg (eds) *International Organizations in Global Environmental Governance* (Abingdon: Routledge), 133–150.

Baber, W. F. and R. V. Bartlett (2009) *Global Democracy and Sustainable Jurisprudence. Deliberative Environmental Law* (Cambridge, MA: MIT Press).

Baber, W. F. and R. V. Bartlett (2013) *Juristic Democracy: A Deliberative Common Law Strategy for Earth System Governance* (Earth System Governance Working Paper 27) (Lund and Amsterdam: The Earth System Governance Project).

Bäckstrand, K., J. Khan, A. Kronsell and E. Lövbrand (eds) (2010) *Environmental Politics and Deliberative Democracy. Examining the Promise of New Modes of Governance* (Cheltenham: Edward Elgar).

Bäckstrand, K., S. Campe, S. Chan, A. Mert and M. Schäfferhof (2012) 'Transnational Public-Private Partnerships', in F. Biermann and P. Pattberg (eds) *Global Environmental Governance Reconsidered* (Cambridge, MA: MIT Press), 123–147.

Bauer, S. (2009) 'The Secretariat of the United Nations Environment Programme: Tangled Up in Blue', in F. Biermann and B. Siebenhüner (eds) *Managers of Global Change. The Influence of International Environmental Bureaucracies* (Cambridge, MA: MIT Press), 169–201.

Bernstein, S. (2005) 'Legitimacy in Global Environmental Governance', *Journal of International Law and International Relations* 1, 1–2, 139–166.

Bernstein, S. and B. Cashore (2007) 'Can Non-state Global Governance Be Legitimate? A Theoretical Framework', *Regulation and Governance* 1, 1–25.

Betsill, M. M. and E. Corell (eds) (2008) *NGO Diplomacy: The Influence of Non-governmental Organizations in International Environmental Negotiations* (Cambridge, MA: The MIT Press).

Betsill, M. M. and H. Bulkeley (2006) 'Cities and the Multilevel Governance of Global Climate Change', *Global Governance* 12, 2, 141–159.

Biermann, F. (2012) 'Greening the United Nations Charter: World Politics in the Anthropocene', *Environment* 54, 3, 6–17.

Biermann, F. and A. Gupta (2011) 'Accountability and Legitimacy in Earth System Governance: A Research Framework', *Ecological Economics* 70, 1856–1864.

Biermann, F. and B. Siebenhüner (eds) (2009) *Managers of Global Change: The Influence of International Environmental Bureaucracies* (Cambridge, MA: MIT Press).

Biermann, F. and P. Pattberg (eds) (2012) *Global Environmental Governance Reconsidered* (Cambridge, MA: MIT Press).

Biermann, F. and S. Bauer (eds) (2005) *A World Environment Organization: Solution or Threat for Effective International Environmental Governance?* (Aldershot, UK: Ashgate).

Biermann, F., K. Abbott, S. Andresen, K. Bäckstrand, S. Bernstein, M. M. Betsill, H. Bulkeley, B. Cashore, J. Clapp, C. Folke, A. Gupta, J. Gupta, P. M. Haas, A. Jordan, N. Kanie, T. Kluvánková-Oravská, L. Lebel, D. Liverman, J. Meadowcroft, R. B. Mitchell, P. Newell, S. Oberthür, L. Olsson, P. Pattberg, R. Sánchez-Rodríguez, H. Schroeder, A. Underdal, S. Camargo Vieira, C. Vogel, O. R. Young, A. Brock and R. Zondervan (2012) 'Navigating the Anthropocene: Improving Earth System Governance', *Science* 335, 6074, 16 March, 1306–1307.

Biermann, F., M. M. Betsill, J. Gupta, N. Kanie, L. Lebel, D. Liverman, H. Schroeder and B. Siebenhüner, with contributions from K. Conca, L. da Costa Ferreira, B. Desai, S. Tay and R. Zondervan (2009) *Earth System Governance: People, Places and the Planet. Science and Implementation Plan of the Earth System Governance Project* (ESG Report 1) (Bonn: The Earth System Governance Project). Available at www.earthsystemgovernance.org

Biermann, F., P. Pattberg, H. van Asselt and F. Zelli (2009) 'The Fragmentation of Global Governance Architectures: A Framework for Analysis', *Global Environmental Politics* 9, 4, 14–40.

Bouteligier, S. (2011) 'Exploring the Agency of Global Environmental Consultancy Firms in Earth System Governance', *International Environmental Agreements: Politics, Law and Economics* 11, 1, 43–61.

Bulkeley, H. and A. Jordan (2012) 'Transnational Environmental Governance: New Findings and Emerging Research Agendas', *Environment and Planning C: Government and Policy* 30, 4, 556–570.

Bulkeley, H., M. J. Hoffmann, S. D. VanDeveer and V. Milledge (2012) 'Transnational Governance Experiments', in F. Biermann and P. Pattberg (eds) *Global Environmental Governance Reconsidered* (Cambridge, MA: MIT Press), 149–171.

Caldwell, L. K. (1984) *International Environmental Policy: Emergence and Dimensions* (Durham, NC: Duke University Press).

Clapp, J. (2009) 'Global Mechanisms for Greening TNCs: Inching Towards Corporate Accountability?' in K. P. Gallagher (eds) *Handbook on Trade and Environment* (Cheltenham: Edward Elgar), 159–170.

Commission on Global Governance (1995) *Our Global Neighbourhood. The Report of the Commission on Global Governance* (Oxford: Oxford University Press).

Crutzen, P. J. and V. Ramanathan (2004) 'Atmospheric Chemistry and Climate in the Anthropocene: Where Are We Heading?', in H.-J. Schellnhuber, P. J.

Crutzen, W. C. Clark, M. Claussen and H. Held (eds) *Earth System Analysis for Sustainability* (Cambridge, MA: MIT Press in cooperation with Dahlem University Press), 266–292.

Dellas, E., P. Pattberg and M. M. Betsill (2011) 'Agency in Earth System Governance: Refining a Research Agenda', *International Environmental Agreements: Politics, Law and Economics* 11, 1, 85–98.

Dingwerth, K. (2005) 'The Democratic Legitimacy of Public-private Rule-making: What Can We Learn from the World Commission on Dams?' *Global Governance* 11, 1, 65–83.

Dingwerth, K. (2007) *The New Transnationalism: Transnational Governance and Democratic Legitimacy* (Basingstoke: Palgrave Macmillan).

Dombrowski, K. (2013) *Filling the Gap? International NGOs as Democratic Links between Local Communities and International Organisations*, Doctoral Dissertation (London: London School of Economics and Political Science).

Dryzek, J. S., A. Bächtiger and K. Milewicz (2011) 'Toward a Deliberative Global Citizens' Assembly', *Global Policy* 2, 1, 33–42.

Dryzek, J. S. and H. Stevenson (2011) 'Global Democracy and Earth System Governance', *Ecological Economics* 70, 11, 1865–1874.

Eckersley, R. (2004) 'The Big Chill: The WTO and Multilateral Environmental Agreements', *Global Environmental Politics* 4, 2, 24–50.

Falk, R. and A. Strauss (2001) 'Toward Global Parliament', *Foreign Affairs* 80, 1, 212–220.

Falkner, R. (2003) 'Private Environmental Governance and International Relations: Exploring the Links', *Global Environmental Politics* 3, 2, 72–87.

Falkner, R., H. Stephan and J. Vogler (2010) 'International Climate Policy after Copenhagen: Towards a "Building Blocks" Approach', *Global Policy* 1, 3, 252–262.

Finkelstein, L. S. (1995) 'What is Global Governance?' *Global Governance* 1, 3, 367–372.

Gulbrandsen, L. H. (2010) *Transnational Environmental Governance. The Emergence and Effects of the Certification of Forests and Fisheries* (Cheltenham: Edward Elgar).

Gupta, A. (2010) 'Transparency in Global Environmental Governance: A Coming of Age?' *Global Environmental Politics* 10, 3, 1–9.

Gupta, A. and M. Mason (eds) (2014) *Transparency in Global Environmental Governance* (Cambridge, MA: MIT Press).

Gupta, A., S. Andresen, F. Biermann and B. Siebenhüner (2012) 'Science Networks', in F. Biermann and P. Pattberg (eds) *Global Environmental Governance Reconsidered* (Cambridge, MA: MIT Press), 69–93.

Gupta, J. (2003) 'The Role of Non-state Actors in International Environmental Affairs', *Heidelberg Journal of International Law* 63, 2, 459–486.

Haas, P. M. (2012) 'The Road from Rio: Why Environmentalism Needs to Come Down from the Summit', *Foreign Affairs*, 16 August.

Hoffmann, M. J. (2011) *Climate Governance at the Crossroads: Experimenting with a Global Response after Kyoto* (Oxford: Oxford University Press).

Ivanova, M. (2010) 'UNEP in Global Environmental Governance: Design, Leadership, Location', *Global Environmental Politics* 10, 1, 30–59.

Jasanoff, S. and M. Long-Martello (eds) (2004) *Earthly Politics: Local and Global in Environmental Governance* (Cambridge, MA: MIT Press).

Johnson, B. (1972) 'The United Nations Institutional Response to Stockholm: A Case Study in the International Politics of Institutional Change', *International Organization* 26, 2, 255–301.

Jordan, A. J. (2008) 'The Governance of Sustainable Development: Taking Stock and Looking Forwards', *Environment and Planning C*, 26, 17–33.

Kanie, N., M. M. Betsill, R. Zondervan, F. Biermann and O. R. Young (2012) 'A Charter Moment: Restructuring Governance for Sustainability', *Public Administration and Development* 32, 292–304.

Kanie, N. and P. M. Haas (eds) (2004) *Emerging Forces in Environmental Governance* (Tokyo: United Nations University Press).

Kennan, G. F. (1970) 'To Prevent a World Wasteland: A Proposal', *Foreign Affairs* 48, 3, 401–413.

Keohane, R. O. and D. G. Victor (2011) 'The Regime Complex for Climate Change', *Perspectives on Politics* 9, 1, 7–23.

Mason, M. R. (2008) 'The Governance of Transnational Environmental Harm: Addressing New Modes of Accountability/Responsibility', *Global Environmental Politics* 8, 3, 8–24.

Mert, A. (2014) *Environmental Governance through Transnational Partnerships: A Discourse Theoretical Study* (Cheltenham: Edward Elgar).

Mitchell, R. B. (2003) 'International Environmental Agreements: A Survey of Their Features, Formation and Effects', *Annual Review of Environment and Resources* 28, 429–461.

Mitchell, R. B. (2012) *International Environmental Agreements Database Project* (Version 2012.1). Available at http://iea.uoregon.edu/

Mitchell, R. B., W. C. Clark, D. W. Cash and N. M. Dickson (eds) (2006) *Global Environmental Assessments: Information and Influence* (Cambridge, MA: MIT Press).

Newell, P. J. (2008) 'Civil Society, Corporate Accountability and the Politics of Climate Change', *Global Environmental Politics* 8, 3, 122–153.

Newell, P., P. Pattberg and H. Schroeder (2012) 'Multiactor Governance and the Environment', *Annual Review of Environment and Resources* 37, 365–387.

Oberthür, S. and O. Schram Stokke (eds) (2011) *Managing Institutional Complexity: Regime Interplay and Global Environmental Change* (Cambridge, MA: MIT Press).

Oberthür, S. and T. Gehring (2005) 'Reforming International Environmental Governance: An Institutional Perspective on Proposals for a World Environment Organization', in F. Biermann and S. Bauer (eds) *A World Environment Organization: Solution or Threat for Effective International Environmental Governance?* (Aldershot, UK: Ashgate), 205–234.

Oberthür, S. and T. Gehring (eds) (2006) *Institutional Interaction in Global Environmental Governance: Synergy and Conflict among International and EU Policies* (Cambridge, MA: The MIT Press).

Okereke, C., H. Bulkeley and H. Schroeder (2009) 'Conceptualizing Climate Governance beyond the International Regime', *Global Environmental Politics* 9, 1, 58–78.

Ostrom, E. (2010) 'Polycentric Systems for Coping with Collective Action and Global Environmental Change', *Global Environmental Change* 20, 4, 550–557.

Overbeek, H., K. Dingwerth, P. Pattberg and D. Compagnon (2010) 'Global Governance: Decline or Maturation of an Academic Concept?' *International Studies Review* 12, 4, 619–642.

Pattberg, P. (2007) *Private Institutions and Global Governance: The New Politics of Environmental Sustainability* (Cheltenham: Edward Elgar).
Pattberg, P. (2012) 'Transnational Environmental Regimes', in F. Biermann and P. Pattberg (eds) *Global Environmental Governance Reconsidered* (Cambridge, MA: MIT Press), 97–121.
Pattberg, P., F. Biermann, S. Chan and A. Mert (eds) (2012) *Public-Private Partnerships for Sustainable Development: Emergence, Influence and Legitimacy* (Cheltenham, UK: Edward Elgar).
Pattberg, P. and J. Stripple (2008) 'Beyond the Public and Private Divide: Remapping Transnational Climate Governance in the 21st Century,' *International Environmental Agreements: Politics, Law and Economics* 8, 4, 367–388.
Rosenau, J. N. (1995) 'Governance in the Twenty-First Century', *Global Governance* 1, 1, 13–43.
Schroeder, H. and H. Lovell (2012) 'The Role of Non-Nation-State Actors and Side Events in the International Climate Negotiations', *Climate Policy* 12, 1, 23–37.
Schroeder, H., M. T. Boykoff and L. Spiers (2012) 'Equity and State Representations in Climate Negotiations', *Nature Climate Change* 2, 834–836.
Skodvin, T. and S. Andresen (2003) 'Nonstate Influence in the International Whaling Commission, 1970–1990', *Global Environmental Politics* 3, 4, 61–86.
South Centre (1996) *For a Strong and Democratic United Nations: A South Perspective on UN Reform* (Geneva: South Centre).
Tamiotti, L. (2001) 'Environmental Organizations: Changing Roles and Functions in Global Politics', *Global Environmental Politics* 1, 1, 56–76.
Tienhaara, K., A. Orsini and R. Falkner (2012) 'Global Corporations', in F. Biermann and P. Pattberg (eds) *Global Environmental Governance Reconsidered* (Cambridge, MA: MIT Press), 45–67.
Van Asselt, H. (2014) *The Fragmentation of Global Climate Governance: Consequences and Management of Regime Interactions* (Cheltenham, UK: Edward Elgar).
Van Kersbergen, K. and F. van Waarden (2004) ' "Governance" as a Bridge Between Disciplines: Cross-disciplinary Inspiration Regarding Shifts in Governance and Problems of Governability, Accountability and Legitimacy', *European Journal of Political Research* 43, 2, 143–171.
Vieira, S. C. (2012) 'From Sustainable Development to Earth System Governance: A View from the South', *Anuário Brasileiro de Direito Internacional/Brazilian Yearbook of International Law/Annuaire Brésilien de Droit International* 2, 11, 167–176.
Vijge, M. J. (2013) 'The Promise of New Institutionalism: Explaining the Absence of a World or United Nations Environment Organization', *International Environmental Agreements: Politics, Law and Economics* 13, 2, 153–176.
Von Moltke, K. (2005) 'Clustering International Environmental Agreements as an Alternative to a World Environment Organization', in F. Biermann and S. Bauer (eds) *A World Environment Organization: Solution or Threat for Effective International Environmental Governance?* (Aldershot, UK: Ashgate), 175–204.
Wapner, P. K. (1996) *Environmental Activism and World Civic Politics* (Albany: State University of New York Press).
Wapner, P. K. (2002) 'Horizontal Politics: Transnational Environmental Activism and Global Cultural Change', *Global Environmental Politics* 2, 2, 37–62.

Young, O. R. (2002) *The Institutional Dimension of Environmental Change: Fit, Interplay and Scale* (Cambridge, MA: MIT Press).

Young, O. R., L. A. King and H. Schroeder (eds) (2008) *Institutions and Environmental Change: Principal Findings, Applications and Research Frontiers* (Cambridge, MA: MIT Press).

Zelli, F. (2007) 'The World Trade Organization: Free Trade and its Environmental Impacts', in K. V. Thai, D. Rahm and J. D. Coggburn (eds) *Handbook of Globalization and the Environment* (London: Taylor and Francis), 177–216.

Zelli, F. (2010) *The Regime Environment of Environmental Regimes: Conceptualizing, Theorizing and Examining Conflicts among International Regimes on Environmental Issues*, Doctoral dissertation (Germany: University of Tübingen).

Zelli, F., A. Gupta and H. van Asselt (2012), 'Horizontal Institutional Interlinkages', in F. Biermann and P. Pattberg (eds) *Global Environmental Governance Reconsidered* (Cambridge, MA: MIT Press), 175–198.

Part III

Frameworks for Evaluating Global Environmental Politics

11
The Effectiveness of International Environmental Regimes: Existing Knowledge, Cutting-edge Themes, and Research Strategies

Oran R. Young

A common observation among those concerned with solving environmental problems and, more generally, with promoting sustainability in human–environment relations is that governance systems work relatively well at the national level but poorly or not at all in efforts to solve international, transnational, and especially global problems (Speth, 2004).[1] While the state is a positive force in managing natural resources and protecting the environment in domestic settings, the anarchic character of international society treated as a society of spatially demarcated sovereign states constitutes a barrier to successful governance at the international level.[2]

Yet both elements of this argument are open to question. Failures to tackle environmental problems effectively, much less to achieve sustainability in human–environment relations, are common not only in societies facing severe problems of poverty and hunger or saddled with the curse of natural resources but also in advanced industrial societies (Collier, 2008). Although efforts to address the grand challenges of climate change, the loss of biological diversity, and the degradation of ecosystem services leave a great deal to be desired, international environmental governance does not present a uniform picture of failure.

Published originally as Oran R. Young, 'Effectiveness of International Environmental Regimes: Existing Knowledge, Cutting-edge Themes, and Research Strategies', *Proceedings of the National Academy of Science USA* 108, 19853–19860. Updated and revised June 2013.

Some international environmental governance systems or, as they are commonly called, international regimes make a difference in the sense that they contribute to solving large-scale problems. Arrangements widely regarded as effective in these terms include the global regime created to protect the stratospheric ozone layer, the governance system applicable to Antarctica, and the multilateral arrangement established to clean up the Rhine River.[3] The lack of efficacy or relative failure of other regimes created to deal with environmental problems is equally evident. Prominent examples include the climate regime, the arrangement created to combat desertification, and some (but not all) of the regional fisheries management regimes. Many regimes fall between these polar categories. They achieve a measure of effectiveness, though it is often hard to place them precisely along a continuum ranging from total failure to clear-cut success. Cases that fit this description include the regime dealing with pollution of the sea from ships, the arrangement dealing with pollution in the North Sea, the regime governing trade in endangered species, and the regime articulated in the Great Lakes Water Quality Agreement. Not surprisingly, some regimes are successful for a time but subsequently decline or even collapse (for example, the regime for North Pacific fur seals), while others are slow to gain traction but become more effective with the passage of time (for example, the transboundary air pollution regime in Europe).

How can we account for this mixed record in efforts to address problems of environmental governance in international society? Are there identifiable factors that contribute to success or cause failure? Can we formulate propositions about such matters that will be of interest both to those responsible for implementing the provisions of environmental regimes and to those engaged in efforts either to strengthen existing arrangements or to create entirely new ones? In this chapter, I address these questions in four steps. The first step involves conceptual and definitional issues; it focuses on clarifying the meaning of effectiveness with regard to environmental regimes. The next step centers on identifying and discussing the most important things we have learned about the determinants of institutional effectiveness. Step three features an exploration of cutting-edge themes or areas ripe for increased attention on the part of researchers going forward. The final step turns to a discussion of the tools available for tackling these themes and recommends strategies likely to produce policy-relevant results. The chapter highlights findings about the determinants of effectiveness in environmental regimes that are relevant to efforts to strengthen existing regimes or to create new ones. The take-home message is one of cautious optimism. There is much we can do to add to our understanding of the effectiveness of international environmental regimes, despite the impacts of some obvious as well as some less apparent limitations on the methods available for pursuing this goal.

What do we mean by effectiveness?

The concept of effectiveness as applied to environmental regimes is complex and subject to a variety of formulations (Mitchell, 2008; Underal, 2008). Arguably, the core concern is the extent to which regimes contribute to solving or mitigating the problems that motivate those who create them (Young, 1999a). But there are other ways of thinking about effectiveness that are both less ambitious and more ambitious than this focus on problem solving. Less ambitious conceptions of effectiveness direct attention to what are known as (i) outputs or the regulations and infrastructure created to move a regime from paper to practice and (ii) outcomes or changes in the behavior of actors relevant to the problem at hand (Young, 1999b). Success in these terms is significant, but it does not guarantee progress in solving the relevant problems. More ambitious conceptions seek to assess the performance of regimes relative not only to the probable course of events in their absence (that is, the no-regime counterfactual) but also to some conception of an ideal outcome known as the collective optimum. The effectiveness of a regime (E) is then measurable as the location of actual performance (AP) on the spectrum ranging between the no-regime counterfactual (NR) and the collective optimum (CO) or:

$$E = \frac{AP - NR}{CO - NR} \quad (1)$$

Normalizing this equation by setting NR equal to 0 and CO equal to 1 produces a way to compare and contrast the effectiveness of different regimes on a common scale that is conceptually attractive, though it is hard to operationalize (Helm and Sprinz, 2000; Hovi et al., 2003a, 2003b; Young, 2003).

Several other aspects of effectiveness deserve notice at the outset. A regime's participants may differ both in the importance they attach to the problem and in the way they frame it for consideration in policy forums. Those who create regimes may harbor unstated goals that differ significantly from those spelled out in constitutive documents. The effectiveness of regimes may vary through time. Some regimes go from strength to strength with the passage of time. Others are relatively ineffective at the outset but gain strength over time or vice versa. Many of those seeking to assess the effectiveness of regimes add other measures of success to the core concern of problem-solving, including some criterion of sustainability or resilience, economic efficiency, various measures of fairness or equity, and one or more considerations embedded in the idea of 'good' governance (Mitchell, 2008).

Evaluating the effectiveness of environmental regimes is a challenging task under the best of circumstances (Underal, 2008). In every case, we want to compare the actual course of events regarding the relevant problem with what would have happened in the no-regime counterfactual. Although this is easier to do with regard to some measures of effectiveness (for example, outputs) than others (for example, problem-solving), documenting the consequences resulting from the creation and operation of a regime is always demanding. Additionally, regimes invariably operate in complex settings in which a variety of other forces are at work. Separating the signal attributable to the operation of a regime from the noise associated with a variety of other forces at work at the same time is a difficult task. I will discuss tools available to those endeavoring to address these issues in the last section of this chapter. Suffice it to say for now that some differences of opinion regarding the effectiveness of regimes are more apparent than real, in the sense that they are artifacts of the definitions of effectiveness selected or the procedures used to evaluate effectiveness rather than substantive disagreements about the actual performance of specific regimes.

What do we know about effectiveness?

Scientists understandably focus on cutting-edge questions that constitute the frontiers of research in their areas of interest, a practice that directs attention to issues that we do not understand or at least do not understand well. But in this discussion of the current state of knowledge regarding the effectiveness of environmental regimes, it is appropriate to begin with an account of what we have learned so far about this subject. I address this topic under three headings: general findings about effectiveness, findings about specific determinants of success, and findings about institutional interplay.

Three distinct bodies of evidence deserve attention in this assessment: qualitative case studies typically carried out by analysts trained as political scientists (Haas et al., 1993; Parson, 2003; Skjaerseth, 2000; Stokke, 2012; Wettestad, 2002), quantitative case studies most often produced by analysts with a background in economics (Bernauer and Singfried, 2008; Finus and Tjøtta, 2003; Murdoch and Sandler, 1997; Murdoch et al., 1997; Ringquist and Kostadinova, 2005), and quantitative analyses that seek to develop generalizations about effectiveness, drawing on evidence from sizable universes of cases (Breitmeier et al., 2006, 2011; Miles et al., 2002). The conclusions emerging from these bodies of evidence overlap, but they are not entirely compatible. Those who have carried out the qualitative case studies, perhaps reflecting a positive attitude toward political institutions common among political scientists, tend to find evidence of the significance of regimes in addressing environmental problems. The quantitative case studies, arguably reflecting skeptical attitudes toward governance systems common among economists, typically raise doubts about the roles regimes play. The large-N studies have sought to move beyond this divide, endeavoring to discriminate among cases in which regimes matter a lot or a little and seeking to identify the determinants of success and failure.

General findings

Though it may be a source of frustration to those hoping for simple generalizations regarding the determinants of effectiveness, differences in the findings flowing from the three bodies of evidence are understandable. In virtually every case, a regime constitutes only one of a number of distinct but interacting forces influencing the course of human–environment relations. What is notable is that there are some general findings about the effectiveness of environmental regimes arising from the research carried out so far. In this subsection, I comment on what seems to me to be the most important of these findings.

Some regimes matter in the sense that they make a (sometimes sizable) difference not only in terms of outputs and outcomes but also in terms of solving the problems leading to their creation. It is easy to overestimate the success of environmental regimes. Quantitative case studies, rooted in a rational-choice paradigm, have suggested that key actors may reduce emissions of ozone-depleting substances or airborne pollutants voluntarily, that non-regime factors may account for as much or more of the success in dealing with water pollution as the operation of the relevant regimes, and that actual outcomes fall short of the collective optimum in most cases. Yet in-depth qualitative case studies, making extensive use of procedures like process tracing and thick description, have concluded that regimes have contributed to the development of new knowledge and new social practices that have played important roles in dealing with long-range transboundary air pollution in Europe (Wettestad,

2002), the depletion of the stratospheric ozone layer (Parson, 2003), the control of pollution in the North Sea (Skjaerseth, 2000), and the management of commercial fisheries in the Barents Sea (Stokke, 2012). In an effort to reconcile these findings, several teams of researchers have created databases containing sufficiently large numbers of cases to allow for the development of empirical generalizations about the effectiveness of environmental regimes. Miles et al. (2002), drawing on a dataset including 37 cases, report that 50 percent of these regimes produced behavioral changes and 35 percent played a significant role in terms of problem-solving (Underal, 2008: 59). Breitmeier et al. (2006), employing a dataset encompassing 172 cases, report that in situations where problems improved slightly or considerably, regimes had a 'significant' or 'very strong' influence 52 percent of the time (Underal, 2008: 59).[4] Environmental regimes can make a difference. But they do not always work, and they never operate in a vacuum devoid of other causal forces (Breitmeier et al., 2011).

The anarchic character of international society is not always an obstacle to building the capacity of regimes to contribute to problem-solving. Many observers regard the absence of a government at the international level as a severe impediment to the establishment of effective regimes, primarily because it rules out the use of enforcement mechanisms of the sort that states employ to induce their subjects to comply with systems of rules and regulations. While this is certainly a concern in some cases, it does not loom large in situations where compliance on the part of most members of the group is unnecessary, the parties to environmental agreements have no incentive to cheat, factors other than sanctions in the ordinary sense provide subjects with good reasons to comply, or various forms of private or hybrid governance are able to exert pressure on subjects to comply (Barrett, 2007; Dai, 2007; Young, 1999a). There is no basis for complacency here when it comes to dealing with the great issues of our times, such as climate change and the loss of biological diversity. But neither is there a basis for dismissing the capacity of regimes to contribute to solving a range of problems.

Regime design is often a more significant determinant of effectiveness than some measure of whether the problem is benign (that is, easy to solve) or malign (that is, hard to solve). Poorly designed regimes can produce disappointing results even in cases where problems are straightforward and relatively easy to solve; well-designed regimes can produce positive results even in dealing with problems that are widely regarded as malign. This has given rise to a stream of research on what has become known as the issue of fit (Galaz et al., 2008), together with a growing interest in institutional diagnostics (Ostrom, 2007; Young, 2002a, 2008). Whereas the effort to conserve Atlantic tunas among generally friendly states has produced poor results, leading states were able to join forces to launch a successful regime for Antarctica during the height of the cold war.

A sizable proportion of the success of environmental regimes is attributable to activities that are not regulatory in the ordinary sense. There is a strong tendency to think of regimes in regulatory terms. The promulgation and implementation of prescriptive regulations setting forth prohibitions, requirements, and permissions are important functions of many regimes (Chayes and Chayes, 1995). But these institutional arrangements regularly perform other functions as well (Young, 1999a). Regimes may perform procedural functions (for example, setting total allowable catches in fisheries on an annual basis or establishing and adjusting phase-out schedules for ozone-depleting substances), and they often oversee programmatic activities (for example, carrying out remedial action plans aimed at alleviating the effects of pollution in lakes or marine systems). Often overlooked is the function of regimes in generating knowledge about the problems to be solved and contributing to a shared understanding of the issues at stake among participating actors (Breitmeier et al., 2006).

Environmental regimes are dynamic in the sense that they change continually after their initial formation. Once established, institutional arrangements do not remain static over time. Environmental regimes wax and wane in terms of their capacity to solve problems. Some take on roles or are brought to bear in efforts to address problems that were not on the agenda at the time of their creation. It is possible to identify a number of patterns that constitute common pathways of institutional development and that generally take the form of emergent properties rather than planned developments (Young, 2010). Some regimes (for example, the ozone regime) go from strength to strength. Others (for example, the Antarctic Treaty System) develop by fits and starts in a pattern of punctuated equilibrium. Still others (for example, the climate regime) run into roadblocks that give rise to a pattern of arrested development.

The success of environmental regimes is highly sensitive to contextual factors. Context matters as a determinant of the effectiveness of regimes. An arrangement that works perfectly well in one setting may fall flat in another setting. It is always important to think about scope conditions in assessing propositions about the effectiveness of environmental regimes. Some of the most notable features of the ozone regime, for instance, are unworkable in addressing the problem of climate change.

This explains the importance of the propositions that we must go 'beyond panaceas' in devising regimes to address real-world problems (Ostrom, 2007) and that it is essential to adopt a diagnostic approach in efforts to design regimes crafted to solve specific problems (Ostrom, 2007; Young, 2002a, 2008).

Specific findings

Beyond these general findings, regime analysis has generated a variety of more specific propositions about the effectiveness of environmental regimes. Some of these findings are negative, in the sense that they disconfirm popular notions about requirements for success. Others are positive, pointing to factors that are commonly associated with success.

Active participation on the part of a single dominant actor (commonly known in regime analysis as a hegemon) is not a necessary condition for success in solving international environmental problems. Dominant actors are important especially when they value a regime's products more than the cost of supplying them, making the relevant social system what is known as a privileged group (Olson, 1965). But the absence of an engaged hegemon does not spell failure in this realm. What does seem important is the existence of a coalition of influential actors prepared to take the lead in jump-starting a regime at the outset and to provide an extra push at critical junctures along the road to success (Schelling, 1978).

Success in the implementation of international regimes is likely to thrive on the establishment and maintenance of maximum winning coalitions rather than minimum winning coalitions. Regimes are public goods, though individual members of the group of subjects may value them differently. In the extreme, some may regard them as public bads. The fact that their products are non-rival makes it desirable to maximize the size of coalitions supporting regimes rather than to form minimum winning coalitions of the sort common in domestic legislative settings (Riker, 1962). Given the existence of the temptation to free ride, leading actors will have an incentive to make participation attractive to others rather than to minimize the number of those entitled to a share of the joint gains. This is particularly so where regimes require ongoing implementation on the part of individual members.

The maintenance of feelings of fairness and legitimacy is important to effectiveness, especially in cases where success requires active participation on the part of the members of the group over time. Neorealist perspectives suggest that both the formation and the implementation of regimes are about power – perhaps including soft power as well as hard power – all the way down (Mearsheimer, 1994/1995; Strange, 1982). The role of power is not only important in such settings; it is also a topic requiring more intensive analysis on the part of those interested in international regimes. But this does not eliminate the role of considerations of fairness and legitimacy (Coicaud and Warner, 2001; Franck, 1990). Given the underdeveloped character of enforcement procedures at the international level, it is hard to elicit compliance on an ongoing basis from actors that do not accept a regime's prohibitions and requirements as fair and legitimate.

Casting arrangements in the form of legally binding conventions or treaties does not ensure higher levels of compliance on the part of subjects. Many analysts assume that the 'normative pull' associated with legally binding arrangements will have a positive effect on compliance. But the available evidence does not support this proposition (Breitmeier et al., 2006). Although hard-law arrangements may be desirable for other reasons, there is often a price to be paid for pursuing such arrangements in terms of both the willingness of key actors to join regimes and the depth of the substantive provisions adopted.

Arrangements featuring private governance and hybrid systems encompassing both public and private elements can solve some types of environmental problems. It is easy to exaggerate growth in the role of non-state actors (for example, multinational corporations, environmental NGOs) at the international level as well as the emergence of global civil society (Kaldor, 2003). But while nation states remain core actors in international society, other actors are gaining ground. This opens up opportunities to solve problems through the development of hybrid systems (for example, the system for classifying and insuring ships) and even private regimes (for example, the certification regimes for sustainable wood products and fish) rather than limiting the roles of non-state actors to efforts to influence the operations of intergovernmental regimes (Delmas and Young, 2009).

Multiple pathways can lead to success in efforts to solve many environmental problems. It is generally a mistake to assume that there is one true path that must be identified and followed in efforts to solve specific environmental problems. Alternative solutions may vary in terms of other considerations, such as fairness or various notions of 'good' governance. But what systems theorists call equifinality is a common phenomenon in the realm of environmental governance. This proposition applies with particular force to the selection of policy instruments (for example, incentive systems vs. command-and-control regulations).

Findings about institutional interplay

Environmental regimes often interact both with one another and with regimes operating in other areas like trade and finance. The growth of interest in what is now known as institutional interplay is a recent development fueled by the observation that the number of distinct regimes operative in international society has grown rapidly in recent decades (Brown Weiss, 1993; Underal and Young, 2004). A simple point of departure in thinking about interplay, pioneered by the long-term project on the Institutional Dimensions of Global Environmental Change, features two primary distinctions: one between horizontal and vertical interactions and the other between functional (or unintended) and political interactions (Young et al., 1999/2005). Because much of the responsibility for implementing the provisions of international regimes falls to their individual members, it is essential in thinking about effectiveness to consider vertical interplay (often known as multi-level governance) as well as interactions among distinct institutional arrangements operative at the international level (Enderlein et al., 2010). Similarly, there is an important distinction between interplay that is largely unintended and often unforeseen and interplay involving intentional moves on the part of actors desiring either to manage interplay in order to promote problem solving or to exploit interplay in order to advance their individual interests (Young, 2002a). Others have moved on from this point of departure. Particularly important in this regard are Stokke's (2001, 2011, 2012) account of the mechanisms of cognition, obligation, and utility maximization as determinants of the effects of interplay on problem solving and Raustiala and Victor's (2004) concept of institutional complexes as loosely coupled sets of arrangements operating in a single issue area (Orsini et al., 2013). Although the study of institutional interplay is a central concern at the domestic level, it constitutes a relatively new area of research at the international level. Nevertheless, some findings are already emerging from analyses of such matters (Oberthür and Gehring, 2006; Oberthür and Stokke, 2011).

Institutional interplay is just as likely to produce positive or even synergistic results as it is to lead to interference between or among regimes. This stream of analysis arose from a concern that tensions or even open conflict between or among distinct regimes would become an increasingly prominent feature of the institutional landscape in international society (Brown, 1993). The logic underlying this concern is simple. As the number and variety of regimes operating in a given social space grow, the overlaps between and among them will increase. Since these overlaps are often unintended and commonly unforeseen in nature, it seems reasonable to expect that tensions will ensue (Lovejoy and Hannah, 2005). But the research done so far on institutional interplay fails to confirm this expectation. Interactions may generate tensions. But institutional interplay often produces positive results and may even prove synergistic, as in the case of the regulation of substances under the ozone regime that are also greenhouse gases (Oberthür and Gehring, 2006; Oberthür, 2009).

There is generally scope for resolving actual or potential conflicts between regimes through negotiations leading to mutual accommodation rather than by subordinating one regime to the other. For the most part, resolving such conflicts is not a matter of applying legal doctrines involving criteria like specificity and temporal sequencing to determine which regime should take precedence in the event of conflict between distinct arrangements. Rather, it is a matter of negotiating workable compromises that allow the regimes in question to operate effectively without undue interference in each other's domains (Oberthür and Stokke, 2011). The most striking examples involve interplay between the global trade regime and a variety of multilateral environmental agreements involving the use of trade restrictions as a policy instrument (for example, the regimes dealing with endangered species, hazardous wastes, the protection of the stratospheric ozone layer, and climate change) (Jinnah, 2010; Faude and Gehring, unpublished). The central challenge is to work out a modus vivendi allowing individual regimes to make progress toward solving the problems motivating their creation.

Regime complexes offer a way forward in situations that do not lend themselves to the creation of a single integrated governance system. Many issue areas (for example, climate, biodiversity, marine pollution) feature networks of distinct regimes or 'loosely coupled set[s] of specific regimes' (Keohane and Victor, 2011: 7) that grow up over time in the absence of an overall blueprint. Such complexes may range along a continuum from comprehensive and integrated governance systems for entire issue areas to total fragmentation (Keohane and Victor, 2011). Regime complexes offer the advantage of being more flexible across issues and adaptable over time than more tightly coupled governance systems. They may be easier to create than fully integrated systems and more resilient in the face of stresses occurring at the international or global level today. On the other hand, some observers have noted the prominence of fragmentation in such settings; they take a less positive view of the phenomenon of regime complexes (Zelli and van Asselt, 2013).

What are the cutting-edge issues regarding regime effectiveness?

Taken together, these findings derived from hundreds of individual studies have significant implications for policy. It is worth bearing in mind, for instance, that not all regimes are regulatory in character, that legally binding arrangements are not always preferable to softer arrangements, that institutional interaction sometimes produces synergistic results, and that regime complexes may prove more successful than fully integrated regimes. Above all, it is essential to understand the problem of fit and, as a result, to discard hopes for panaceas and to sharpen the skills needed to engage in institutional diagnostics (Young, 2013).

At the same time, there is much more we can learn about the effectiveness of environmental regimes that will be of interest to policymakers. This section identifies a series of topics that constitute cutting-edge concerns in this field (Biermann et al., 2009; Young et al., 2008). There is no need to forge a consensus regarding the precise content of the research agenda. But it is helpful to get a sense of where we are headed regarding research on the effectiveness of environmental regimes.

Deep structure

Environmental regimes are specialized arrangements embedded in and reflecting the deep structure of international society (Underal, 2008). To be effective such arrangements must be broadly compatible with the essential features of the prevailing deep structure (Conca, 2006). There is little point, for instance, in creating a regime complex for climate that requires the imposition of fundamental restrictions on the sovereignty of member states or the creation of enforcement mechanisms that rely on severe sanctions to elicit compliance with the regime's rules. Yet it is easy to carry this line of thinking too far. The deep structure of international society is not static (Buzan, 2004). The familiar power structure of the postwar era is shifting dramatically. We need to recognize the growing importance of non-state actors and the emergence of global civil society and to think about the implications of these developments for perspectives built on the assumption that environmental governance is largely a matter of intergovernmental relations (Betsill et al., 2007; Biermann and Pattberg, 2012; Wapner, 1997). So long as the normative gap is not too great, the development of innovative regimes can play a role in driving the evolution of the deep structure of international society. The hopes of those who foresee the rise of a liberal environmental order rest on this line of thinking (Bernstein and Pauly, 2007). What we need to know in this realm is more about the constraints and opportunities associated with deep structure as they pertain to the operation of governance systems for specific problems like climate change and the protection of biodiversity.

Problem structure

It is intuitively appealing to adopt the view that some environmental problems are harder to solve than others or, to use the terminology of the Miles et al.'s (2002) team, that we can locate specific problems on a benign-malign spectrum. Climate change is certainly a more challenging problem than the depletion of the stratospheric ozone layer. But what exactly are the factors that make environmental problems harder or easier to solve, and can we devise a metric for assessing problems in these terms (Young, 1999a)? Underdal (2010) argues that environmental problems are hard to solve to the extent that they (i) are 'long-term policy problems with time lags between policy measures...and effects,' (ii) 'are embedded in very complex systems' clouded by uncertainties, and (iii) 'involve global collective goods' not subject to single best effort solutions. These factors do pose important challenges for those seeking to solve environmental problems. They go some way, for example, toward explaining why it is so hard to come to grips with the problem of climate change. Yet there is considerable evidence to suggest that solutions to seemingly easy or benign problems can prove elusive and that groups sometimes succeed in banding together to make serious efforts to tackle seemingly hard or malign problems. What we need to know here is whether we are doomed to suffer the consequences of hard problems like climate change or we can come up with innovative strategies to address such problems, given the emergence of effective leadership and the will to collaborate on the part of key actors.

Power

The role of power as a determinant of regime effectiveness is complex and contested, especially if we construe power to encompass soft power as well as hard power, cognitive power as well as structural power, and issue-specific power as well as general power (Nye, 2011). Critics of regime analysis have often dismissed institutions as epiphenomena that reflect underlying distributions of power and that change as these distributions shift (Strange, 1982). Those studying regimes sometimes seem to ignore or at least to marginalize the role of power as a determinant of the capacity of these arrangements to solve problems. How can we come to terms with these diverging perspectives on the role of power?

Regimes are embedded in overarching political orders, and they reflect the general principles of political discourses dominant at the time of their creation. But this does not mean that they are of no significance in their own right, especially when treated as intervening forces that form links between the underlying drivers of human behavior and the outcomes flowing from human-environment interactions (Krasner, 1983; Young, 2002b). What we need to know here is how to think about the role of power as a driving force in world affairs that does not blind us to the significance of other forces.

Breadth vs. depth

Because participation in international environmental regimes is voluntary, there is a tendency to settle for arrangements that are shallow in terms of substance in order to make them palatable to all relevant actors. This is what Underdal (2002) and others have described as the law of the least ambitious program. The logic of those who advocate going forward even when commitments are shallow is that it is important to get the ball rolling and that institutional evolution will lead to a deepening of commitments over time. This is an intuitively appealing argument, and examples like the regime for the protection of stratospheric ozone and (to a lesser extent) the regime dealing with long-range transboundary air pollution in Europe suggest that such a dynamic occurs under some conditions. But there is no reason to assume that such evolutionary processes will occur in all cases (Barrett, 2003). The contrast between the regime for stratospheric ozone and the climate regime in these terms is striking. Not only has the stratospheric ozone regime proved effective in reducing drastically the production and consumption of ozone depleting substances, it has also proven more effective in reducing emissions of greenhouse gases than the climate regime itself (Velders et al., 2007). What we need to know in this case centers on scope conditions. Under what conditions is it realistic to expect institutional evolution to work as a mechanism for deepening commitments in a manner required to ensure success in efforts to solve problems?

Compliance

We already know a lot about the sources of compliance (Mitchell, 1994; Raustiala and Slaughter, 2002; Young, 1979). Because many observers regard the lack of compliance mechanisms as the Achilles heel of international governance, however, this subject remains on the list of research priorities. The absence of a government in the ordinary sense in international society makes it hard to use sanctions – graduated or otherwise – effectively as a means of persuading or compelling those subject to a regime's rules to comply with their obligations. But this is not a fatal flaw (Young, 1999a). In single best effort situations where one or a few actors can solve the problem once and for all, compliance is not a critical issue (Barrett, 2007). Compliance is not a concern with regard to regimes that are not fundamentally regulatory in character. Even in regulatory settings, a management approach is sometimes more effective than an enforcement approach as a means of maximizing compliant behavior on the part of a regime's subjects (Chayes and Chayes, 1995). Other factors, such as the extent to which subjects have engaged actively in the process of regime creation and the extent to which they feel that a regime constitutes a fair deal, can make a big difference in inducing actors to comply with a regime's rules and regulatory measures. What we need to know here is more about the sources of compliance (Victor et al., 1998). Because the emphasis at the international level must be on governance without government for the foreseeable future, we have a particular need to deepen our understanding of mechanisms that can produce compliant behavior in the absence of the sorts of sanctions we generally associate with the idea of enforcement (Skjaerseth et al., 2006).

Fairness and legitimacy

Despite the finding reported in the preceding section, there is substantial variation in the views that analysts have expressed regarding the roles of fairness and legitimacy as determinants of the effectiveness of international regimes (Young). Those who follow the logic of consequences (March and Olsen, 1998) and frame issues in collective-action terms have a tendency to dismiss or downplay the role of fairness, equity, and other normative concerns in thinking about the success or failure of environmental regimes (Victor, 2011). Those who think in terms of the logic of appropriateness (March and Olsen, 1998) and approach issues in social-practice terms, by contrast, are more receptive to the idea that such considerations are important determinants of effectiveness (Franck, 1990, 1995). This divergence is not peculiar to the analysis of governance systems or regimes. It mirrors a larger and ongoing debate about the role of normative considerations as driving forces in global society. We do not need to adopt one or the other of these views in analyzing the effectiveness of international environmental regimes. But we do need an understanding of the conditions under which fairness and legitimacy are significant forces in this realm (Bernstein and Pauly, 2007; Bernstein, 2005). This knowledge will have important implications for those designing regimes to address specific environmental problems, such as climate change or the loss of biological diversity.

Policy instruments

The ideas of those who espouse incentive mechanisms, such as tradable catch shares or carbon taxes, in contrast to the more traditional mechanisms we generally lump under the heading of command-and-control regulations have dominated the discussion of policy instruments for several decades (Cole, 2002). Clearly, the emphasis on such mechanisms has been salutary. Incentive mechanisms can alleviate the dynamic giving rise to the tragedy of the commons; they also can give subjects reason to focus on innovation on an ongoing basis. Yet it would be unfortunate if this were to lead to a situation in which one set of tools dominates our thinking about governance to the exclusion of others. There are important cases (for example, climate change) in which it is difficult to make calculations regarding both the costs of leaving the problem unattended and the costs of taking effective action to alleviate the problem. There are also cases in which we have good reasons to override the use of discount rates of the sort commonly considered in conjunction with incentive mechanisms. It may make good sense in such cases to use command-and-control measures in place of or as a supplement to incentive mechanisms. Maintaining a well-stocked toolkit is clearly a good idea (Zaelke et al., 2005). What we need to know in this regard is more about the conditions under which specific policy instruments are likely to prove effective and how to make use of diagnostic procedures to bring this knowledge to bear on specific cases (Young, 2008).

Interplay management

Institutional interplay is on the rise. Whatever the attractions of creating comprehensive and integrated governance systems to address problems like climate change and the loss of biological diversity, we must prepare for a world that features rising levels of interplay between and among distinct regimes. As Keohane and Victor (2011) observe in their analysis of efforts to address climate change, regime complexes or loosely coupled sets of specific regimes or regime elements dealing with broad and complex issues like climate change may well prove advantageous in terms of flexibility across issues and adaptability over time. But those who emphasis the consequences of fragmentation have a point as well (Zelli and van Asselt, 2013). The implication of these observations is that we must shift our attention from intensive studies of individual regimes to more expansive accounts of institutional interactions and especially regime complexes. Long familiar in domestic systems, this is a perspective that is relatively new at the level of international society. What we need to know here is more about the conditions leading to synergy rather than interference in institutional interactions and the conditions under which regime complexes produce flexibility and adaptability rather than chaos and confusion (Oberthür and Stokke, 2011).

Non-linear processes

As we move deeper into a world of human-dominated ecosystems and the new planetary era now widely characterized as the Anthropocene, the need to improve our understanding of thresholds and tipping points together with the events triggering non-linear changes has become urgent (Carpenter et al., 2009; Lenton et al., 2008). Non-linear changes are typically irreversible, sometimes abrupt, and often nasty from the perspective of human welfare. This makes it important not only to devise procedures to provide early warning regarding the onset of such changes but also to devise governance systems able to adjust nimbly to the impacts of these changes. The trick is to create governance systems that are resilient in the sense that they have the staying power to be effective combined with the adaptability to adjust quickly to changing circumstances. A regime that changes too readily in the face of modest stresses will not be effective. But a regime that is too rigid in the sense that it is unresponsive to major changes in the socio-ecological environment will be vulnerable to forces leading to institutional collapse in a world in which non-linear changes are common. What we need is a major step forward in our understanding of how to structure governance systems to maximize resilience while, at the same time, including procedures allowing for timely adjustments of the sort needed to maintain a good fit between socio-ecological conditions and institutional arrangements (Boyd and Folke, 2011; Galaz et al., 2010; Young, 2012).

Scale

Scale in this context is a matter of the generalizability of findings regarding the effectiveness of governance systems across levels of social organization (Bernauer and Siegfried, 2008). To what extent do findings about issues like avoiding the tragedy of the commons derived from analyses of smallscale or local cases apply to comparable issues at the international level and vice versa (Cash et al., 2006; Dietz et al., 2003; Keohane and Ostrom, 1995)? There are clear parallels between smallscale systems and largescale or global systems with regard to the need to produce governance without government (Young, 2005). But there are also differences between these settings. Although many analysts use the term 'international community' in discussing global issues, for example, there are major differences between local and global systems with regard to what is meant by the idea of community. What we need to know here is more about the limits of generalizability across levels of social organization regarding factors that determine the effectiveness of governance systems.

What are the most promising research strategies in this field?

What tools are available to those desiring to tackle these themes? How can we use these tools to greatest effect to deepen our understanding of the determinants of regime effectiveness and to generate conclusions that will prove helpful to those responsible for designing and administering these arrangements? The proper response to these questions is to think in terms of a methodological portfolio or toolkit containing a collection of distinct but complementary modes of analysis and to urge those seeking to understand the effectiveness of regimes to employ multiple methods whenever possible (Delmas and Young, 2009; Underal and Young, 2004; Young et al., 2006). Taking this proposition as a point of departure, several specific observations about strategies for advancing understanding of the effectiveness of international environmental regimes come into focus.

As in other fields of study, finding ways to combine quantitative and qualitative methods is a priority in studies of effectiveness. Quantitative procedures produce measures of association but are limited in terms of their capacity to reveal the causal mechanisms underlying the relationships identified. Theoretical case studies, by contrast, can probe the causal forces at work in specific situations but do not produce results that are easy to generalize (Young, 1999b). The central flow of research on environmental regimes consists of studies that provide in-depth analyses of individual regimes or of a handful of regimes examined from a common perspective. There is every reason to continue to nurture this flow of research (Andresen and Wettestad, 2004; Andresen et al., 2012). But a priority is to build up the stock of large-N quantitative studies to facilitate triangulation in efforts to enhance our understanding of the determinants of effectiveness (Mitchell, 2004). Research of this type has already yielded insights regarding the importance of knowledge production as a source of regime effectiveness, the role of 'pushers' versus 'laggards' in making regimes effective, and the occurrence of synergy in contrast to interference in situations involving institutional interplay (Breitmeier et al., 2011; Oberthür and Gehring, 2006). Small universes of cases impose limits on what is possible in this realm. But they do not rule out progress in applying quantitative methods to develop empirical generalizations about effectiveness (Poteete et al., 2010).

Another priority is to devise methods that can shed light on the role of complex causation as a determinant of effectiveness. Complex causation occurs when clusters of causal forces interact with one another in ways that make it difficult to pull them apart through the use of normal statistical procedures (Young, 2002b; Young et al., 2006). What is needed to illuminate situations of this kind are methods that direct attention to (i) conjunctural causation (for example, Ragin's qualitative comparative analysis or QCA) (Regin, 1987, 2000; Stokke, 2004), (ii) emergent properties of complex systems (for example, simulations employing agent-based models) (Poteete et al., 2010), and (iii) recurrent relationships that become apparent in examining large numbers of comparable case studies (for example, meta-analysis) (Rudel, 2005). Whereas reductionist methods are especially useful in separating out the effects of individual variables and assigning weights to them as distinct factors in accounting for outcomes of interest, methods focusing on causal clusters concentrate on identifying combinations of interacting forces that together constitute necessary or, more often, sufficient conditions to produce results like the success of environmental regimes in solving problems (Young et al., 2006).[5] We need to pursue both pathways in examining the sources of effectiveness.

Formal modeling is another important tool in research on regime effectiveness. Modeling of this sort is based on a strategy of abstracting away many factors to highlight the core logic of social relationships in contrast to developing directly testable hypotheses. A particularly productive effort of this sort focuses on understanding dilemmas of collective action, like the tragedy of the commons and free ridership in the supply of public goods, as outcomes of interdependent decision-making in which participants select strategies that seem rational in individualistic terms but that lead to socially undesirable outcomes (Barrett, 2007; Schelling, 1978). The facts that the dynamic of the tragedy of the commons can be represented in terms of the game-theoretic construct known as prisoner's dilemma and that the choice generally labeled 'defect' constitutes what is known as a dominant strategy for each participant do not allow us to predict that actors relying on common property arrangements to manage the use of common pool resources are bound to come to grief. Taking this concern as a point of departure, analyses of real-world cases make it clear that the tragedy does not always occur (Ostrom et al., 2002). But using the prisoner's dilemma to model such interactions has sharpened our understanding of the issues involved. Similar remarks are in order regarding the role of privileged groups, coalitions of pushers, and burden-sharing arrangements in situations involving the supply of things like clean air that can be construed as public goods (Olson, 1965; Schelling, 1978).

Research carried out by individual scholars will always play an important role in this field. But research teams and networks are becoming more important as we endeavor to use this toolkit to increase understanding of the effectiveness of international environmental regimes. The work of Miles et al. (2002), Breitmeier et al. (2006), and Breitmeier et al. (2011) resulted from the efforts of integrated teams. More recently, researchers have focused on the development of larger and looser networks of individuals working independently but adhering to a common science plan. The project on the Institutional Dimensions of Global Environment Change (1998–2007) and the Earth System Governance project (2009–) (Biermann et al., 2009), both core projects of the International Human Dimensions Programme on Global Environmental Change, exemplify this strategy.

Moving toward Pasteur's quadrant

Research on the determinants of effectiveness in international environmental regimes constitutes a young field. But it has already generated results of interest both to practitioners charged with administering regimes dealing with specific problems and to analysts seeking to understand the nature of governance, particularly in social settings where there is no government in the ordinary sense of the term. My own experience has convinced me that many rewards flow from a strategy of working back and forth between the worlds of analysis and praxis. It would be naïve to suppose that this line of research can reveal simple solutions to the great issues of our times, like controlling climate change and preventing the loss of biological diversity. But it would be equally inappropriate to dismiss the role of environmental regimes because they do not provide us with simple solutions to such overarching concerns. The way forward in efforts to enhance our understanding of the determinants of effectiveness is to make use of a suite of complementary modes of analysis. When the results converge, our confidence in the relevant findings rises. When they diverge, we are presented with puzzles of the sort on which science thrives. With persistence and a certain amount of good fortune, we will succeed in producing results that are of interest to analysts and practitioners alike and that, as a result, land us squarely in the domain of Pasteur's Quadrant (Clark, 2007; Stokes, 1997).

Acknowledgments

I thank Bill Clark, Ron Mitchell, Matthew Stilwell, Arild Underdal, Durwood Zaelke, and three anonymous reviewers for helpful comments on earlier drafts of this chapter.

Notes

1. A more radical view asserts that the existing approach to environmental governance, which accepts the dominant role of the state, emphasizes distinct initiatives at different levels of social organization, and focuses on intergovernmental agreements at the international level, is fundamentally flawed and bound to fail (Park, et al., 2008).
2. More generally, the premise that anarchy leads to collective-action problems is the point of departure for neoliberal approaches to international environmental politics (Paterson, this book).
3. There are dissenters even in these cases. Some see evidence that countries have reduced emissions of chlorofluorocarbons voluntarily and point to the fact that healing the ozone layer will take decades (Murdoch, et al., 1997). Others emphasize the role of non-regime factors in the effort to clean up the Rhine and comment on the slow pace of negotiations in this case (Bernauer, 1996; Bernauer and Moser, 1996).
4. In both studies, 'cases' are defined in such a way that they include regime elements or components rather than discrete regimes. The climate regime, for example, encompasses three components in the Breitmeier, Young, and Zürn study (2006).
5. The recent work of Stokke on the fisheries of the Barents Sea demonstrates the potential of QCA in addressing such issues (Stokke, 2012). New developments in QCA make it possible to apply this method to situations involving up to 100–200 cases.

Annotated list of readings

Andresen, S., E. L. Boasson and G. Hønneland (eds) (2012) *International Environmental Agreements: An Introduction* (New York: Routledge).

An overview of qualitative studies of international environmental regimes.

Breitmeier, H., A. Underdal and O. R. Young (2011) 'The Effectiveness of International Environmental Regimes: Comparing Findings from Quantitative Research', *International Studies Review,* 13, 1–27.

A prominent example of quantitative research on international environmental regimes.

Mitchell, R. B. (2008) 'Evaluating the Performance of Environmental Institutions: What to Evaluate and How to Evaluate It?' in O. R. Young, L. A. King and H. Schroeder (eds) *Institutions and Environmental Change: Principal Findings, Applications, and Research Frontiers* (Cambridge: MIT Press), 79–114.

A sophisticated overview of ways to think about evaluating effectiveness.

Underdal, A. (2008) 'Determining the Causal Significance of Institutions: Accomplishments and Challenges' in O. R. Young, L. A. King and H. Schroeder (eds) *Institutions and Environmental Change: Principal Findings, Applications, and Research Frontiers* (Cambridge: MIT Press), 49–78.

A careful assessment of the ways in which regimes matter.

Young, O. R. (2013) *On Environmental Governance: Sustainability, Efficiency, and Equity* (Boulder: Paradigm Publishers).

An account stressing the impact of the 'Great Acceleration' and the challenges of developing effective regimes in the Anthropocene.

Works cited

Andresen, S. and J. Wettestad (2004) 'Case Studies of the Effectiveness of International Environmental Regimes' in *Regime Consequences*, ed. Underdal, A. and O. R. Young (Dordrecht: Kluwer Academic Publishers), 27–48.

Andresen, S., E. L. Boassum and G. Hønneland (eds) (2012) *International Environmental Agreements: An Introduction* (New York: Routledge).

Barrett, S. (2003) *Environment and Statecraft: The Strategy of Environmental Treaty Making* (New York: Oxford University Press).

Barrett, S. (2007) *Why Cooperate? The Incentive to Supply Collective Goods* (Oxford: Oxford University Press).

Bernauer, T. (1996) 'Protecting the Rhine River against Chloride Pollution' in *Institutions for Environmental Aid,* ed. Keohane, R. O. and M. L. Levy (Cambridge: MIT Press), 201–232.

Bernauer, T. and P. Moser (1996) 'Reducing Pollution of the River Rhine: The Influence of International Cooperation' *Journal of Environment and Development* 5, 391–417.

Bernauer, T. and T. Siegfried (2008) 'Compliance and Performance in International Water Agreements: The Case of the Naryn/Syr Darya Basin' *Global Governance* 14, 479–501.

Bernstein, S. (2005) 'Legitimacy in Global Environmental Governance' *Journal of International Law and International Relations* 1, 139–166.

Bernstein, S. and L. W. Pauly (eds) (2007) *Global Liberalism and Political Order: Towards a New Grand Compromise* (Albany: State University of New York Press).

Betsill, M. and E. Corell (eds) (2007) *NGO Diplomacy: The Influence of Nongovernmental Organizations in International Environmental Negotiations* (Cambridge: MIT Press).

Biermann, F. and P. Pattberg (eds) (2012) *Global Environmental Governance Reconsidered* (Cambridge: MIT Press).

Biermann, F., M. M. Betsill, J. Gupta, N. Kanie, L. Lebel, D. Liverman, H. Schroeder and B. Siebenhüner with contributions from K. Conca, L. da Costa Ferreira,

B. Desai, S. Tay and R. Zondervan (2009) *Earth System Governance: People, Places, and the Planet*, IHDP Report No. 20 (Bonn: IHDP).

Boyd, E. and C. Folke (2011) *Adapting Institutions: Governance, Complexity, and Social-Ecological Systems* (Cambridge: Cambridge University Press).

Breitmeier, H., A. Underdal and O. R. Young (2011) 'The Effectiveness of International Environmental Regimes: Comparing and Contrasting Findings from Quantitative Research' *International Studies Review* 13, 579–605.

Breitmeier, H., O. R. Young and M. Zürn (2006) *Analyzing International Regimes: From Case Study to Database* (Cambridge: MIT Press).

Brown Weiss, E. (1993) 'International Environmental Issues and the Ermergence of a New World Order' *Georgetown Law Journal* 81, 675–710

Buzan, B (2004) *From International to World Society* (Cambridge: Cambridge University Press).

Carpenter, S. R., H. A. Mooney, J. Agard, D. Capistrano, R. S. DeFries, S. Díaz, T. Dietz, A. K. Duraiappah, A. Oteng-Yeboah, H. M. Pereira, C. Perrings, W. V. Reid, J. Sarukhan, R. J. Scholes and A. Whyte (2009) 'Science for Managing Ecosystem Services: Beyond the Millennium Ecosystem Assessment' *Proceedings of the National Academy of Sciences USA* 106, 1305–1312.

Cash, D. W. et al. (2006) 'Scale and Cross-Scale Dynamics: Governance and Information in a Multilevel World' *Ecology and Society* 11, 181–192.

Chayes, A. and A. H. Chayes (1995) *The New Sovereignty: Compliance with International Regulatory Agreements* (Cambridge: Harvard University Press).

Clark, W. C. (2007) 'Sustainability Science: A Room of its Own' *Proceedings of the National Academy of Sciences USA*, 104, 1737–1738.

Coicaud, J. M. and D. Warner (eds) (2001) *Ethics and International Affairs: Extent and Limits* (Tokyo: UNU Press).

Cole, D. H. (2002) *Pollution and Property: Comparing Ownership Institutions for Environmental Protection* (Cambridge: Cambrdige University Press).

Collier, P. (2008) *The Bottom Billion: Why the Poorest Countries are Failing and What can be Done about it* (Oxford: Oxford University Press).

Conca, K. (2006) *Governing Water: Contentious Transnational Politics and Global Institution Building* (Cambridge: MIT).

Dai, X. (2007) *International Institutions and National Policies* (Cambridge: Cambridge University Press).

Delmas, M. and O. R. Young (eds) (2009) *Governance for the Environment: New Perspectives* (Cambridge: Cambridge University Press).

Dietz, T., E. Ostrom and P. C. Stern (2003) 'The Struggle to Govern the Commons' *Science* 302, 1907–1912.

Enderlein, H., S. Wälti and M. Zürn (eds) (2010) *Handbook on Multi-Level Governance* (Cheltenham: Edward Elgar).

Faude, B. and T. Gehring (unpublished) 'Institutional Ecology and the Evolution of Broader Structures in Regime Complexes: The Trade and Environment Overlap' (available from thomas.gehring@sowi.uni-bamberg.de).

Finus, M. and S. Tjøtta (2003) 'The Oslo Protocol on Sulfur Reducation: The Great Leap Forward?' *Journal of Public Economics* 87, 2031–2048.

Franck, T. M. (1990) *The Power of Legitimacy among Nations* (New York: Oxford University Press).

Franck, T. M. (1995) *Fairness in International Law and Institutions* (New York: Oxford University Press).

Galaz, V., A. Duit, K. Eckerbereg and J. Ebbesson (eds) (2010) Governance, Complexitry, and Resilience. A special issue of *Global Environmental Change,* 20 (3).

Galaz, V., P. Olsson, T. Hahn, C. Folke and U. Svedin (2008) 'The Problem of Fit among Biophysical Systems, Environmental and Resource Regimes, and Broader Governance Systems: Insights and Emerging Challenges' in *Institutions and Environmental Change,* ed. Young, O. R., L. King and H. Schroeder (Cambridge: MIT Press), 147–186.

Haas, P. M., R. O. Keohane and M. L. Levy (eds) (1993) *Institutions for the Earth: Sources of Effective International Environmental Protection* (Cambridge: MIT Press).

Helm, C. and D. Sprinz (2000) 'Measuring the Effectiveness of International Environmental Regimes' *Journal of Conflict Resolution* 44, 639–652.

Hovi, J., D. Sprinz and A. Underdal (2003a) 'The Oslo-Potsdam Solution to Measuring Regime Effectiveness: Critique, Response, and the Road Ahead' *Global Environmental Politics* 3, 74–96.

Hovi, J., D. Sprinz and A. Underdal (2003b) 'Regime Effectiveness and the Oslo-Potsdam Solution: A Rejoinder to Oran Young' *Global Environmental Politics* 3, 105–107.

Jinnah, S. (2010) 'Overlap Management in the World Trade Organization: Secretariat Influence on Trade-Environment Politics' *Global Environmental Politics* 10, 54–79.

Kaldor, M. (2003) *Global Civil Society: An Answer to War* (Cambridge: Polity Press).

Keohane, R. O. and D. G. Victor (2011) 'The Regime Complex for Climate Change' *Perspectives on Politics* 9, 7–23.

Keohane, R. O. and E. Ostrom (eds) (1995) *Local Commons and Global Interdependence* (London: Sage Publications).

Krasner, S. D. (1983) 'Structural Causes and Regime Consequences: Regimes as Intervening Variables' in *International Regimes,* ed. Krasner, S. D. (Ithaca: Cornell University Press), 1–21.

Lenton, T. M., H. Held, E. Kriegler, J. W. Hall, W. Lucht, S. Rahmstorf and H. J. Schellnhuber (2008) 'Tipping Elements in the Earth's Climate System' *Proceedings of the National Academy of Sciences USA,* 105, 1786–1793.

Lovejoy, T. and L. Hannah (eds) (2005) *Climate Change and Biodiversity* (New Haven: Yale University Press).

March, J. G. and J. P. Olsen (1998) 'The Institutional Dynamics of International Political Orders' *International Organization* 52, 943–969.

Mearsheimer, J. J. (1994/1995) 'The False Promise of International Institutions' *International Security* 19, 5–49.

Miles, E., S. Andresen, E.M Carlin, J. B. Skjaerseth, A. Underdal and J. Wettestad (2002) *Environmental Regime Effectiveness: Confronting Theory with Evidence* (Cambridge: MIT Press).

Mitchell, R. (2008) 'Evaluating the Performance of Environmental Institutions: What to Evaluate and How to Evaluate It?' in *Institutions and Environmental Change,* ed. Young, O. R., L. King and H. Schroeder (Cambridge: MIT Press), 79–114.

Mitchell, R. B. (1994) *International Oil Pollution at Sea: Environmental Policy and Treaty Compliance* (Cambridge: MIT Press).

Mitchell, R. B. (2004) 'A Quantitative Approach to Evaluating International Environmental Regimes' in *Regime Consequences,* ed. Underdal, A. and O. R. Young (Dordrecht: Kluwer Academic Publishers), 121–149.

Murdoch, J. C. and T. Sandler (1997) 'The Voluntary Provision of a Public Good: The Case of Reduced CFC Emissions and the Montreal Protocol' *Journal of Public Economics* 63, 331–349.

Murdoch, J. S., T. Sandler and K. Sargent (1997) 'A Tale of Two Collectives: Sulphur versus Nitrogen Oxide Emissions Reduction in Europe' *Economica* 64, 281–301.

Nye, J. S. (2011) *The Future of Power* (New York: Public Affairs).

Oberthür, S. (2009) 'Interplay Management: Enhancing Environmental Policy Integration among International Institutions' *International Environmental Agreements* 9, 371–391.

Oberthür, S. and O. S. Stokke (eds) (2011) *Institutional Interaction and Global Environmental Change* (Cambridge: MIT Press).

Oberthür, S. and T. Gehring (eds) (2006) *Institutional Interaction in Global Environmental Governance: Synergy and Conflict among International and EU Policies* (Cambridge: MIT).

Olson, M. (1965) *The Logic of Collective Action* (Cambridge: Harvard University Press).

Orsini, A., J-F. Morin and O. R. Young (2013) 'Regime Complexes: A Buzz, A Boom or a Boost for Global Governance' *Global Governance* 19, 27–39.

Ostrom, E. (2007) 'A Diagnostic Approach for Going beyond Panaceas' *Proceedings of the National Academy of Sciences USA* 104, 15181–15187.

Ostrom, E., T. Dietz, N. Dolsak, P. C. Stern, S. Stonich and E. U. Weber (eds) (2002) *The Drama of the Commons* (Washington, D.C.: National Academy Press).

Park, J., K. Conca and M. Finger (2008) *The Crisis of Global Environmental Politics: Towards a New Political Economy of Sustainability* (London: Routledge).

Parson, E. A. (2003) *Protecting the Ozone Layer: Science and Strategy* (New York: Oxford University Press).

Poteete, A. R., M. A. Janssen and E. Ostrom (2010) *Working Together: Collective Action, the Commons, and Multiple Methods in Practice* (Princeton: Princeton University Press).

Ragin, C. C. (1987) *The Comparative Method* (Berkeley: University of California Press).

Ragin, C. C. (2000) *Fuzzy-Set Social Science* (Chicago: University of Chicago Press).

Raustiala, K. and A. Slaughter (2002) 'International Law, International Relations and Compliance' in *Handbook of International Relations,* ed. Carlsnaes, W., T. Risse and B. A. Simmons (London: Sage Publications), 538–558.

Raustiala, K. and D. G. Victor (2004) 'The Regime Complex for Plant Genetic Resources' *International Organization* 55, 277–309.

Riker, W. (1962) *The Theory of Political Coalitions* (New Haven: Yale University Press).

Ringquist, E. J. and T. Kostadinova (2005) 'Assessing the Effectiveness of International Environmental Agreements: The Case of the 1985 Helsinki Protocol' *American Journal of Political Science* 49, 86–102.

Rudel, T. K. (2005) *Tropical Forests: Regional Paths of Destruction and Regeneration in the Late 20th Century* (New York: W. W. Norton).

Schelling, T. C. (1978) *Micromotives and Macrobehavior* (New York: W. W. Norton).
Skjaerseth, J. B. (2000) *North Sea Cooperation: Linking International and Domestic Pollution Control* (Manchester: Manchester University Press).
Skjaerseth, J. B., O. S. Stokke and J. Wettestad (2006) 'Soft Law, Hard Law, and Effective Implementation of International Environmental Norms' *Global Environmental Politics* 6, 104–120.
Speth, J. G. (2004) *Red Sky at Morning: America and the Crisis of Global Government* (New York: Oxford University Press).
Stokes, D. (1997) *Pasteur's Quadrant: Basic Science and Technological Innovations* (Washington, DC: Brookings Institution).
Stokke, O. S. (2001) *The Interplay of International Regimes: Putting Effectiveness Theory to Work, Report No. 14* (Lysaker, Norway: Nansen Institute).
Stokke, O. S. (2004) 'Boolean Analysis, Mechanisms, and the Study of Regime Effectiveness' in *Regime Consequences*, ed. Underdal, A. and O. R. Young (Dordrecht: Kluwer Academic Publishers), 87–120.
Stokke, O. S. (2011) 'Interplay Management: Niche Selection, and Arctic Environmental Governance' in *Institutional Interaction and Global Environmental Change*, ed. Oberthür, S. and O. S. Stokke (Cambridge: MIT Press), 143–170.
Stokke, O. S. (2012*) Disaggregating International Regimes: A New Approach to Evaluation and Comparison* (Cambridge: MIT Press).
Strange, S. (1982) '*Cave! hic dragones:* A Critique of Regime Analysis' *International Organization* 36, 479–497.
Underdal, A. (2002) 'One Question: Two Answers' in *International Regime Effectiveness*, ed. Miles, E. L., S. Andresen, E. M. Carlin, J. B. Skjaerseth, A. Underdal and J. Wettestad (Cambridge: MIT Press), 3–45.
Underdal, A. (2008) 'Determining the Causal Significance of Institutions: Accomplishments and Challenges' in *Institutions and Environmental Change*, ed. Young, O. R., L. King and H. Schroeder (Cambridge: MIT Press), 49–78.
Underdal, A. (2010) 'Complexity and Challenges of Long-term Environmental Governance' *Global Environmental Change* 20, 386–393.
Underdal, A. and O. R. Young (eds) (2004) *Regime Consequences: Methodological Challenges and Research Strategies* (Dordrecht: Kluwer Academic Publishers).
Velders, G. J. M., S. O. Andersen, J. S. Daniel, D. W. Fahey and M. McFarland (2007) 'The Importance of the Montreal Protocol in Protecting Climate' *Proceedings of the National Academy of Sciences USA* 104, 4814–4819.
Victor, D. G. (2011) *Global Warming Gridlock: Creating More Effective Strategies for Protecting the Planet* (Cambridge: Cambridge University Press).
Victor, D. G., K. Raustiala and E. B. Skolnikoff (eds) (1998) *The Implementation and Effectiveness of International Environmental Commitments* (Cambridge: MIT Press).
Wapner, P. (1997) 'Governance in Global Civil Society' in *Global Governance*, ed. Young, O. R. (Cambridge: MIT Press), 65–84.
Wettestad, J. (2002) *Clearing the Air – European Advances in Tackling Acid Rain and Atmospheric Pollution* (Aldershot: Ashgate).
Young, O. R. (2014) 'Does Fairness Matter in International Environmental Governance? Creating an Effective and Equitable Climate Regime' in *Towards a New Climate Agreement*, ed. Cherry, T., J. Hovi and D. McEvoy (London: Routledge).

Young, O. R. (1979) *Compliance and Public Authority: A Theory with International Applications* (Baltimore: Johns Hopkins University Press).
Young, O. R. (1999a) *Governance in World Affairs* (Ithaca: Cornell University Press).
Young, O. R. (ed.) (1999b) *The Effectiveness of International Environmental Regimes: Causal Connections and Behavioral Mechanisms* (Cambridge: MIT Press).
Young, O. R. (2002a) *The Institutional Dimensions of Environmental Change: Fit, Interplay, and Scale* (Cambridge: MIT Press).
Young, O. R. (2002b) 'Are Institutions Intervening Variables or Basic Causal Forces? Causal Clusters vs Causal Chains in International Society' in *Millennium Reflections on International Studies*, ed. Brecher, M. and F. Harvey (Ann Arbor: University of Michigan Press), 176–191.
Young, O. R. (2003) 'Determining Regime Effectiveness: A Comment on the Oslo-Potsdam Solution' *Global Environmental Politics* 3, 97–104.
Young, O. R. (2005) 'Why Is There No Unified Theory of Environmental Governance?' in *Handbook of Global Environmental Politics*, ed. Dauvergne, P. (Cheltenham: Edward Elgar), 170–184.
Young, O. R. (2008) 'Building Regimes for Socioecological Systems: Institutional Diagnostics' in *Institutions and Environmental Change*, ed. Young, O. R., L. King and H. Schroeder (Cambridge: MIT Press), 115–144.
Young, O. R. (2010) *Institutional Dynamics: Emergent Patterns in International Environmental Governance* (Cambridge: MIT Press).
Young, O. R. (2012) 'Navigating the Sustainability Transition' in *Global Environmental Commons*, ed. Brousseau, E., T. Dedeurwaedere, P. A. Jouvet and M. Willinger (Oxford: Oxford University Press), 80–101.
Young, O. R. (2013) 'Sugaring Off: Enduring Insights from Long-term Research on Environmental Governance' *International Environmental Agreements* 13, 87–105.
Young, O. R., A. Agrawal, L. A. King, P. H. Sand, A. Underdal, and M. Wasson (1999/2005) *Institutional Dimensions of Global Environmental Change* IHDP Reports 9 and 16 (Bonn: IHDP).
Young, O. R. et al. (2006) 'A Portfolio Approach to Analyzing Complex Human-Environment Interactions: Institutions and Land Use' *Ecology and Society* 11 (article 31).
Young, O. R., L. King and H. Schroeder (eds) (2008) *Institutions and Environmental Change: Principal Findings, Applications, and Research Frontiers* (Cambridge: MIT Press).
Zaelke, Z., D. Kaniaru and E. Kružiková (eds) (2005) *Making Law Work: Environmental Compliance and Sustainable Development*, 2 volumes (London: Cameron May).
Zelli, F. and H. van Asselt (2013) 'Introduction. The Institutional Fragmentation of Global Environmental Governance: Causes, Consequences, and Responses' *Global Environmental Politics* 13, 1–13.

12
Sustainable Development: The Institutionalization of a Contested Policy Concept

Sander Happaerts and Hans Bruyninckx

Very few concepts have made such a fast and pervasive career in policy discourses as sustainable development. Since its introduction as a guiding policy principle with the Brundtland Report in 1987 and the Rio Conference in 1992, it has been accepted as a framework for policy agendas as widely different as macroeconomic development and the provision of basic healthcare services. It quickly became the central concept in environmental policy, economic planning, spatial planning, and development policy, at all levels of policymaking. Outside of government and policymaking, it has also been a defining concept for non-governmental organizations (NGOs) of different types, of business associations, labor unions, and even churches. Sustainable development has achieved an enormous reach in terms of its use as a framework for desired or intended societal action. Yet, at the same time, the concept remains contested at different levels. Critics point to the vagueness of the concept, the level of aggregation that is not adapted for pragmatic policymaking, the Northern bias, its voluntaristic and unrealistic view on the role of economic dynamics, or even the marginalization of the environmental dimension.

This chapter will first provide an overview of the conceptual history of sustainable development and its basic content. Next, the main elements of debate that have crystallized in the 25 years that the concept has been used by policymakers and other actors will be discussed. Several examples of the way in which sustainable development is being implemented in actual policy processes will be given by looking at the institutionalization of sustainable development at different levels of governance and by different actors. Finally, we discuss some recent developments in the debate about sustainable development, as new concepts are emerging in political and academic circles.

The conceptual history of sustainable development

Although sustainable development as a concept has become important only since the publication of the Brundtland Report, *Our Common Future*, in 1987 (WCED, 1987) and the United Nations Conference on Environment and Development (UNCED), the so-called Rio Conference in 1992, it is clearly embedded in a number of currents that have existed much longer.

The early scenario builders have shaped our thinking about the interaction between human systems of production and consumption, population dynamics, and the fundamental environmental and natural resource basis on which our society is dependent (see Stevis, this book). In their influential *Limits to Growth* report, the Club of Rome (1971) projects predictions about resource scarcity and pollution into the 21st century. Their conclusion was as simple as sobering: our exploitation of natural resources and its negative side effects are not tenable – or sustainable – in the long run.

Another origin of the sustainable development concept can be found in the developmentalist literature and a number of critical international reports on the enormous differences between dynamics in rich and in poor countries, such as the Tinbergen Report (1970) and the Brandt Commission Report (1977). They explained the differences between North and South primarily through the fundamental imbalances in the global economic system of trade and production, and described the unacceptable and dangerous continuation of those differences.

An important step forward in the international policy debate occurred in the early 1970s. Building upon earlier work (Almeida, 1972; Sachs, 1970), the United Nations Conference on the Human Environment (UNCHE), the first global summit on the environment held in Stockholm in 1972, formulated the first internationally recognized and carefully developed link between environmental problems and poverty. In the aftermath of Stockholm, we see the emergence of a growing literature on international environment and development issues, which increasingly put the emphasis on the connections between economic dimensions of North–South relations and their impact on the environment (see Stevis, this book). To underline that reasoning, during the second half of the 1980s and the 1990s a number of global environmental issues were discovered and placed high on the international agenda. The depletion of the ozone layer, global warming, and the loss of biodiversity undeniably demonstrated that this global interconnectedness could no longer be ignored. It was also increasingly recognized in that period that solutions to those problems could only be formulated at the global level (Caldwell, 1990; Haas et al., 1993; Hurrell and Kingsbury, 1992).

This global dimension is strongly present in the Brundtland Report of the World Commission on Environment and Development (WCED, led by Gro Harlem Brundtland), which was established by the United Nations in 1983 after it became clear that international environmental and developmental policies were not leading to satisfactory results (Chasek and Wagner, 2012). Although the obvious differentiations between North and South were made by the Commission in terms of impacts, capacities, responsibilities, and so on, the underlying message was that we had entered a period of global problems that require global solutions; hence the title, *Our Common Future*. It advanced the view that environmental challenges lie at the heart of economic development, social problems and even international peace and security. Accordingly, its major lesson was that environmental concerns need to be integrated in economic policy and in mainstream decision-making (Runnalls, 2008). The Brundtland Commission introduced and defined sustainable development as 'development that meets the needs of the present, without compromising the ability of future generations to meet their own needs' (WCED, 1987: 43). Since the publication of the Report, the term has been increasingly used in international literature, negotiations, and policymaking.

Regardless of the strong differences on a number of issues, it was hard to be against the basic ideas behind sustainable development as such. It became the central concept around which the debates were organized during the United Nations Conference on Environment and Development, held in Rio de Janeiro (Brazil) in 1992. The main documents of the Rio Conference, the *Rio Declaration* and *Agenda 21*, further defined sustainable development and gave it a more policy-oriented content (UNCED, 1992a, 1992b). The emphasis became the 'balancing' or integration of environmental, economic and social goals: a stable economy should be able to produce enough welfare for everybody, and to distribute the benefits and the costs in a much more equitable way, without endangering the environment on which the whole system is based. *Agenda 21* formulated goals and implementation mechanisms referring to institutional, economic, and other changes that are deemed necessary to bring about the turn toward a more sustainable society.

Both *Our Common Future* and *Agenda 21* mention several policy principles for sustainable development (see also IISD, 2013). Their application is meant to achieve a political operationalization of the meta-goals of sustainable development. A number of them seem to have reached (at least in their theoretical dimension) consensual status. They include integration, equity, intergenerational solidarity, internalization, and participatory policymaking. An absolute core principle is the necessity to integrate different policies. Horizontal *Policy Integration* is defined as recognition of the linkages between different policy domains and the need to approach them together (Lafferty and Hovden, 2003). Vertical integration refers to the need to come to better policy coherence between different levels of policymaking and implementation, for example, subnational, national, and international (Berger and Steurer, 2008; Happaerts, 2012a). *Equity* forms the strong normative foundation for the social dimension of sustainable development (Ikeme, 2003). Production and consumption have to be based on a more equitable distribution of costs and benefits, both within Northern and Southern countries and between them (Agyeman et al., 2003). *Intergenerational Solidarity* refers to the – until now – often absent long-term planning that is necessary to come to fundamental changes in our society. It will be increasingly necessary to take the next generations into account when we make decisions, which was a totally new notion in politics when the Brundtland Report advanced it in 1987. The *Internalization of Social and Environmental Costs* is another key principle (Bartelmus, 1994). It has become increasingly clear that the market price of goods does not fully reflect cost elements such as environmental damage during the complete production cycle, from the extraction of resources and energy production to the problem of dealing with the waste at the end of the consumption cycle. Finally, *Participatory Policymaking* involves both instrumental and normative hypotheses (Hemmati, 2002). On the one hand, more participation by stakeholders is believed to result in better policymaking and implementation. On the other hand, a participatory society is believed to be a better society, as it fundamentally recognizes the role of citizens and social groups for the legitimacy of policymaking processes.

Since the Rio Conference, those principles have been accepted as guidelines for international policy debates. Actors have committed to them and have adopted myriad programs and changes in order to further the sustainable development agenda. In 2002, they met again in Johannesburg (South Africa) for the Rio + 10 Conference known as the World Summit on Sustainable Development (WSSD). The main issue on the agenda was the lack of strong implementation that was generally observed in the decade since UNCED. States and other actors discussed better strategies to push forward the common agenda. It had become clear that implementing the multifaceted concept was far more difficult and required much more political commitment than was generally admitted (Nierynck et al., 2003). Held in a far less optimistic international atmosphere than UNCED, debates in Johannesburg had become more political and confrontational. In the process leading up to Johannesburg, the South was reluctant to hold a new summit with an explicit environmental agenda without a strong focus on development issues, and it had become difficult to come to global strategies on biodiversity and climate change. The international community continued to try to address these issues at the Rio + 20 conference in 2012 (see below).

Academic debates about sustainable development as a policy concept

The Brundtland and Rio definitions can be considered as meta-concepts, which capture an integrated, holistic vision of the future (O'Toole, 2004). The translation of sustainable development as a policy concept has proven to be very difficult, both conceptually and in its implementation. Since its conception, several critiques have developed.

The vagueness of the concept

The broad use of sustainable development has led some to claim that *sustainable* has become an adjective that can be placed in front of nearly anything. One of the criticisms about the concept is that it is vague and means something different to all actors in the debate (Blühdorn, 2007; Hajer and Versteeg, 2005). That is probably correct, although it is, to a certain extent, not surprising and perhaps intended (Daly, 1991; Spangenberg, 2004). Like other political terms such as 'freedom' or 'democracy', sustainable development is an essentially contested concept (Connelly, 2007). Contested concepts have two levels of meaning. A first level of meaning is commonly accepted but rather vague, like the Brundtland definition and three dimensional representation of sustainable development are. The semantic battle takes place at the second level of meaning, where the contested concept has to be interpreted in practice and translated into concrete actions. The argument over the second level of meaning reflects an array of different conceptualizations of sustainable development, some of which contradict each other. Sustainable development thus becomes a 'legitimating concept': through multiple interpretations, it justifies diverging or even opposing ambitions (Gendron and Revéret, 2000). One instance of such opposing ambitions is between industrial elites, who would never abandon the premise of economic growth, and environmental movements, which strive for fundamentally new priorities and decision-making criteria. While those two actors applied totally different discourses in the past, they are now both embracing the concept of sustainable development. The battle between them is fought with regard to the exact interpretation of the concept (Gendron, 2006).

Environmental sustainability or broader interpretations?

Most actors defend a broad definition and emphasize that the integration of social, economic, and environmental goals is the central difference between traditional environmental policies and sustainable development (Zaccai, 2002). However, the policy translation of sustainable development often has a strong environmental bias, which has several explanations. First, there clearly exists an ecological essentialism in its foundation. The ecosystem is seen as an essential precondition for human functioning in its social and economic dimensions. It is obvious to many authors that the environmental dimension forms the foundation for the other two (Gendron, 2005; Zaccai, 2011), and for that reason they denounce a three-pillar visualization or an idea of balancing one dimension against another (Kemp et al., 2005). A second explanation is that environmental groups have from the start been the strongest proponents of the sustainability concept, and have hence had a very significant impact on the debates. Third, many of the policy-oriented translations of sustainable development are based on some form of environmental policy integration. That could explain why, in many countries, sustainable development was put onto the political agenda but never achieved a central status, as environment ministries (which often have the competency over sustainable development) are usually still considered 'junior' departments.

A number of countries and actors have, however, chosen the more holistic interpretation of sustainable development. Some countries in Europe (for example, Belgium, France, and the Netherlands) have included very strong social and economic goals in their national sustainable development plans. But developing nations are probably the strongest proponents of the social and economic dimensions of sustainable development. They put *Human Development* centrally in the whole enterprise (Mestrum, 2003), meaning that basic economic welfare and social development in terms of education, healthcare, and access to services (such as sanitation and waste management) are the central elements and the real basis for sustainable development programs.

The impact of the concept on the North–South debate

Although the Brundtland Report and the preparation of UNCED placed a heavy emphasis on the North–South dimension of sustainable development, the concept increasingly became an element of debate between North and South (Najam, 2011). That debate is often narrowed down to a development versus environment debate. Yet it is much more complex and refers to both development and environmental dynamics in industrialized and in developing countries, and to the connection between those two (Faber and McCarthy, 2003; McLaren, 2003). In that sense, sustainable development can be interpreted as an essential concept in the globalization debate.

Indeed, some of the harshest criticisms of sustainable development are based on either the fact that is the nth concept coming from Northern intellectuals trying to capture global inequalities, or that it is reaffirming precisely those power structures that underlie the issues for which it claims to be a cure (Faber and McCarthym, 2003; Lélé, 1991; Najam, 2011). Those critiques question both the fundamental analysis that is behind the use of the concept and the sincerity of the real agenda behind its use.

Conceptualizations of sustainable development by authors from the South usually go in one of two directions. Some approach the issue as closely connected to structural elements of the global economy and the impacts it has on socioeconomic conditions and subsequently also the environment (Najam, 2011; Napolitano, 2013; Stevis and Assetto, 2001). Sustainable development then refers to fundamental changes in international economic parameters. The other approach is much more linked to poverty as a pervasive phenomenon in the South (Mestrum, 2003), leading to recommendations in the sphere of basic needs. Furthermore, developmentalist approaches to sustainability usually place a larger emphasis on the bottom-up or communitarian approach because of a better cultural fit, or because of weakness of state institutions (Arunachalam et al., 2007; Fisher, 1993).

Is there really much beyond the discourse?

Ever since the Johannesburg Summit, a growing number of critics have been claiming that the implemented actions until now are very limited in comparison to the challenges or that sustainable development is staying at the discourse level altogether. One part of the critique is that real political will for social change toward sustainable development is largely absent (Van Ypersele, 2003). Another fundamental criticism is that the capacity to really implement changes is not made available, referring to the limited amount of funds that industrialized countries have effectively made available to countries of the South in the context of multilateral environmental agreements. For critics, fundamental change of trade mechanisms, debt relief schemes, and financial transactions will need to be discussed from the perspective of sustainable development (Petrella, 2003).

A number of authors that have analyzed governmental policies for sustainable development claim that they are characterized by symbolic politics. That means that they have low impact effectiveness but high politico-strategic effectiveness (Happaerts, 2012b; Newig, 2007). Their impact effectiveness is low because they do not solve any problems in the 'real' world (Szerszynski, 2007). Many policies result in transversal instruments such as sustainable development strategies and interdepartmental working groups, but have less direct policy measures in specific sectors. They often have a low administrative relevance, receive little agenda attention, and mostly stay off the political radar. Paradoxically, sustainable development does attract a high degree of declaratory commitment, as political discourses are usually filled with all sorts of references to 'sustainability'. Since the policies have low impact effectiveness, that discourse is characterized as lip service, window-dressing, or empty rhetoric, and some authors explicitly oppose it to effective action (Baker, 2007). Yet the policies do meet a determined, politico-strategic goal. They succeed in displaying to a broad public the commitment of political officials, and they remove a pressing issue from the political agenda. In that sense, policies can become highly cosmetic (Meadowcroft, 2007).

In short, although sustainable development has been widely accepted as a relevant and useful concept, there is much debate on its meaning, applicability, and impact. Some of that discussion is about definitional issues and interpretations, but it is clear that there is also a more fundamental debate about some of the key elements of sustainable development, which touch upon basic features of our system of production and consumption.

Academic debates and research on the institutionalization and practices of sustainable development

In the following pages we will look at several dimensions of the institutionalization of sustainable development at different levels of governance. Moreover, we look at academic debates about the actors involved in changes toward sustainability. In conclusion, we discuss the types of knowledge claims that have been made regarding sustainable development.

Sustainable development and global governance

One of the central issues in the sustainable development literature is about the necessity for, and the feasibility of, a functioning system of global governance for sustainable development. The necessity of such a system is defended because reconciling global sustainability with the current economic forces of globalization will require some sort of governance regime (see also Biermann, this book). A number of academic research agendas concerned with global governance for sustainable development have converged into a program on earth system governance, which is defined as

> the interrelated and increasingly integrated system of formal and informal rules, rule-making systems, and actor-networks at all levels of human society (from local to global) that are set up to steer societies towards preventing, mitigating, and adapting to global and local environmental change and, in particular, earth system transformation, within the normative context of sustainable development.
> (Biermann et al., 2009)

The answer to the global governance for sustainability question can be approached in different ways. Institutionalists, who believe in the potential of incremental institutional adaptation for problem-solving in the direction of sustainable development, tend to start from the United Nations as an institutional anchoring point (Pallemaerts, 2003; Paterson, this book).[1] A number of organizational steps, which form a sort of skeleton of a global sustainable development governance regime, have been taken. For example, after the Rio conference, the UN Commission for Sustainable Development (CSD) was founded to oversee the implementation of *Agenda 21*. Also, numerous global conferences on partial themes of sustainable development have been held, such as on the role of women (Fourth World Conference on Women, 1995, Beijing), on sustainable housing (6th World Urban Forum, 2012, Naples), on issues related to indigenous people (World Conference on Indigenous Peoples, 2014, New York), and on numerous other issues. Certain global conventions are labeled as post-Rio regimes because they illustrate that the sustainability paradigm has entered global policymaking dynamics. The Convention to Combat Desertification (1994), for instance, incorporates a number of innovative elements, such as discourses of participatory policymaking and implementation, decentralization as a fundamental policy goal, and the use of local knowledge as an explicit 'good'. The sustainable development concept formed the overarching umbrella for those discourses, which represented at that point a very new dimension in international environmental policymaking (Bruyninckx, 2004).

Other approaches are of a more structuralist nature and emphasize the essentially unsustainable character of the global political economy. In that view, the embedded imbalances between economic performance and social and environmental consequences and the unequal distribution of wealth between and within North and South are key characteristics of the current system of production and consumption and are reflected in the institutional outcomes at the level of global governance (Zaccai, 2002; Paterson, this book). Those social and material foundations of the current system prevent social change in the direction of sustainable development (Petrella, 2003). From a structuralist perspective, the current functioning of global governance is inadequate and incremental changes will be unable to bring about the social transformations required to come to a sustainable society. The debate on fundamental changes in the UN system (or on multilateral reform more broadly) can be placed in this context. Attention should go to financial and economic institutions such as the World Bank, the International Monetary Fund (IMF), the Group of 8 (G8), the World Trade Organization (WTO), and the World Economic Forum (WEF) as promoters of unsustainable development (Clapp, this book). Those debates are recently seen in a new light, as emerging powerhouses such as China, India, and Brazil are blurring the lines between North and South, and claiming a more significant role in multilateral institutions (Vihma, 2011).

Regional organizations and governance for sustainable development

Regional institutions have been put forward as one of the elements in multi-level governance arrangements on sustainable development. The EU, which has always been a promoter of global sustainable development (Van den Brande, 2012), can be considered as a special case. It is the only international organization that has competencies in all relevant policy areas important to sustainable development and with a strong impact on the policies of its member states. That includes agriculture, transport, trade relations with countries in the South, and so on. Sustainable development has also been added as a central objective in the EU's founding treaty.

The EU developed its own Sustainable Development Strategy (EUSDS) in preparation for the Johannesburg Summit (European Commission, 2002). Elaborating goals in themes such as clean energy and the management of natural resources, the strategy was considered to be a necessary environmental add-on to the Lisbon process, the EU's main socioeconomic strategy between 2000 and 2010. However, the EUSDS never received much political attention and was rather disconnected from the Lisbon process. When Europe 2020, the EU's new socioeconomic strategy, was launched in 2010, an effort was made to include sustainable development concerns directly into Europe 2020. That strategy now aims at 'smart, sustainable and inclusive growth' (European Commission, 2010). In response to some criticisms on the Lisbon process, Europe 2020 gives a more prominent place to social and environmental goals and indicators. One of the priorities that the European Commission puts forward is resource efficiency (European Commission, 2011). While environmental concerns – managing natural resources more responsibly – are at the core of that policy, the Commission frames it as a predominantly economic issue (Happaerts and Bruyninckx, 2013). In a continent that depends on imports of many commodities, achieving more economic growth with fewer resources is presented as a strategy for innovation, jobs and competitiveness. That illustrates some of the common points of the EU's sustainable development agenda since the 1990s, namely environmental policy integration and the decoupling of economic growth from material input, energy use, and environmental degradation. A more recent phenomenon in the EU, which can also be traced back to the core principles of sustainable development, is the adoption of 2050 as a long-term time horizon. That not only happens in the EU's resource efficiency policy, but also with regard to climate change, energy, and transport.

Although the inclusion of sustainable development in the high-level Europe 2020 strategy can be regarded as a step forward, Pisano and colleagues (2013) show that this high-level socioeconomic strategy has a narrow interpretation of sustainable development and tackles only a limited number of issues. Moreover, it is unclear how the integration of different strategies will play out at the EU level. Especially in light of the global financial and economic crisis, and the sovereign debt crisis in the European monetary union, it is unpredictable how the priorities of European leaders will evolve.

The state and the institutionalization of sustainable development

One of the most lively and interesting debates of recent years in the academic literature on globalization has been on environmental governance and the role of the state. The central idea of governance is that the nation-state is not the exclusive actor anymore, but that societal functions and processes are performed and implemented in different ways, by a variety of actors and at different levels (Biermann, this book; Rosenau, 2005). Sustainable development has given rise to innovative forms of multi-actor governance, as we explain below. However, most authors also agree that a key role in sustainable development governance is still put aside for governments (Jordan, 2008; Meadowcroft, 2008). While governments are increasingly dependent on the cooperation of civil society or private actors and have lost some of their powers due to globalization and economic and political integration, most multi-actor interactions (for example, participative processes, government business cooperation, and so on) rely on a strong center that is often still provided by governments. Moreover, in many parts of the world citizens continue to look at governments as the major catalysts of societal change (Kemp et al., 2005).

The active role of government can be illustrated by looking at the most visible state-based practices in sustainable development policymaking. A large number of countries are implementing national sustainable development policies, sometimes leading to legislative or even constitutional initiatives on sustainable development. In many countries sustainable development is recognized as a policy field at the level of ministerial competence and in a large number of countries sustainable development agencies have been set up (Meadowcroft, 2007). All of this has its impact on bureaucratic functioning, for instance with regard to interagency cooperation or long-term strategic planning. But, sustainable development has also led to conflict within state bureaucracies as the allocation and redistribution of responsibilities, influence, and budgets have been feeding turf battles.

Sustainable development policy processes have changed the institutional aspects of policy participation (Frickel and Davidson, 2004; Niestroy, 2005). In almost all countries, national advisory bodies or councils are functioning. This has increased the opportunities of environmental and developmental NGOs and actors to influence governmental policies (at least in principle) (Betsill, this book). In some countries with strong neocorporatist traditions, the advisory landscape has been redrawn. Whereas labor unions and employers' organizations have been the preferred negotiating partners of the government, sustainable development added other dimensions to socioeconomic policymaking and shifted the debate in different ways. That has meant that traditional bodies have sometimes added the environmental theme to their agenda, or have been enlarged with environmental groups. Sometimes new bodies incorporating the traditional social partners have formed the new arena to discuss issues.

The debate on the local dimension: Decentralizing sustainability

A recurring debate in the academic literature goes back to the 'small is beautiful' debate of the 1970s (Schumacher, 1973). After the example of states, and sometimes as a reaction to the lack of decisive national action, subnational governments such as regions or provinces have taken steps to institutionalize sustainable development (Bruyninckx et al., 2012). Another interesting evolution has been the application of sustainable development at the local or city level, including in the South (Qi et al., 2008). For instance, *Local Agenda 21* initiatives have spread surprisingly fast to all countries and very different types of municipalities (Kern et al., 2007).

In the spirit of 'think globally, act locally', a number of those local initiatives also explicitly include a more global dimension, either through evaluating their own local contribution to reaching international policy goals, such as the reduction of greenhouse gas emissions, or because they have (if they are in the North) a clear connection with the South. Although *Local Agenda 21* activities are driven by local dynamics, significant transnational networks and structures have been developed (for example, Local Governments for Sustainability (ICLEI)) with an emphasis on developing countries, or the Network of Regional Governments for Sustainable Development (nrg4SD)). Those networks provide communities with local approaches to issues, practical toolkits or instruments, and good practices. They also create a space for interaction and discussion and often fulfil a lobbying function for local and subnational governments in global governance processes (Betsill and Bulkeley, 2004; Happaerts et al., 2011). Academic attention has included studies the role of cities on the global stage. That role is significantly altered as a result of globalization, as especially mega-cities or global cities have become strategic sites of global environmental governance, concentrating knowledge, infrastructure and institutions vital to the function of transnational actors (Bouteligier, 2012).

Debates on stakeholders and actors

Few elements of the sustainable development discourse have produced such a large social science literature as participation. As mentioned earlier, one of the strong suggestions in the academic literature is that sustainable development initiatives of various types illustrate the shift from traditional government to governance arrangements that may or may not include the state as key point of reference (Jordan, 2008; Mol et al., 2005). If we regard sustainable development as a process of social change, rather than a policy process, that makes sense. In the absence of traditional government, and based on interpretations of the participatory dimensions of sustainable development, a number of innovative new networks on specific environmental issues have emerged that are closely linked to sustainable development practices (Hemmati, 2002). Indeed, economic, social, and environmental actors have created networks that influence production and consumption processes in such areas as tropical forest products, agricultural products, and energy consumption. Through labeling networks for example, sustainable production and consumption are promoted (Cashore and Bernstein, 2004). The fact that state institutions only play a marginal or even negligible role in some of these schemes demonstrates that sustainable development does have a viable existence outside of formal state politics. We will illustrate this by emphasizing the various roles played by stakeholders in this sort of arrangements.

Environmental and development NGOs have been among the earliest and most enthusiastic supporters of the sustainability concept (Zaccai, 2002; Agyeman et al., 2003). They have used it to emphasize their older ideas on the essential nature of environmental protection, on participation in policymaking, on the need for solidarity between North and South, and maybe even more, they have found elements in the sustainability discourse in support of more structural changes in our system of production and consumption. In that vein, they support the global solidarity movement, which defends the interests of the South's victims of globalization and developed-world policies, and the movement for sustainable livelihoods, which tries to create sustenance opportunities in the South that offer alternatives to current development processes (Kates et al., 2005).

NGOs have gradually gained a more visible position in international negotiations on sustainable development (Betsill, this book). Where they were only marginally represented in the first real side conference at the Stockholm Conference in 1972, they were represented with literally tens of thousands at Rio and Johannesburg. They now have received observer status or speaking rights in many post-Rio conventions and have also started their own more independent processes of international negotiation on a more sustainable global society, such as the annual World Social Forum. A new sort of NGOs, often in the form of umbrella organizations, has emerged to represent sustainable development as such, and multiple new alliances between NGOs have formed, bringing some to speak of the formation of a global citizenship as a counter-force to liberal globalization (Attfield, 2002).

In the business world we have witnessed a number of interpretations of sustainable development (Zadek, 2001; Kates et al., 2005). They often link the concept to technological innovation, in an attempt to stress technocratic solutions to environmental problems (Sagoff, 2000), or they emphasize that, regardless of environmental or social concerns, welfare is dependent on 'sustainable' growth by businesses to guarantee long-term employment and profit. More comprehensive translations of sustainable development into the business community are based on concepts such as triple bottom-line management, integrated business management, stakeholder management, corporate social responsibility, and so on (Capron and Quairel-Lanoizelée, 2004). As actors in a social context, companies have social and economic responsibilities toward the community in which they operate (Mol et al., 2005). Companies have come together at the global level, in the World Business Council for Sustainable Development and in other joint initiatives, to underscore their commitment to being a responsible global player.

Critical voices have correctly pointed out that the number of companies that have really incorporated this sort of new approach is still rather limited. In addition, in times of economic crisis, sustainable development is easily considered as a threat to the bottom line of the company, namely profit. So the critique is that sustainable entrepreneurship seems to be a sort of 'luxury' in times when things are going well. Others talk about a fundamental trend and claim that global scrutiny by NGOs and other interest groups has changed the environment in which companies operate so fundamentally that they are taking public opinion seriously into account and have made significant changes (Mol et al., 2005).

Workers' organizations or labor unions have been rather slow to adopt sustainable development wholeheartedly as a concept that could further their claims (Kjaergaard and Westphalen, 2001). They have been especially hesitant because of the rather dominant position of environmental elements in the discourse. Unions have taken a very ambivalent position toward environmental issues in general. If it referred to workers' health and safety, they have supported them, but as soon as there is reference to more general environmental issues associated with certain industrial sectors, such as the petrochemical or the energy sector, they regard environmental issues as a potential threat to employment. This history of mixed feelings partially explains the lukewarm acceptance of sustainable development as a concept (Bruyninckx, 2002). In more recent debates such as the green economy (see below), labor unions tie in with their tradition of advocating social rights and fairness, by emphasizing the need for a 'just transition' (Räthzel and Uzzell, 2013; Rosemberg, 2010; Stevis, 2011).

The role of knowledge and instruments

A specific literature has emerged about the knowledge requirements for a more sustainable society (Lövbrand, this book). The more practical translation of this debate has been reflected in an approach on 'instruments for sustainable development' (Damon and Sterner, 2012). At a more fundamental level, authors urge for a new framework to look at global society and a completely new ability to face the enormous complexity of challenges in the sphere of sustainable development (Capra, 2002; Homer-Dixon, 2000; Urry, 2003).

In order to answer some of the major calls for new knowledge-based approaches, global networks on partial issues of sustainable development have formed, which could be described as epistemic communities (Haas, 1992). The Intergovernmental Panel on Climate Change (IPCC), to which thousands of scientists from different disciplines worldwide contribute, is probably the most well-known example. Regardless of the existence of such epistemic communities, critical questions can be asked about the role of scientific knowledge in international negotiations. In several international processes, such as climate change, it is clear that the gap is widening between the sense of urgency proclaimed by scientists and the slow pace of political negotiations.

An interesting part of the knowledge and instruments issue is the potential of local or indigenous knowledge (Corell, 1999). The idea that hard science constituted a Western-biased approach to sustainable development lived strongly among some. Local knowledge is supposed to be more authentic and more adapted to local demands (Bruyninckx, 2004).

Another theme in the knowledge for sustainable development debate is linked to the development of sustainability indicators. Recommended by *Agenda 21*, indicators are widely regarded as one of the essential policy tools for sustainable development (Hák et al., 2007; Kates et al., 2005). Practitioners and scientists alike have engaged in indicator development in the 1990s and the 2000s. One of the most well-known exercises is the Ecological Footprint (Wackernagel and Rees, 1996).

Sustainability indicators have become the standard instrument to monitor progress toward sustainable development (Steurer and Hametner, 2010). Some feel that the proliferation of indicator sets is due to the fact that many institutions find work on indicators less threatening than actually intervening for change (Kemp et al., 2005). More recently, the attention has shifted from developing specific sets of sustainability indicators toward replacing or complementing GDP as the common indicator of economic welfare. In the 'beyond GDP' debate, efforts are made to include environmental and social indicators in a more comprehensive measure of welfare or well-being (Fleurbaey, 2009).

Looking at the knowledge challenges for the future, Dedeurwaerdere (2013) argues that 'sustainability science' needs to address two issues. First, scientists not only have to provide models of the complex systems underlying sustainable development but also look at required changes in core values and worldviews of individual and collective actions. Second, scientists have to contribute to the removal of practical and institutional barriers for sustainable development. Those two issues require an interdisciplinary (or transdisciplinary) perspective, combining not only all scientific disciplines but also extra-scientific stakeholder expertise, and a shift away from a value-neutral toward an ethical stance on sustainability (Dedeurwaerdere, 2013).

In this part of the chapter we have looked at different processes that form part of the gradual institutionalization of sustainable development, and we have situated them in ongoing academic debates. It has become clear that many aspects of sustainable development are linked to essential discussions in the social sciences in general and international relations more specifically.

The downturn of sustainable development and the emergence of new debates

Rio + 20

If sustainable development experienced a fast and steady institutionalization at all levels and in all spheres of global society in the 1990s and early 2000s, several signs point toward a downturn in recent years. The political attention to sustainable development has visibly faded. The negative tendency is marked by a decrease of attention to political initiatives aimed at sustainable development, which pushes some scholars to announce the *'fin de règne'* of the concept (Zaccai, 2011). The Rio + 20 summit was another expression of the diminishing enthusiasm.

Despite initial blockage by the North, the United Nations agreed to hold a third sustainable development summit in Rio de Janeiro in June 2012, the UN Conference on Sustainable Development (UNCSD), nicknamed 'Rio + 20'. With nearly 44,000 participants, it was the largest global summit ever. The number of state leaders that were present (79), however, was much lower than in 1992 (117) or 2002 (104). In stark contrast to the initial Rio summit, the pessimism at Rio + 20 was so great that the mere fact of having an outcome document was viewed by many governments as a success, while the Conference should actually be seen as a failure (Biermann, 2012).

Two themes were central: the green economy within the context of sustainable development and poverty eradication (see next section), and the institutional framework for sustainable development. Regarding the latter issue, results remained largely below the expectations of some, who hoped, for instance, for a breakthrough in the debate on a World Environmental Organization or, at least, a major upgrade of the status of UNEP (Biermann et al., 2012). The most noteworthy results of Rio + 20 are the replacement of CSD with a high-level political forum, and the formulation of Sustainable Development Goals. While most observers agree that the CSD has never exceeded the status of a talk shop (Chasek, 2007), it is doubtful whether the new forum will really lift sustainable development discussions to a more political level within the UN system. As for the proposed Sustainable Development Goals, they should build on the Millennium Development Goals after 2015. A high-level group is tasked with the formulation of the goals. Sustainable development experts urge for them to target the life-support systems of the planet, which are the boundaries of societal development and the source of welfare for current and future generations (Griggs et al., 2013).

We see three complementary explanations for the current downturn of sustainable development. The first is the less favorable international context compared to 1992, when the end of the Cold War was followed by a general feeling of optimism about negotiating global solutions to environmental (and other) problems. At present, the world is struggling with the effects of a severe financial crisis that erupted in 2008, and an unprecedented global economic crisis that followed. As a consequence, priorities are even more so than before aimed at the short term, at domestic politics and at economic growth. The second and related reason is the overall pessimistic atmosphere in multilateral environmental governance in recent years (with the failure of the Copenhagen

Conference in 2009 and the ensuing deadlock in climate talks as an emblematic negative milestone). The third explanation relates to a common feeling of disappointment with the actual impact of the institutionalization of sustainable development. To an increasing extent, the optimism about the fast and pervasive rise of sustainable development in policy discourses is replaced by the sobering reality of symbolic politics in some cases and the overall lack of improvement in environmental and social issues.

'It's the *green* economy, stupid!'

The choice of the green economy as one of the main themes of Rio + 20 denotes a consensus among states to put the economy at the center of attention in times of global economic crisis. It was also in line with the headway that this new concept was making in recent years, for instance with the publication of UNEP's *Green Economy Report* (UNEP, 2011). In that Report, the concept is presented as a strategy to achieve sustainable development and poverty eradication by investing 2 percent of global GDP in ten key sectors, such as energy, water, and agriculture. Public and private actors should both engage in the transition to a low-carbon, resource-efficient, and socially inclusive economy. The challenges that UNEP puts forward do not differ much from those described by the Brundtland Report in 1987, seeing that the core problems remain the same. But the issues are now framed much more as economic problems. For instance, a large emphasis is put on job creation, and investments occupy a significant place among the policy instruments that are promoted.

The shifting of global attention away from sustainable development toward a concept such as green economy entails both opportunities and risks. As the concept seems to have a certain appeal for governments, because it justifies their intuitive priority for economic development, it can have a less threatening character for some than sustainable development did. Optimistically, it could thus be more easily integrated into governments' main economic policies and lead to concrete improvements there, and therefore has more chance to exceed the status of symbolic politics. On the other hand, the concept bears the risk of neglecting some of the more essential elements of sustainable development, such as the social dimension or North–South equity, if it is too narrowly interpreted as a more efficient and cleaner economy (Onestini, 2012). Another risk is that a purely instrumental view on the environment is adopted as a 'capital stock' which offers resources and ecosystem services, and which absorbs waste. In that sense, the focus on a green economy could be a serious setback in the sustainable development debate.

A fundamental solution for the persistent problems of sustainable development

After several decades of international environmental politics and 25 years of debate about sustainable development, there is a clear discrepancy between institutional progress on the one hand and the lack of environmental effectiveness on the other. To explain that discrepancy, a growing group of scholars approach sustainable development through the transition perspective. That is a relatively recent strand in the literature that finds its origins in systems and complexity thinking, innovation and technology science, among others. In transition theory, the issues of sustainable development are thought of as persistent problems. Such problems are particularly difficult to steer, because they are complex (as they involve multiple scales, actors, and levels of governance), interdependent (think, for example, of the various links between climate change, transport, and energy), uncertain (for example, with regard to tipping points and causality chains), and deeply embedded in the fabric of society (Loorbach, 2007). Most essentially, they are linked to our dominant patterns of production and consumption. To solve the persistent problems of sustainability, transition scholars maintain that regular policy and market solutions have proven to be insufficient, and that incremental institutional steps rather reinforce than reduce the problems and the structures underpinning them (Frantzeskaki and Loorbach, 2008). Instead, innovations at systemic level are needed. That refers to the concept of socio-technical systems, which consist of the dominant structures, cultures, and practices that have emerged to fulfil the major societal functions (such as food, housing, energy, and mobility) (Rotmans and Loorbach, 2010). In that line of thinking, sustainability transitions are understood as fundamental changes in dominant structures, practices, technologies, policies, lifestyles, and thinking, in order to come to real system innovations (Kemp and Rotmans, 2005). Other strategies, which perpetuate the incumbent – unsustainable – systems, are considered to lead to a lock-in of those systems and are an obstacle for sustainable development (van der Brugge and Rotmans, 2007).

Although the assumptions underlying transition theory are built on a fundamental critique of current modes of development, this thinking has enjoyed a firm resonance in international discourse in recent years (Happaerts and Bruyninckx, 2013). It relates strongly to debates about degrowth, a concept that advocates the maximization of well-being in combination with the contraction of production and consumption (Jackson, 2011). Assuming that the resonance of transition thinking will be mostly discursive, it can still be seen as an interesting evolution. As the consensus among scientists is growing about the need for far-reaching changes in our production and consumption patterns if we want to achieve a sustainable society, the inclusion in international discourse of a concept that advocates such fundamental transitions is indeed significant.

Conclusion

Sustainable development has quickly conquered the discourse in a number of very important fields of policymaking such as environmental policy, development policy, spatial planning, and so on. It has done so in a surprisingly pervasive fashion and at all levels of governance. In addition, the concept is used by all sorts of social actors in highly varied contexts in both developed or industrialized countries and developing or industrializing countries. It remains, however, a very contested concept. We discussed its vagueness, the distinction between holistic versus more ecological interpretations of sustainable development, and the different critiques that the concept provokes in North–South debates.

We have also given numerous examples of the institutional consequences of sustainable development at all sorts of policymaking levels. The importance of those processes of institutionalization is that they embed sustainable development in concrete practices, involving (networks of) actors and giving a certain permanence in behavioral patterns at the policy level. Nevertheless, after 25 years of institutionalization, some initiatives have proven to be rather symbolic, and an overall pessimistic international atmosphere has contributed to a diminishing enthusiasm for the concept of sustainable development.

As new concepts such as the green economy are emerging at the policy level, attention in academic debates is moving increasingly toward systemic approaches that advocate fundamental transitions in our modes of production and consumption, lifestyles and thinking. Future research should focus especially on how such approaches fit within global environmental politics and could be introduced into international policy and decision-making. Such a perspective is specifically relevant as current policy approaches inspired by the green economy have a tendency to ignore some of the key characteristics of the sustainable development paradigm.

Note

1. Strong proponents of 'institutionalism' are found among regime theorists, neo-institutionalists, and idealists (Gupta, 2002).

Annotated bibliography

Jordan, A. (2008) 'The governance of sustainable development: taking stock and looking forwards', *Environment and Planning C: Government and Policy* 26, 17–33.

This article makes a valuable attempt to take stock of the literature that tried to combine the concept of governance with sustainable development. Jordan shows that existing studies mostly adopt either a normative or an empirical perspective, and he recommends researchers to further explore the causal relations between certain types of governance and the specific outcomes that they bring about.

Loorbach, D. (2007) *Transition Management. New Mode of Governance for Sustainable Development* (Utrecht: International Books).

This book is one of the key references of Transition Management, the most well-known operational approach within the literature on transitions. Transition processes that are modeled after this approach have been put into practice at the (sub)national level in the Netherlands and Belgium, and in local communities around the world, where Transition Management shows potential in terms of mobilizing stakeholders and formulating a common long-term vision for sustainability.

Meadowcroft, J. (2008) 'Who is in Charge here? Governance for Sustainable Development in a Complex World', In J. Newig, J.-P. Voß and J. Monstadt (eds), *Governance for Sustainable Development. Coping with Ambivalence, Uncertainty and Distributed Power* (London and New York: Routledge).

This chapter argues that the state retains a pivotal role in steering societies toward sustainable development, despite the tendency that existed in the early 2000s to dismiss the role of national governments within the governance paradigm. He pleads to give more attention to what states do in the context of distributed power, and to focus on their possibilities to reorient societal development toward sustainability.

Works cited

Agyeman, J., R. D. Bullard and B. Evans (2003) *Just Sustainabilities in an Unequal World* (Cambridge, MA: MIT Press).

Almeida, M. (1972) 'Environment and Development: The Founex Report' (United Nations Conference on the Human Environment).

Arunachalam, M., S. Lawrence, M. Kelly and J. Locke (2007) 'A Communitarian Approach to Constructing Accountability and Strategies for Sustainable Development', *Issues in Social and Environmental Accounting* 1, 2, 217–242.

Attfield, R. (2002) 'Global Citizenship and the Global Environment', In N. Dower and J. Williams (eds), *Global Citizenship: A Critical Introduction* (New York: Routledge).

Baker, S. (2007) 'Sustainable Development as Symbolic Commitment: Declaratory Politics and the Seductive Appeal of Ecological Modernisation in the European Union', *Environmental Politics* 16, 2, 297–317.

Barney, G. (1982) *The Global 2000 Report to the President: Entering the 21st Century* (New York: Penguin Books).

Bartelmus, P. (1994) *Environment, Growth and Development: The Concepts and Strategies of Sustainability* (London: Routledge Publishing).

Berger, G. and R. Steurer (2008) 'National Sustainable Development Strategies in EU Member States: The Regional Dimension', In S. Baker and K. Eckerberg (eds), *In Pursuit of Sustainable Development. New Governance Practices at the Sub-national Level in Europe* (London and New York: Routledge).

Betsill, M. M. and H. Bulkeley (2004) 'Transnational Networks and Global Environmental Governance: The Cities for Climate Protection Program', *International Studies Quarterly* 48, 471–493.

Biermann, F. (2012) 'Curtain Down and Nothing Settled. Global Sustainability Governance after the "Rio+20" Earth Summit', Earth System Governance Working Paper No. 26, (Lund & Amsterdam: Earth System Governance Project).

Biermann, F., M. M. Betsill, J. Gupta, N. Kanie, L. Lebel, D. Liverman, H. Schroeder, B. Siebenhüner, K. Conca, L. da Costa Ferreira, B. Desai, S. Tay and R. Zondervan (2009) *Earth System Governance: People, Places and the Planet* (Science and Implementation Plan of the Earth System Governance Project) (Bonn: IHDP: The Earth System Governance Project).

Biermann F., K. Abbott, S. Andresen, K. Bäckstrand, S. Bernstein, M. M. Betsill, H. Bulkeley, B. Cashore, J. Clapp, C. Folke, A. Gupta, J. Gupta, P. M. Haas, A. Jordan, N. Kanie, T. Kluvankova-Oravska, L. Lebel, D. Liverman, J. Meadowcroft, R. B. Mitchell, P. Newell, S. D. Oberthür, L. Olsson, P. H. Pattberg, R. Sanchez-Rodriguez, H. Schroeder, A. Underdal, S. Carmago Vieira, C. Vogel, O. R. Young, A. Brock and R. Zondervan (2012) 'Navigating the Anthropocene: Improving Earth System Governance', *Science* 335, 13061307.

Blühdorn, I. (2007) 'Sustaining the Unsustainable: Symbolic Politics and the Politics of Simulation', *Environmental Politics* 16, 2, 251–275.

Bouteligier, S. (2012) *Cities, Networks, and Global Environmental Governance. Spaces of Innovation, Places of Leadership* (New York and Abingdon: Routledge).

Brandt Commission (1977) *Report of the Independent Commission for International Developmental Issues* (the 'North-South Commission') (World Bank).

Bruyninckx, H. (2002) *Towards a Social Pact in Sustainable Matters* (Brussels: DWTC Publication).

Bruyninckx, H. (2004) 'The Convention to Combat Desertification and the Role of Innovative Policy Making Discourses: The Case of Burkina Faso', *Global Environmental Politics* 4, 3, 107–127.

Bruyninckx, H., S. Happaerts and K. Van den Brande (eds) (2012) *Sustainable Development and Subnational Governments: Policy-Making and Multi-Level Interactions*. (Basingstoke: Palgrave Macmillan).

Caldwell, L. (1990) *International Environmental Policy: Emergence and Dimensions* (Durham: Duke University Press).

Capra, F. (2002) *The Hidden Connections: A Science for Sustainable Living* (New York: HarperCollins).

Capron, M. and F. Quairel-Lanoizelée (2004) *Mythes et réalités de l'entreprise responsable* (Paris: La Découverte).

Cashore, B. and S. Bernstein (2004) 'Non-State Global Governance: Is Forest Certification a Legitimate Alternative to a Global Forest Convention?', In J. Kirton and M. Trebilcock (eds), *Hard Choices, Soft Law: Combining Trade, Environment, and Social Cohesion in Global Governance* (Aldershot: Ashgate).

Chasek, P. (2007) 'U.S. Policy in the UN Environmental Arena: Powerful Laggard or Constructive Leader?', *International Environmental Agreements: Politics, Law and Economics* 7, 363–387.

Chasek, P. S. and L. M. Wagner (2012) 'An Insider's Guide to Multilateral Environmental Negotiations since the Earth Summit', In P. S. Chasek and L. M. Wagner (eds), *The Roads from Rio. Lessons Learned from Twenty Years of Multilateral Environmental Negotiations* (New York and London: Routledge).

Club of Rome (1971) *Limits to Growth*, First report of the Club of Rome.

Connelly, S. (2007) 'Mapping Sustainable Development as a Contested Concept', *Local Environment* 12, 3, 259–278.

Correll, E. (1999) *The Negotiable Desert: Expert Knowledge in the Negotiations on the Convention to Combat Desertification*, PhD thesis, Linköping Studies in Arts and Science, 191.

Daly, H. E. (1991) 'Operational Principles for Sustainable Development', *Earth Ethics*, Summer 1991, 6–7.

Damon, M. and T. Sterner (2012) 'Policy Instruments for Sustainable Development at Rio + 20', *The Journal of Environment and Development* 21, 2, 143–151.

Dedeurwaerdere, T. (2013) 'Sustainability Science for Strong Sustainability' (Louvain-la-Neuve: Université catholique de Louvain).

European Commission (2002) *A European Union Strategy for Sustainable Development* (Luxembourg: Office for Official Publications of the European Communities).

European Commission (2010) *Europe 2020. A Strategy for Smart, Sustainable and Inclusive Growth* (Brussels: European Commission).

European Commission (2011) *Roadmap to a Resource Efficient Europe* (COM(2011) 571 final) (Brussels: European Commission).

Faber, D. and D. McCarthy (2003) 'Neo-liberalism, Globalization and the Struggle for Ecological Democracy: Linking Sustainability and Environmental Justice', In J. Agyeman, R. D. Bullard and B. Evans (eds), *Just Sustainabilities in an Unequal World* (Cambridge: MIT Press).

Fisher, J. (1993) *The Road From Rio: Sustainable Development and the Nongovernmental Movement in the Third World* (Westport, CT: Praeger Publishing).

Fleurbaey, M. (2009) 'Beyond GDP: The Quest for a Measure of Social Welfare', *Journal of Economic Literature* 47, 4, 1029–1075.

Frantzeskaki, N. and D. Loorbach (2008) 'Infrastructures in Transition. Role and Response of Infrastructures in Societal Transitions.' Paper read at the First International Conference on Infrastructure Systems and Services: Building Networks for a Brighter Future (INFRA) 10–12 November, at Rotterdam.

Frickel, S. and D. J. Davidson (2004) 'Understanding Environmental Governance: A Critical Review', *Organization and Environment* 17, 4, 471492.

Gendron, C. (2005) 'Le Québec à l'ère du développement durable', *Options politiques/Policy Options* July–August 2005, 20–25.

Gendron, C. (2006) *Le développement durable comme compromis. La modernisation écologique de l'économie à l'ère de la mondialisation* (Québec: Presses de l'Université du Québec).

Gendron, C. and J. Revéret (2000) 'Le développement durable', *Économies et Sociétés* 37, 111–124.
Griggs, D., M. Stafford-Smith, O. Gaffney, J. Rockström, M. C. Öhman, P. Shyamsundar, W. Steffen, G. Glaser, N. Kanie and I. Noble (2013) 'Sustainable Development Goals for People and Planet', *Nature* 495, 305–307.
Gupta, J. (2002) 'Global Sustainable Development Governance: Institutional Challenges from a Theoretical Perspective', *International Environmental Agreements: Politics, Law and Economics* 2, 361–388.
Haas, P. M. (ed.) (1992) *Knowledge, Power, and International Policy Coordination* (Columbia: University of South Carolina Press).
Haas, P., R. Keohane and M. Levy (1993) *Institutions for the Earth: Sources of Effective International Environmental Protection* (Cambridge, MA: MIT Press).
Hajer, M. and W. Versteeg (2005) 'A Decade of Discourse Analysis of Environmental Politics: Achievements, Challenges, Perspectives', *Journal of Environmental Policy and Planning* 7, 3, 175–184.
Hák, T., B. Moldan and A. Lyon Dahl (eds) (2007) *Sustainability Indicators, A Scientific Assessment* (Washington, Covelo and London: Island Press).
Happaerts, S. (2012a) 'Does Autonomy Matter? Subnational Governments and the Challenge of Vertical Policy Integration for Sustainable Development: A Comparative Analysis of Quebec, Flanders, North Rhine-Westphalia and North Holland', *Canadian Journal of Political Science* 45, 1, 141–161.
Happaerts, S. (2012b) 'Sustainable Development and Subnational Governments: Going Beyond Symbolic Politics?', *Environmental Development* 4, 2–17.
Happaerts, S. and H. Bruyninckx (2013) 'The Discourse and Practice of Transitions in Global Governance. An Exploration of the International Transition towards Sustainable Materials Management', Paper read at the ISA Annual Convention, 3–6 April, at San Francisco.
Happaerts, S., K. Van den Brande and H. Bruyninckx (2011) 'Subnational Governments in Transnational Networks for Sustainable Development', *International Environmental Agreements: Politics, Law and Economics* 11, 4, 321–339.
Hemmati, M. (2002) *Multi-stakeholder Processes for Governance and Sustainability: Beyond Deadlock and Conflict* (London: Earthscan Publications Ltd).
Homer-Dixon, T. (2000) *The Ingenuity Gap* (Random House).
Ikeme, J. (2003) 'Equity, Environmental Justice and Sustainability: Incomplete Approaches in Climate Change Politics', *Global Environmental Change* 13, 3, 195–206.
International Institute for Sustainable Development (IISD) (2013) *Sustainable Development Principles*. Available from http://www.iisd.org/sd/principle.asp.
International Union for the Conservation of Nature and Natural Resources (IUCN) (1980) *The World Conservation Strategy* (Gland, Switzerland: IUCN).
Jackson, T. (2011) 'Societal Transformations for a Sustainable Economy', *Natural Resources Forum* 35, 165–174.
Jordan, A. (2008) 'The Governance of Sustainable Development: Taking Stock and Looking Forwards', *Environment and Planning C: Government and Policy* 26, 17–33.
Kates, R., T. Parris and A. Leiserowitz (2005) 'What is Sustainable Development? Goals, Indicators, Values and Practice', *Environment* 47, 3, 821.
Kemp, R. and J. Rotmans (2005) 'The Management of the Co-evolution of Technical, Environmental and Social Systems', In M. Weber and J. Hemmelskamp

(eds), *Towards Environmental Innovation Systems* (Heidelberg: Springer Berling).

Kemp, R., S. Parto and R. B. Gibson (2005) 'Governance for Sustainable Development: Moving from Theory to Practice', *International Journal of Sustainable Development* 8, 1/2, 12–30.

Kern, K., C. Koll and M. Schophaus (2007) 'The Diffusion of Local Agenda 21 in Germany: Comparing the German Federal States', *Environmental Politics* 16, 4, 604–624.

Kjaergaard, C. and S. Westphalen (2001) *From Collective Bargaining to Social Partnerships: New Roles of the Social Partners in Europe* (Copenhagen: The Copenhagen Centre Publications).

Lafferty, W. M. and E. Hovden (2003) 'Environmental Policy Integration: Towards an Analytical Framework', *Environmental Politics* 12, 3, 1–22.

Lélé, S. (1991) 'Sustainable Development: A Critical Review', *World Development* 19, 6, 607–621.

Loorbach, D. (2007) *Transition Management. New Mode of Governance for Sustainable Development* (Utrecht: International Books).

McLaren, D. (2003) 'Environmental Space, Equity and Ecological Debt', In J. Agyeman, R. D. Bullard and B. Evans (eds), *Just Sustainabilities in an Unequal World* (Cambridge: MIT Press).

Meadowcroft, J. (2007) 'National Sustainable Development Strategies: Features, Challenges and Reflexivity' *European Environment* 17, 152–163.

Meadowcroft, J. (2008) 'Who is in Charge here? Governance for Sustainable Development in a Complex World', In J. Newig, J.-P. Voß and J. Monstadt (eds), *Governance for Sustainable Development. Coping with Ambivalence, Uncertainty and Distributed Power* (London and New York: Routledge).

Mestrum, F. (2003) 'Poverty Reduction and Sustainable Development', *Environment, Development and Sustainability* 5, 1–2, 41–61.

Mol, A., F. Buttel and G. Spaargaren (2005) *Governing Environmental Flows* (Cambridge: MIT Press).

Najam, A. (2011) 'The View from the South: Developing Countries in Global Environmental Politics', In R. S. Axelrod, S. D. VanDeveer and D. L. Downie (eds), *The Global Environment: Institutions, Law, and Policy* (Washington: CQ Press).

Napolitano, J. (2013) 'Development, Sustainability and International Politics', In L. Meuleman (ed.), *Transgovernance. Advancing Sustainability Governance* (Berlin and Heidelberg: Springer).

Newig, J. (2007) 'Symbolic Environmental Legislation and Societal Self-deception', *Environmental Politics* 16, 2, 276–296.

Nierynck, E., A. Van Overschelde, T. Bauler, E. Zaccai, L. Hens and M. Pallemaerts (eds) (2003) *Making Globalisation Sustainable: The Johannesburg Summit on Sustainable Development and Beyond* (Brussels: VUB University Press).

Niestroy, I. (2005) *Sustaining Sustainability. A Benchmark Study on National Strategies towards Sustainable Development and the Impact of Councils in Nine EU Member States*, EEAC series, Background study no. 2 (Utrecht: Lemma).

Onestini, M. (2012) 'Latin America and the Winding Road to Rio + 20: From Sustainable Development to Green Economy Discourse' *The Journal of Environment & Development* 21, 1, 32–35.

O'Toole, L. J. Jr. (2004) 'Implementation Theory and the Challenge of Sustainable Development: The Transformative Role of Learning', In W. M. Lafferty (ed.), *Governance for Sustainable Development. The Challenge of Adapting Form to Function* (Cheltenham and Northampton: Edward Elgar).

Pallemaerts, M. (2003) 'Is Multilateralism the Future? Sustainable Development or Globalisation as a Comprehensive Vision of the Future of Humanity', *Environment, Development and Sustainability* 5, 275–295.

Pearce, D. (1999) 'Economic Analysis of Global Environmental Issues: Global Warming, Stratospheric Ozone and Biodiversity', In J. C. J. M. van den Bergh (ed.), *Handbook of Environmental and Resource Economics* (Cheltenham and Northampton: Edward Elgar).

Petrella, R. (2003) *Sustainable Development in a Globalizing World*. In Nierynck, E., A. Van Overschelde, T. Bauler, E. Zaccai, L. Hens and M. Pallemaerts (eds), *Making Globalisation Sustainable: The Johannesburg Summit on Sustainable Development and Beyond* (Brussels: VUB University Press).

Pisano, U., A. Endl and G. Berger (2013) *The Future of the EU SDS in Light of the Rio+20 Outcomes* (ESDN Quarterly Report No. 28) (Vienna: European Sustainable Development Network).

Qi, Y., L. Ma, H. Zhang and H. Li (2008) 'Translating a Global Issue Into Local Priority. China's Local Government Response to Climate Change', *The Journal of Environment and Development* 17, 4, 379–400.

Räthzel, N. and D. Uzzell (eds) (2013) *Trade Unions in the Green Economy. Working for the Environment* (New York: Routledge).

Rosemberg A. (2010) 'Building a Just Transition: The Linkages between Climate Change and Employment', *International Journal of Labour Research* 2, 2, 125–162.

Rosenau, J. N. (2005) 'Strong Demand, Huge Supply: Governance in an Emerging Epoch', In I. Bache and M. Flinders (eds), *Multi-Level Governance* (Oxford: Oxford University Press).

Rotmans, J. and D. Loorbach (2010) 'Towards a Better Understanding of Transitions and Their Governance: A Systemic and Reflexive Approac', In J. Grin, J. Rotmans and J. Schot (eds), *Transitions to Sustainable Development. New Directions in the Study of Long Term Transformative Change* (New York: Routledge).

Runnalls, D. (2008) *Why Aren't We There Yet?* (Commentary) (Winnipeg: International Institute for Sustainable Development).

Sachs, I. (1970) 'Environmental Concern and Development Planning', *International Conciliation* 39, 72.

Sagoff, M. (2000) 'Can Technology Make the World Safe for Development? The Environment in the Age of Information', In L. Keekok, A. Holland and D. McNeill (eds), *Global Sustainable Development in the 21st Century* (Edinburgh: Edinburgh University Press).

Spangenberg, J. H. (2004) 'Reconciling Sustainability and Growth: Criteria, Indicators, Policies', *Sustainable Development* 12, 74–86.

Steurer, R. and M. Hametner (2010) 'Objectives and Indicators in Sustainable Development Strategies: Similarities and Variances across Europe', *Sustainable Development* (DOI: 10.1002/sd.501).

Stevis D. (2011) 'Unions and the Environment: Pathways to Global Labour Environmentalism', *Working USA* 14, 2, 145–159.

Stevis, D. and V. Assetto (2001) *The International Political Economy of the Environment: Critical Perspectives* (Boulder, CO: Lynne Rienner Publishers).
Swart, R. J., P. Raskin and J. Robinson (2003) 'The Problem of the Future: Sustainability and Scenario Analysis', *Global Environmental Change* 14, 137146.
Szerszynski, B. (2007) 'The Post-ecologist Condition: Irony as Symptom and Cure', *Environmental Politics* 16, 2, 337–355.
Tinbergen, J. (1970) *Report of the Tinbergen Commission* (New York: UN Press).
United Nations Conference on Environment and Development (UNCED) (1992a) *Agenda 21: Programme of Action for Sustainable Development* (United Nations Department of Public Information).
United Nations Conference on Environment and Development (UNCED) (1992b) *The Rio Declaration on Environment and Development* (New York: United Nations).
United Nations Environment Programme (UNEP) (2011) 'Towards a Green Economy: Pathways to Sustainable Development and Poverty Eradication' (UNEP).
Urry, J. (2003) *Global Complexity* (London: Polity Press).
Van den Brande, K. (2012) 'The European Union in the Commission on Sustainable Development', In J. Wouters, H. Bruyninckx, S. Basu and S. Schunz (eds), *The European Union and Multilateral Governance: Assessing EU Participation in United Nations Human Rights and Environmental Fora* (Basingstoke: Palgrave Macmillan).
van der Brugge, R. and J. Rotmans (2007) 'Towards Transition Management of European Water Resources', *Water Resource Management* 21, 249–267.
Van Ypersele, J. (2003) *The 2002 Johannesburg Summit and Global Warming*, In Nierynck, E., A. Van Overschelde, T. Bauler, E. Zaccai, L. Hens and M. Pallemaerts (eds), *Making Globalisation Sustainable: The Johannesburg Summit on Sustainable Development and Beyond* (Brussels: VUB University Press).
Vihma, A. (2011) 'India and the Global Climate Governance: Between Principles and Pragmatism', *The Journal of Environment and Development* 20, 1, 69–94.
Wackernagel, M. and W. E. Rees (1996) *Our Ecological Footprint. Reducing Human Impact on the Earth* (Gabriola Island: New Society Publishers).
World Commission on Environment and Development (WCED) (1987) *Our Common Future* (New York and Oxford: Oxford University Press).
Zaccai, E. (2002) *Le Développement Durable: Dynamique et constitution d'un projet* (Brussels: P.I.E. Peter Lang).
Zaccai, E. (2011) *25 ans de développement durable, et après?* (Paris: Presses Universitaires de France).
Zadek, S. (2001) *The Civil Corporation: The New Economy of Corporate Citizenship* (London: Earthscan Publications, Ltd).

13
Environmental and Ecological Justice

Chukwumerije Okereke and Mark Charlesworth

Introduction

Justice is a central theme in international environmental politics (IEP). In fact, one of the most distinctive contributions of IEP to the broader International Relations (IR) scholarship and discourse is arguably the firm insertion and elevation of questions of distributive justice. It is broadly accepted that justice occupies a central position in Western moral political philosophy. Hume (1975) described justice as the most important virtue of social relations and political institutions. St Augustine (1467/2003: 139) considered that the very legitimacy of a state lay on its claim to do justice. Rawls (1971) argued that any political institution deserves to be abolished if found to be unjust, because 'justice', he says, 'is the first virtue of social institutions' (3) and the 'rights secured by justice are not subject to political bargaining or the calculus of social interest' (5). Aristotle (1874/1998) regarded justice as co-extensive with virtue and therefore as the greatest of all virtues. 'In justice', he said, 'every virtue is comprehended'.

Despite the centrality of justice in political institutions and its potency as a 'tool for social political mobilization' (Okereke, 2008a: 32), IR scholarship has traditionally (and even now to a degree) failed to accord adequate concern to questions of justice. The conventional wisdom is that justice, to the measure that it can, should be focused on relations among people within national political boundaries rather than rigorously applied to IR (Caney, 2001).

However, from the contestations about fair distribution of putative proceeds from the exploitation of mineral resources in the seabed in the UN Third Law of the Sea (1967–1983), through disputes about how to balance between concerns for ecology and development (Stockholm and Rio), to current debates about how to share the burden of climate change, issues of equity and justice have certainly become one of the most contentious elements in the global sustainable development discourse. Yet, while the quest for environmental justice has played a central role in defining the character of international environmental rule-making over the last 30 years, these agitations have been generally less successful in upturning the fundamental structures and relationships that engender ecological degradation and environmental injustice.

There are at least three main reasons why justice is a very contentious and 'slippery' subject, especially in the context of IR. First, there are several dimensions of justice and the relationship between these aspects and frameworks for organizing the different dimensions are not always clear (Dobson, 1998). Notable aspects include distributional justice, which focuses on outcomes; procedural justice, which emphasizes how decisions are made; and compensatory justice, which is concerned with how to calculate and offset the effect of historical injustice. Other dimensions include intergenerational justice, which concerns justice between present and future generations, and interspecies justice, which deals with justice between human and non-human beings. Furthermore, there are gender, individual, community, national, and international dimensions of justice (Beck, 2008; Caney, 2005; Clark, 2008; Dobson, 2005; Page, 2011; Roberts and Parks, 2007; Skinner, 2011). The multiplicity of visions of an ecologically just world and the diverse array of voices and emphasis often raise the legitimate question of whether in fact it is 'possible to identify a single paradigm of environmental justice based on a common set of arguments' (Conca et al., 1995: 279). In fact, as we shall see below there are those who suggest making a distinction between environmental and ecological justice.

A second and related challenge is that despite the elegant definition proposed by Aristotle (1874/1998) – 'justice is giving to each their due' – it is still very difficult, even with the best effort, to decide what justice should look like in practice. The simple reason is that there are many different ways of deciding what is due to people, which correspond roughly to different conceptions of justice. Notable formulations include (1) advocacy of *equality* of resource, of income, or of opportunity; (2) desert – for example, distribution on the basis of effort and/or merit; (3) need – distribution on the basis of need; (4) fairness – for example, behind Rawls' veil of ignorance or following 'The Golden Rule', which has versions in all societies, including 'do unto others as you would have them do unto you'; (5) property rights; and (6) utility or welfare maximization which emphasizes distribution to achieve the greatest happiness for the greatest number of people. Furthermore, not only does each of these conceptions of justice command decent ancestry and strong adherents, making any reconciliation difficult; most are developed in the context of national politics and transposing the core arguments to the international context is fraught with conceptual and practical difficulties. Hence, in many cases, any statement to the effect that one is interested in environmental justice immediately invites a riposte, 'which justice?' (see Dobson, 1998: 5).

A third major challenge is that IR, both in theory and practice, has developed in ways that tend to endorse the centrality of anarchy and self-interest. This development, alongside the weakness of institutions for addressing cross-border justice, often conspires to dampen even the best efforts to insert equity and justice in global (environmental) politics (Hurrell, 2001).

This chapter traces the evolution of key debates on international environmental justice and suggests important research frontiers. Although practical actions are yet to catch up with rhetoric, it is impressive that questions of ethics and justice can no longer be regarded as external or marginal concerns in global environmental governance (Biermann, this book).

The next section gives an orientation to questions of justice, the environment, and IR, before the following sections look in some detail at more specific facets of these questions; specifically, justice and deep ecology, social ecology, 'empty-belly' environmentalism, intergenerational justice and procedural justice, before climate change is used as a case study to illustrate some of these aspects.

Environmentalism, justice, and IR

Ideas about justice in one form or another have always been a feature of international politics. A world vision of justice and peace is at the heart of idealism as an IR theory (Bull, 1983; Viotti and Kauppi, 2011) and the establishment of key international institutions such as the United Nations and the International Labour Organization (ILO) (Pogge, 2001). Also the quest for justice underpins much of the anti-slavery, anti-colonization, and anti-apartheid movements in the 19th and 20th centuries, respectively (Bull, 1983; Nelson, 1974). However, to a large extent these concerns were muffled by the two World Wars and the dominance of realism which emphasizes power and relative self-interest as the key principles in IR. Moreover, the claims for justice as expressed in these movements were more about rights to self-determination than distributional in nature (Beitz, 1999; Nagel, 2005).

Moral theories of IR were revived in the 1970s through the works of English School scholars like Hedley Bull, Barry Buzan, and Martin Wright, as well as through the writings of international political theorists such as Charles Beitz and Chris Brown. However, these theoretical endeavors quickly succumbed to revolutionary changes in global economic structures (Beitz, 1999: 515) and the dominant influence of liberal institutionalism, which emphasized the centrality of national economic interests, interdependence, and regime complexes in IR (Keohane, 1984; Viotti and Kauppi, 2011).

But while popular Western IR theorizing was dominated by power and self-interested assumptions, the vision of a more solidarist international society that incorporates adequate concerns for social justice remained strong in countries from the global South who had in the early 1960s set up the Non-aligned Movement and the Group of 77 (G77) to protest against prevailing global political systems of domination and to promote a more equitable international order (Willetts, 1978; Geldart and Lyon, 1980; Salter et al., 1993). It is therefore not surprising that as environmental issues gained ascendancy as a major topic for international cooperation in the late 1960s and early 1970s, developing countries wasted no opportunity to articulate 'clear and coherent reasons why the North stood to gain by discriminating in their favour' (Parks and Roberts, 2006: 331) regarding 'arguments about how responses to global environmental problems can be legitimised' (Paterson, this book).

Although, as stated, concern for a just and more equitable international order has been a dominant theme in developing countries' approach to IR, they enjoyed more success with environmental issues for at least three reasons (cf. Okereke, 2008b). First, environmental problems provoked a unique sense of urgency for international cooperation because they highlighted the close interdependence of humankind as well as our dependence on a single natural world regardless of national boundaries and geographies (Sachs, 1993, 1999). Critically, they highlighted the limited nature of political boundaries as causes and receptors of environmental pollution. Second, environmental issues created a heightened sense of vulnerability among the rich countries, and with that a sense of urgent need for global action (Beck, 1992, 1999). Third, scholarly works like *Silent Spring* (Carson, 1962) and *Limits to Growth* (Meadows, et al., 1972) plus major environmental disasters, such as the Bhopal gas-leak tragedy in India which killed over 3,000 people and the

Chernobyl nuclear power plant disaster which affected millions, created a favorable 'normative temper' (Okereke, 2008b) for questions of justice in the international realm.

A particularly important set of events that helped to amplify and create a favorable 'moment' for International Environmental Justice (IEJ) was the anti-toxic waste dump movement in the United States (Bullard, 1983; Bryant and Mohai, 1992). The anti-toxic campaigns which started as a protest to block the dumping of contaminated soil in a poor and black-dominated area in North Carolina (Schlosberg, 2007: 47) soon broadened to include unequal exposure by class, race, and ethnicity. In addition to inspiring a wave of literature on the concept of environmental justice, the anti-toxic movement also produced emotive terms such as 'environmental racism', 'environmental genocide', 'toxic racism', and other vocabularies which provided important rhetorical devices and mobilization tools in the campaign for international environmental justice. In fact, from this time on, IEJ and questions of environmental justice at national levels were viewed as conceptually linked and resultant of the same systems of domination operating across different scales (Byrne et al., 2007; Conca et al., 1995; Elliott, 1998; Okereke, 2011).

As noted, environmental justice is far from being a unified or unifying concept. Right from the outset, the notion of environmentalism and its relationship with social justice has always been a hotly contested subject. It was not long after environmentalism became mainstream that various 'shades' of green movements or environmentalism were identified in scholarly literature and public discourse. To give just a limited view, O'Riordan in 1977 distinguished between 'conservative ecocentric' and 'liberal ecocentric' ideologies, and again in 1981 he made a distinction between 'ecocentric' and 'technocentic' modes of environmentalism. Sandbach (1980) differentiates between 'functionalist' and 'Marxist' environmentalism. Norwegian philosopher Arne Naess (1973) espoused a deep ecology philosophy which he set against the so-called shallow ecology. Dobson's (1998) typology distilled four different types of environmentalism, including 'very weak sustainability', 'weak sustainability', 'strong sustainability', and 'very strong sustainability'. For Dobson, these four categories correspond to four environmental philosophies: Cornucopia, Accommodating, Communalism, and Deep Ecology. A provocative treatment by Guha and Martinez-Alier (1997) distinguishes between the 'full-stomach' environmentalism of the North and the 'empty-belly' environmentalism of the South. Paterson (2000) writing from an IR perspective separates between six categories, including liberal environmentalism, realism, ecoauthoritarianism, ecosocialism, social ecology, and deep ecology.

Critically, these shades of green do not simply reflect varying emphasis on what is the most important aspect of environmentalism; rather many reflect fundamental (and sometimes irreconcilable) differences about worldview, and about the cause and solutions for global environmental problems (Dobson, 1998). In the following sections, we will trace the evolution of these ideas and their impact on IEP. A historical account is necessary because an understanding of how things came to be is often essential in comprehending the options for, and barriers against, achieving desired change (see also Stevis, this book). As Pepper (1984) puts it,

> if we seek for the future the kind of real social and environmental changes which much of the standard environmental literature calls for,... then we must develop an historical perspective of how we and others have arrived at our present set of attitudes and understand what material changes will be needed to help foster a new one. (3)

For the sake of simplicity, we will limit ourselves to four broad categorizations and prevalent ideologies – all of which are mainly concerned with distributional aspects of justice. These are deep ecology, social ecology, 'empty-belly' environmentalism, and intergenerational justice. Subsequently, we will briefly discuss procedural justice, which is one distinct form of justice that has enjoyed a rise in recent times. Next, we will use the case of climate change to provide more practical illustration before making some concluding remarks.

Deep ecology, justice, and IR

The notion of deep ecology as a concept and philosophical tradition has its roots in the life and works of Norwegian philosopher Arne Naess. In a seminal article in 1973, Naess criticized what he called the 'shallow range ecological movement', which he said was the prevailing ecological moment of the time. He argued that although powerful and influential, shallow ecology was very limited and in some cases outright dangerous. It was limited because it has as its central focus the fight against pollution and resource depletion, with the central motif being mainly the preservation of the health and affluence of people from the industrialized countries of the world. The shallowness of the movement lies, he argues, in its failure to grasp the deeper and intimate connection between man and nature. For Naess, shallow ecology can be outright dangerous because while purporting to recognize the importance of ecological thinking, policies inspired by this approach are not holistic but often result in displacing environmental harm, increasing the prices of life's necessities, increasing class divisions, and privileging economic growth over nature conservation. In short, then, this type of ecology is nothing but another form of utilitarian pragmatism by Western governments and businesses. Naess argued for a movement away from shallow ecology to deep ecology.

Deep ecology, according to Naess, recognizes the intimate interdependence between humankind and their ecology. The central claim is that humans are part of nature and not separate from it. Equally central is the notion that nature has an intrinsic or inherent value that is not based on its instrumental utility to humans. In fact, it advocates the doctrine of 'biospherical egalitarianism' (Naess, 1973: 95), which suggests that human beings and living nature each have equal right to life. Deep ecology rejects the domination of humans over nature to satisfy human wants, but rather presses the claim that humans have equivalent or at least near equivalent worth to nature. Following, it is suggested that a 'deep-seated respect [for nature], or even veneration' (Naess, 1973: 95) is required preferably from all, and most certainly from ecologists. Naess argues that deep ecology warrants a deeper recognition of the complexity of the world and a search for whole person and whole system solutions.

Although Naess is often linked to Buddhism because of his profession of Gandhian non-violence, it is in fact the case that his philosophy has a long and rich Western ancestry. In particular, his idea of deep ecology has close parallels with the writing and philosophy of many 20th-century Western ecologists, especially Aldo Leopold, the close friend and adviser to US President Theodore Roosevelt (Bowler, 1992). Leopold had in his book, *A Sand County Almanac* (2001), advocated a 'Land Ethic', which is based on the claim that humanity and all other beings are aspects of a single unfolding reality/community. Deep ecologists would tend to support Leopold's 'Land Ethic': 'a thing is right when it tends to preserve the integrity, stability and beauty of the biotic community. It is wrong when it tends otherwise' (Leopold, 2001: 189). According to Bowler (1992), Leopold's views and his influence on Roosevelt were very much instrumental to the establishment of the first wave of nature and game reserves in the United States and other parts of Europe.

What, though, is the connection between deep ecology and justice? To be sure, Naess did not use the term 'justice' in his seminal article setting out the key tenets of deep ecology. In fact, the term 'ecological justice' was apparently coined by Low and Gleeson as recently as 1998, though Conca et al. (1995) indicated the term. However, it is obvious that the deep ecology movement as elucidated by Naess has strong implications for justice both nationally and internationally. The rejection of human supremacy over nature or a domineering posture against nature is certainly at the heart of the idea of interspecies justice, which has been a key dimension of environmental ethics (Agar, 1997; Alder and Wilkinson, 1999; Brennan, 2003). A number of different broad arguments are used by philosophers to advance the claim for non-anthropocentric environmental ethics and to assert that non-human things are supposed to be included as subjects for justice (Baxter, 2005; Schlosberg, 2007). The first is the consequentialist approach, which draws mainly from utilitarianism and suggests that non-human things, especially those that can experience pain and pleasure, should be included in the community of justice (Regan, 1987; Singer, 1975). The notion that non-human things have intrinsic value and are deserving of justice consideration is the philosophical backbone for the animal rights movement, which has had considerable influence in relevant legislation and praxis in the Western democracies. The second is the deontological approach, the basic argument of which is that it is morally perverse for humans to destroy or dominate what they have not and cannot create (Alder and Wilkinson, 1999). Another strand is to argue that it is the whole biotic community (rather than individual species) that constitutes the community of justice (Callicott, 1989; cf. Leopold, 2001). The third approach is virtue, which emphasizes the role of ethical behavior such as moderation and prudence (Aristotle, 1874; MacIntyre, 1984). While deep ecology's main contribution to equity is seen to lie in the realm of interspecies justice, it should be pointed out that the philosophy also has implications for social justice at national and international levels. For example, Naess clearly identified egalitarianism and classlessness as key concerns for deep ecology. He wrote about the need to encourage diversity globally and promote an '[a]nti-class posture' both within and between nations. Moreover, his preference for local autonomy and decentralization both for production and, to the extent possible, for adequate political coordination has far-reaching implications for current globalized political and economic structures (see Byrne et al., 2007). It is hard to argue that deep ecology does not embody a concern for justice.

Nevertheless, deep ecology has been criticized precisely for failing to give adequate consideration to social justice at both the national and international levels. The most popular critique is arguably the one launched by Bookchin (1987) in his work, 'Social Ecology versus Deep Ecology: A Challenge for the Ecology Movement'. Here, Bookchin criticizes deep ecology for paying too much attention to justice for nature while ignoring issues of social justice. He argues that deep ecology is in error for failing to recognize that ecological degradation was ultimately a reflection and manifestation of unequal and hierarchical social relations.

Another main strand of criticism which comes mostly from scholars from the South is the accusation that deep ecology stems from Western romanticism of nature and fails to adequately address questions of social justice, especially between rich and poor countries. Southern scholars point out that international environmental cooperation has not developed with adequate concern for distributive justice (Kiss, 1985; Okereke, 2008a). They argue that the focus has been more on conservation of endangered non-human species and less on distributive justice. Here it is frequently pointed out that the first set of international environmental agreements such as the Antarctic Treaty System (1959), the Ramsar Convention on Wetlands (1971), the London Convention on Dumping at Sea (1972), and the Convention on the Conservation of Migratory Species of Wild Animals (1979) have all been about conservation and less about human welfare. The argument is that attention to conservation masks the asymmetrical distribution of environmental benefits and burdens across nations. It is fair to say that while in principle deep ecology has radical implications for justice at national and international levels, the attention of the key proponents has been more on justice between human beings and nature. Deep ecology has important influence in the emergence and philosophy of some of the most globally influential non-governmental environmental organizations like WWF, IUCN, and Greenpeace (Bowler, 1992). These organizations had tremendous influence in bringing many environmental issues into the international agenda and some even took positions that had important consequence for justice (Betsill, this book). However, these organizations have been criticized for not giving adequate attention to social justice, especially in relation to their support of programs such as debt for nature swaps, which tend to privilege biodiversity conservation over human welfare (Hamlin, 1989; Knicley, 2012). Deep ecology sentiments are still strong in the discourse and international environmental policymaking circles. Organizations such as Greenpeace and Friends of the Earth frequently attend important international conferences on environment to campaign for nature-centered views on policymaking. However, it is fair to say that for the most part their philosophy has shifted significantly from deep ecology sentiments to more of 'shallow ecology' or what Steven Bernstein (2001) calls 'liberal environmentalism'.

Social ecology, justice, and IR

Social ecology presses the claim that ecological problems are rooted in social problems and especially the social relations of hierarchy and domination (see also Detraz, this book). Founded by green author and activist Murray Bookchin, the philosophy espouses the belief that it is impossible to address the world's ecological problems unless one fully recognizes that these problems are symptomatic of unequal social relations. Although critical of all forms of hierarchical relations (for example, gender, race), social ecology singles out class-based capitalism as the ultimate and most ecologically destructive form of hierarchical relations. Bookchin argues that capitalism pits man against man and results in the conversion of community relations to impersonal market relations where people are basically treated as commodities. The same logic results in various aspects of nature being treated as commodities that need to be exploited for instrumental reasons, especially market profit. Social ecology presses the claim for the need for more community, equality, and diversity. Like deep ecology, it argues for humanity to better recognize its relationship with nature, that is, recognize its place within as opposed to outside nature. Social ecology is very critical of solutions to ecological problems rooted in market transactions, such as green or ethical consumerism. It is suggested that such schemes mask the fundamental changes in structure of production that are required to achieve lasting ecological integrity. It argues for more decentralization and for democratic participation. Bookchin envisioned a post-capitalist society characterized by greater diversity, natural abundance, and participatory democracy, all of which he says will be made possible by technology.

As noted, social ecology is critical of deep ecology for focusing attention on humans' domination of nature and less on humans' domination over one another. Yet these traditions have far more in common than first meets the eye. Both believe in decentralization; both are against instrumental approaches to environmental conservation; both are critical of capitalism; and both believe in the need for greater diversity and equality. However, despite giving more explicit emphasis to unequal social relations in its thesis, social ecology does not dwell to any appreciable degree on international environmental justice. Bookchin's focus, like those after him, has been at the national level. Hence, it is subject to the same critique that Southern scholars make against deep ecology.

There is evidence that the philosophy of social ecology had at least indirect effect in promoting the quest for international environmental justice. For one, it certainly provided philosophical impetus for the campaign against environmental injustice that swept through the United States in the 1980s. However, given perhaps its focus on the national level, the influence of social ecology as a philosophy on IEP has been very limited.

'Empty-belly' environmentalism, justice, and IR

By far the strongest source of agitation for international environmental justice has come from what Guha and Martinez-Alier (1997) describe as the 'empty-belly' environmentalism of developing country citizens and their political leaders. This environmental 'movement' has thrown up all sorts of difficult questions and made a serious impact on the direction and character of IR over recent years. As a result of this movement, it has now become almost unthinkable that any serious account of international relations will ignore the moral and ethical dimensions of international relations. Unlike deep and social ecology, both of which have as their focus social and ecological relations at the national level, the focus of 'empty-belly' environmentalism is squarely at the international level and in short the relationship between developed and developing countries. Although a coherent philosophical treatment is still lacking, agitation for international environmental justice by the South is based closely on three thematic rationales. These include the following.

Unequal access to natural resources between the North and South

Current and historical political economy structures mean that resources tend to be more readily available to Northern citizens and companies than those in countries with lower per capita income – typically in the South. Examples include energy, minerals, food, medicine, and clean water. The reasons for this situation are complex but certainly include historical factors such as colonialism and associated domination, exploitation, and abuses of power.

The historical overuse of 'global commons' by the North

It is often asserted (for example, WCED, 1987) that the atmosphere, deep oceans, and geosynchronous orbital space required for telecommunications satellites should be considered global commons that are not and cannot be owned by individuals or nations. In each case in the last few hundred years, Western nations have made more use of these resources than other nations with the vast majority of the benefits accruing. The clearest example is the emission of greenhouse gases (GHGs) such as carbon dioxide into the atmosphere, to which we return later in the chapter. Other examples include industrial whaling, fishing, exploitation of biological resources, and waste dumping in the oceans.

Externalization of environmental risk by the North

It is generally agreed that the risks associated with climate change caused by GHGs will fall principally on the poor in the South. A similar transfer of environmental risks is the movement of hazardous waste from North to South, which continues despite the Basel Convention (Okereke, 2008a).

In addition to the above, emphasis on the quality of participation in international environmental decision-making has grown in recent years, leading many commentators to focus on procedural equity as a separate dimension of justice which deserves close attention in the context of IEP. We provide a brief treatment later in the chapter.

Intergenerational justice and environmental relations

The fourth dimension of distributional justice that has exerted important influence on the discourse of IEP is intergenerational justice. This is justice between present and future (as well as past) generations. The Brundtland Report defines sustainable development as development that 'meets the needs of the present without compromising the ability of future generations to meet their own needs' (WCED, 1987: 53). This definition is an explicit endorsement that the idea of intergenerational justice is central to the notion of global sustainability. A number of popular maxims in the sustainable community also support the idea of intergenerational justice. One example is the famous Native American saying: 'Treat the earth well: it was not given to you by your parents, it was loaned to you by your children.'

However, while intergenerational environmental justice may have a powerful intuitive appeal, providing a robust philosophical justification of why unborn people should be considered a legitimate subject of justice is not easy (Page, 2007: 54). Perhaps the best-known philosophical argument for attempting to establish the claim of intergenerational environmental justice is the indirect reciprocity thesis, which owes much to the work of Brian Barry (Barry, 1989; see Page 2007; Gosseries, 2006, 2008). Here the main argument is that the current generation owes something to the next generation because it received something from the previous one. In other words, because we received something from our parents we are obligated to pass the same 'in return' to our children. On this frame, justice also demands that the current generation must pass on to the next capital at least equivalent to what it inherited from the previous one, so that later generations should never 'be left worse off' (Barry, 1989: 159). One serious problem with this formulation however relates to the exact content of the capital that ought to be bequeathed to future generations – that is, whether it refers to natural resources, man-made resources, welfare, or social and technological capabilities. A related controversy is about whether these resources are commensurable and perfectly substitutable. In other words, is it fair for the present generation to exhaust resource X and compensate a future generation by leaving an abundance of resource Y? Some others reject the notion that future generations have rights and can be a legitimate community of justice on the ground that people yet unborn do not have identity and 'cannot now, therefore, be the present bearer or subject of anything including rights' (De George, 1981: 161).

Gosseries (2006, 2008) and Page (2006, 2007) among others have provided extensive treatments to show that considerations of justice across generations are implicated in important environmental policies, such as defining the level of a global cap on GHG emissions, deciding an emission reduction trajectory over time, justifying the preservation of biodiversity, conservation of species, setting a moratorium on whaling, and so on. This may well be the case. In fact as stated, leaving behind a decent planet for future generations constitutes a central tenet and impetus for environmental sustainability. However, it is less clear that this intuition is best cast in terms of justice or that such considerations are overly significant in international environmental decision-making quarters.

Procedural environmental justice and IR

The question of who has the power to participate and shape decisions has become increasingly crucial in global environmental governance literature (Auer, 2002; Betsill, this book; Bulkeley and Mol, 2003; Clapp, 1998; Newig and Fritsch, 2009; Najam, 2005).

Although many comprehensive theories of justice include procedure and outcome (Miller, 1999; Nozick, 1974; Rawls, 1971), there is a growing tendency by commentators to give specific attention to procedural justice as an independent aspect of fairness in environmental politics (Adger et al., 2003; Newig and Fritsch, 2009; Paavola, 2005; Shrader-Frechette, 2002). Ceva (2012) for one argues that fairness of governance systems and corresponding equity in participation should be a legitimate focus of justice and moral scrutiny independent of the outcomes of such systems. He contends that questions about the principles and fairness of participation remain expedient even where a governance system appears to produce just decisions and outcomes. Barry (1965) stressed the need to distinguish between fairness in the rules of participation and 'background fairness' where emphasis is on making sure that all parties have equal opportunity to participate under the set rules. This distinction is very crucial as will become evident below.

The United Nations and many international environmental regimes have elaborate rules of procedure which are often rigorously debated, suggesting states are mindful of the need for procedural justice. However, much of the focus appears to be on rules of process with little clarification of intended outcomes. Moreover, the huge asymmetry in numbers, technical, language, and informational capabilities, among other factors, all but makes a joke of this formal attempt to secure procedural justice defined narrowly as being present in meetings (see Najam, 2005; Roberts and Parks, 2007). Table 13.1 presents a head count of the number of delegates from select developed and developing countries (based on comparable populations) that attended the annual UN climate change meetings.

One quickly sees that even when huge technical advantages are discounted, the sheer numerical advantage of industrialized countries in these meetings leaves poor countries absolutely little or no chance to exert significant influence in decision-making. It should be noted that considerations of procedural justice in IEP extend beyond the balance of power and participation between developed and developing countries. Concerns have been raised about the extent to which states truly represent the various segments, interests, and communities that constitute them (Bäckstrand, 2003; Wapner, 1995, 1996). The lack of representativeness of states in IEP has also been voiced with particular reference to the interests of indigenous communities (Bäckstrand, 2006; Dove, 2006; Lane and Corbet, 2005). Other accounts have highlighted the need for gender balance in both national and international environmental governance (Bretherton, 2003; Bäckstrand, 2004; Detraz, this book). In fact, one aspect of procedural injustice often identified in the literature is that while individuals, communities, and states alike have the ability

Table 13.1 Headcount of delegates from various countries who attended annual UN climate change meetings

	Chad 11.2m	Germany 81.8	Ethiopia 82.1m	UK 62.4m	DR of Congo 65.9	Brazil 190m	Nigeria 160m	Canada 34m	Algeria 36m	Japan 127m
2000	2	75	5	41	2	66	15	81	8	69
2001	2	56	3	37	2	40	19	46	8	98
2002	2	54	3	43	2	30	8	54	6	73
2003	1	62	0	38	2	55	13	66	14	76
2004	1	46	2	47	6	207	18	71	13	81
2005	1	48	2	83	7	34	9	371	11	70
2006	1	45	0	40	3	15	7	48	1	39
2007	5	101	2	64	9	196	31	61	8	75
2008	2	57	2	42	2	17	11	33	2	54
2009	2	31	7	22	7	34	27	24	11	55
2010	10	110	28	75	58	736	83	93	27	135

Notes: Lead Author's head count from UNFCCC attendance register from unfccc.int/resource/docs.

to affect or be affected by environmental challenges, the legitimacy to decide solutions remains almost exclusively the preserve of nation-states (or more specifically, a few state-representing elites). In defense of the need to democratize global environmental spaces, some have suggested formal space should be created for transnational NGOs and subnational organizations such as cities to have greater say in global environmental decision-making (Bäckstrand, 2004, 2006; Cashore, 2002; Wapner, 2006, 2007; see also Betsill, this book).

At the same time some have questioned the rationale for opening up international institutions to community groups, NGOs, and the like when these entities might themselves be in dire need of democratization (Lehr-Lehnardt, 2005; Jordan and Van Tuijl, 2000). Still others have questioned the justice of according equal vote to countries like the United States and Chad despite the huge difference in their population (Leech, 2002). There are no easy answers. However, questions about how best to democratize or achieve greater procedural justice in global environmental decision-making systems remains a priority in both theoretical and empirical terms (Farber, 2007; Tol and Verheyen, 2004).

Case study: Climate change

Climate change is a very good example for illustrating the role of justice in IEP and in particular the potential and limits of justice in global environmental governance. Although, as noted, demands for justice in global environmental rule-making have a fairly long history, climate change has provided the space and platform for the 'loudest' articulation of the need for justice in IEP in both policy and academic circles. In fact, it is fair to say that contestations for justice have been one, if not the most, defining feature of global climate policy since 1992.

There are at least two main reasons that account for the high-profile role of equity (a specific concern of justice) and justice (a broad term that typically has legal, political, and moral aspects) in global climate policy. The first is that climate change is implicated in fundamental developmental activities of all nations; thus the rules and policies adopted (or not) in addressing the challenge are bound to have serious consequences for the economy of nations. The second and related point is that climate change, with its massive negative impacts on natural, human, and social systems, has the ability to fundamentally alter the development path or options of many countries, especially those in the poorer regions of the world. In fact, for many countries in sub-Saharan Africa as well as several small island countries, climate change is an existential threat – holding out the possibility of forcing the entire country to relocate to new political territories (Biermann and Boas, 2010; Myers, 2002). These reasons are further amplified because of the huge asymmetry in contribution, impact, and decision-making abilities within and between countries.

In terms of impact, the vast proportion of the negative effects of climate change will be borne by the world populations that have least contributed to the problem. The implication is that climate change is essentially a case of the rich imposing their risks on the poor. Beyond that current decision-making processes and power structures mean that rich countries have advantages over poorer countries in the rather convoluted and complex process of deciding how to tackle the problem of climate change. This creates a distinct possibility that those responsible for the problem can use their decision powers to approve rules that further exacerbate their advantages.

In extant scholarship, several dimensions of justice in the context of climate change have been articulated (Adger, et al., 2006; Gardiner, 2004; Grasso, 2007; Page, 2007; Okereke and Schroeder, 2009). This ranges from two broad categories – justice in mitigation and adaptation – to 'ten layers' of justice as proposed by Parks and Roberts (2006). In our view, there are at least five practical justice challenges to which any global climate policy must respond. These include the following:

(i) Justice in mitigation: Given the vast asymmetry in contribution to climate change between nations in both historical and current terms, plus the urgent need for drastic emission reductions, as indicated by climate science, how should the burden sharing for climate mitigation be best distributed within and between nations (Adger, 2001; Müller, 2001; Thomas and Twyman, 2005)?

(ii) Justice in adaptation: Given that the vast proportion of the negative impacts of climate change will be on the people who have contributed least to the problem, what is the most just way for distributing the limited finance and capacity for helping vulnerable communities and countries adapt, plus how should the cost for adaption be distributed within and between countries (Adger, 2001; Adger et al., 2006; Thomas and Twyman, 2005)?

(iii) Justice in procedure: Given the huge asymmetry in decision-making power within and between countries, what are just ways of ensuring effective participation in climate negotiation processes? Critically, how can the notion of effective participation be made applicable across scales of geography from communities, through

states, to world regions (Thomas and Twyman, 2005; Adger et al., 2006; Vanderheiden, 2008)?
(iv) Justice in compensation: Given that climate change is already having devastating impacts on the lives and development paths of many countries, especially in the poor South, what are just ways of calculating and providing adequate compensation for the victims of climate change (Farber, 2007; Smit and Skinner, 2002; Tol and Verheyen, 2004)?
(v) Justice in background political structures: Given that the asymmetries implicated in climate change (contribution, impact, ability to participate, and so on) are only symptomatic of broader and deeper structural inequities, what are just ways to address the deep causes of these fundamental inequities, in a bid to create a fairer world (Shue, 1992; Okereke, 2011)?

In practice, these questions have been a regular feature of the climate regime since its inception in 1992. There is every indication that they will continue to dominate the effort to find a replacement for the new climate regime that will replace the Kyoto protocol in 2015.

In light of the above, the question then is what, if any, has been the role of the different environmental justice philosophies/movements in the development of the global climate regime? It is probably fair to say that deep ecology ideas are barely considered in climate policy. It is true that the target of climate stabilization is often discussed in the context of preventing irreversible damage to the natural ecosystem. However, what instrumental benefits humans can get from nature and threats to these benefits have almost exclusively been the concern. Thus any 'justice' – even gratitude as part of right relationship (cf. Barry, 1999: 263) – that might be owed to aspects of nature is barely considered in practice. More social aspects of deep ecology ideas, such as implications for social class, do have some consideration in climate change policy and discussions, if not explicitly labeled as such.

Similarly, social ecology has had little explicit impact on climate policy. There is growing recognition that the climate problem has an intimate link with prevailing class-based capitalist structures and that the most vulnerable to climate change are those at the bottom of the ladder in the world capitalist structure. Despite this, however, the idea that the market and neoclassical development patterns provide the best way of addressing climate change remains extremely strong in policy and academic discourse. The central assumption is that climate represents nothing more than a market failure and that the best way to address it is to internalize the externalities and in so doing correct the inefficiencies that caused the problem. This explains why market instruments such as carbon taxes, emissions trading, the Clean Development Mechanism (CDM), rating and disclosure, joint implementation, and voluntary carbon offsets are by far the dominant instruments for addressing climate change. This practice barely considers the mounting scholarship, casting serious doubt on the efficiency and poor distributional impacts of these instruments (Böhm and Dabhi, 2009, 2011; Lohmann, 2009, 2010).

Predictably, empty-belly ecology has had the most effect on climate policy. Right from the outset, developing countries resisted the attempt by the North to make climate change a matter of narrow discussion about the technical mitigation or sequestration of carbon. In an intense diplomatic effort led mostly by Brazil, China, and India, the South insisted that global efforts to address climate change must be put in the broader context of equitable international development. In fact, they argued that effort to address climate change must be seen as a tool for correcting fundamental injustice in global economic structures (Borione and Ripert, 1994; Dasgupta, 1994). It was through these efforts that concepts like North–South financial and technology transfer, per capita emissions, historical emissions, capacity building, and most notably the common but differentiated responsibility principle became very common vocabularies within the regime's development circles.

Yet, while there is evidence that the empty-belly ecology of the South has led to a very strong insertion of justice concerns in the climate regime, it is fair to say that actual practical policies aiming at achieving greater justice have at best had very marginal impact. The minimal evidence of such impact includes, firstly, the exclusion of the countries with historically low emissions from being required to undertake quantified emission reduction targets in the Kyoto regime. Secondly, there is the establishment of a climate adaption fund to help the Least Developed Countries (LDCs) deal with the negative impact of climate change. Thirdly, there is the CDM which provided opportunities for rich countries to make technological investments in developing countries in return for certified emission reduction (CER) credits. However, North–South transfers in the order of hundreds of billions of dollars originally envisaged from the outset of the global climate negotiations have not materialized (Grubb et al., 1991; Hayes, 1993). Nor has there been any radical change in the world capitalist and neoliberal system, which is responsible for climate destruction and Northern domination.

Summary and conclusion

Normative approaches to IEP have served to bring the 'unavoidability of justice' to mainstream scholarship and the forefront of global environmental policymaking. Concerted diplomatic effort and scholarship on this area have joined forces to overturn the notion that international politics is beyond the pale of morality and subject only to the calculus of egoistic and self-interested state actors. The impact of considerations of justice – deep ecology, social ecology, and most of all empty-belly ecological justice – litters the international environmental policy landscape. Yet, it remains the case that international diplomacy still tends in practice to support gains in wealth and power by those countries, companies, and individuals who already have power. Hence despite the central role of justice in international environmental policymaking and environmental studies scholarship, IR practice has so far failed to accord adequate concerns to questions of justice. As efforts to negotiate a new regime for climate change get under way, one of the interesting questions will be whether the new regime will build on the advances made or whether there will be significant regression.

Each of the facets of justice discussed here will bear sustained efforts to be applied to environmental policy in general and climate policy in particular, as justice questions, if not solutions, are clear in this field. Perhaps the most important are considerations of formal international procedures that address empty-belly environmental justice questions in the context of difficulties predicting the climate system (Charlesworth and Okereke, 2010), in particular, by gathering weight of evidence of the distribution approaches to justice and environmental ethics questions globally. This appears likely to suggest that policy and international agreements should be less focused on imposing maximization of economic growth and more focused on allowing the wealthy to live simply so that the poor and future generations can simply live. There is already evidence to suggest that more people in the world support moderate consumption rather than humans eating ourselves out of heart and home.

Annotated bibliography

Dobson, A. (1998) *Justice and the Environment: Conceptions of Environmental Sustainability and Dimensions of Social Justice* (Oxford: Oxford University Press).

A preeminent work which provides a detailed account of the relationship and trade-offs between different conceptions of environmental justice and various principles of social justice. It is the first book to robustly argue that while environmental sustainability and social justice may both be desirable social goals, these two objectives in reality are not always compatible or mutually reinforcing. A lot depends, at least in part, on *Which* conception of sustainability and to *Which* dimension of justice one subscribes.

Shue, H. (1992) 'The Unavoidability of Justice', in A. Hurrell and B. Kingsbury (eds) *International Politics of the Environment: Actors Interests and Institutions* (Oxford: Clarendon Press).

This is one of the earliest accounts that mounted a robust defense of the centrality of questions of justice in international environmental relations. The central argument is that questions of justice are not external to international environmental negotiations on three grounds. First is that negotiation between states takes place in the context of power asymmetry and conditions of 'background injustice'. The second is that most of the environmental problems that international cooperation attempts to solve have been caused, albeit unintentionally, by the rich and more powerful countries. It follows therefore that avoiding questions of justice will most likely result in poor nations having to bear a greater burden of environmental harm and international cooperation.

Roberts, J. T. and B. Parks (2007) *A Climate of Injustice: Global Inequality, North-South Policies, and Climate Policy* (Cambridge, MA: MIT Press).

This is a rich and well-argued account of the role of inequality between poor and rich countries in global climate negotiations. It contains an extensive analysis of the structural and historical origins of North–South injustice. It is strongly argued that there is little chance of getting a workable global agreement to deal with climate change and a strong argument unless close attention is paid to the fundamental issues of inequality and getting a good balance between conservation and development objectives.

Okereke, C. (2008) *Global Justice and Neoliberal Environmental Governance: Ethics, Sustainable Development and International Co-Operation* (London: Routledge).

This is the first book devoted – not to a moral argument about the importance of justice in global environmental governance – but to actually analyzing and identifying the different conceptions that actually underpin key global environmental agreements. Covering three important environmental regimes, Third UN Law of the Sea (UNCLOS III), Hazardous Waste Regime (Basel convention), and the global Climate Change Agreement (UNFCCC), it is shown that while a variety of conceptions of justice can be found in international environmental agreements, by far the most dominant are conceptions of justice that are consistent with neoliberalism and free market philosophy.

Agarwal, A. and N. Sunita (1991) *Global Warming in an Unequal World: A Case of Environmental Colonialism* (New Delhi: Centre for Science and Environment).

A trenchant critique of the data and argument presented by the World Resource Institute (WRI) which appears to suggest that big developing countries, especially India and China, have comparable historical contributions to global warming with industrialized countries. The authors say that the WRI study and publication was 'an excellent example of *environmental colonialism*' (italics in original) because it seeks to arm developed countries and especially the United States with (manipulated) data, upon which they can continue their goal to blame developing countries for causing global warming and perpetuate existing North–South inequality.

Works cited

Adger, W. N. (2001) 'Scales of Governance and Environmental Justice for Adaptation and Mitigation of Climate Change', *Journal of International Development* 13, 7, 921–931.
Adger, W. N., J. Paavola, S. Huq and M. J. Mace (eds) (2006) *Fairness in Adaptation to Climate Change* (Cambridge, MA: MIT Press).
Agar, N. (1997) 'Biocentrism and the Concept of Life', *Ethics* 108, 1, 147–168.
Alder, J. and D. Wilkinson (1999) *Environmental Law and Ethics (Palgrave Law Masters)* (Basingstoke: Palgrave Macmillan), 167–185. Available at http://www.lead-journal.org/content/09167.pdf.
Aristotle (1874/1998) *Nicomachean Ethics Book IV*, Translated by J. A. K. Thompson (Harmondsworth: Penguin).
Auer, M. R. (2000) 'Who Participates in Global Environmental Governance? Partial Answers from International Relations Theory', *Policy Sciences* 33, 2, 155–180.
Bachram, H. (2005) 'Climate Fraud and Carbon Colonialism: The New Trade in Greenhouse Gases', *Capitalism Nature and Socialism* 15, 4, 5–20.
Bäckstrand, K. (2003) 'Civic Science for Sustainability: Reframing the Role of Experts, Policy-makers and Citizens in Environmental Governance', *Global Environmental Politics* 3, 4, 24–41.
Bäckstrand, K. (2004) 'Scientisation vs. Civic Expertise in Environmental Governance: Eco-feminist, Eco-modern and Post-modern Responses', *Environmental Politics* 13, 4, 695–714.
Bäckstrand, K. (2006) 'Democratizing Global Environmental Governance? Stakeholder Democracy after the World Summit on Sustainable Development', *European Journal of International Relations* 12, 4, 467–498.
Barry, B. (1965) *Political Argument* (London: Routledge and Kegan Paul).
Barry, B. (1989) *Theories of Justice: A Treatise on Social Justice, Vol. 1 (California Series on Social Choice and Political Economy)* (Berkeley and Los Angeles: University of California Press).
Barry, J. (1999) *Rethinking Green Politics: Nature, Virtue and Progress* (London: Sage).
Baxter, B. (2005) *A Theory of Ecological Justice* (Abingdon: Routledge).
Beck, U. (1992) *Risk Society: Towards a New Modernity* (London: Sage).
Beck, U. (1999) *World Risk Society* (Cambridge: Polity Press).
Beck, U. (2010) 'Remapping Social Inequalities in an Age of Climate Change: For a Cosmopolitan Renewal of Sociology', *Global Networks* 10, 2, 165–181.
Beitz, C. R (1999) 'Social and Cosmopolitan Liberalism', *International Affairs* 75, 3, 515–529.
Bernstein, S. (2001) *The Compromise of Liberal Environmentalism* (New York: Columbia University Press).
Bhagwati, J. (1991) *The World Trading System in Risk* (Hemel Hempstead: Harvester Wheatsheaf).
Bhagwati, J. (1993) 'The Case for Free Trade', *Scientific American* 269 (November), 42–49.

Biermann, F. and I. Boas (2010) 'Preparing for a Warmer World: Towards a Global Governance System to Protect Climate Refugees', *Global Environmental Politics* 10, 1, 60–88.

Böhm, S. and S. Dabhi (eds) (2009) *Upsetting the Offset: The Political Economy of Carbon Markets* (London: Mayfly).

Böhm, S. and S. Dabhi (2011) 'Fault Lines in Climate policy: What Role for Carbon Markets?' *Climate Policy* 11, 6, 1389–1392.

Bookchin, M. (1987) 'Social Ecology Versus Deep Ecology: A Challenge for the Ecology Movement', *Green Perspectives: Newsletter of the Green Program Project*, 4–5. Available at http://dwardmac.pitzer.edu/Anarchist_Archives/bookchin/socecovdeepeco.html

Borione, D. and J. Ripert (1994) 'Exercising Common but Differentiated Responsibility', in I. Mintzer, J. A. Leonard and M. J. Chadwick (eds), *Negotiating Climate Change: The Inside Story of the Rio Convention* (Cambridge: Cambridge University Press).

Bowler, P. J. (1992) *The Fontana History of the Environmental Sciences* (London: Fontana Press).

Brennan, A. (2003) 'Chapter 2: Philosophy', in E. Page and J. Proops (eds) *Environmental Thought* (Cheltenham: Edward Elgar), 15–33.

Bretherton, C. (2003) 'Movements, Networks, Hierarchies: A Gender Perspective on Global Environmental Governance', *Global Environmental Politics* 3, 2, 103–119.

Bryant, B. and P. Mohai (eds) (1992) *Race and the Incidence of Environmental Hazards: A Time for Discourse* (Boulder, CO: Westview Press).

Bulkeley, H. and A. P. Mol (2003) 'Participation and Environmental Governance: Consensus, Ambivalence and Debate', *Environmental Values* 12, 2, 143–154.

Bull, H. (1977) *The Anarchical Society: A Study of Order in World Politics* (London: Macmillan).

Bull, H. (1983) 'Justice in International Relations', in A. Hurrell and K. Alderson (eds) (2000) *Hedley Bull on International Society* (Basingstoke: Macmillan).

Bullard, R. D. (1983) *Confronting Environmental Racism: Voices from the Grassroots* (Boston: South End Press).

Buzan, B. and R. Little (2001) 'Why International Relations has Failed as an Intellectual Project and What to do About It', *Millennium Journal of International Studies* 30, 1, 19–39.

Byrne, J., L. Glover and C. Martinez (eds) (2002) *Environmental Justice: Discourses in International Political Economy, Energy and Environmental Policy Series (Book 8)* (New Brunswick, NJ: Transaction Books).

Callicott, J. B. (1989) *In Defense of the Land Ethic: Essays in Environmental Philosophy* (Albany: State University of New York Press).

Caney, S. (2001) 'Review article: International Distributive Justice', *Political Studies* 49, 5, 974–981.

Caney, S. (2005) 'Cosmopolitan Justice, Responsibility, and Global Climate Change', *Leiden Journal of International Law* 18, 4, 747–775.

Carson, R. (1962) *Silent Spring* (Boston, MA: Houghton Mifflin).

Cashore, B. (2002) 'Legitimacy and the Privatization of Environmental Governance: How Non – state Market – Driven (NSMD) Governance Systems Gain Rule – Making Authority', *Governance* 15, 4, 503–529.

Ceva, E. (2012) 'Beyond Legitimacy. Can Proceduralism Say Anything Relevant About Justice?', *Critical Review of International Social and Political Philosophy* 15, 2, 183–200.
Charlesworth, M. and C. Okereke (2010) 'Policy Responses to Rapid Climate Change: An Epistemological Critique of Dominant Approaches', *Global Environmental Change* 20, 1, 121–129.
Clark, N. (2008) 'Aboriginal Cosmopolitanism', *International Journal of Urban and Regional Research* 32, 3, 737–744.
Conca, K., M. Alberty and G. Dabelko (eds) *Green Plant Blue, Environmental Politics from Stockholm to Rio* (Boulder: Westview Press).
Clapp, J. (1998) 'The Privatization of Global Environmental Governance: ISO 14000 and the Developing World', *Global Governance* 4, 3, 295–316.
Dasgupta, C. (1994) 'The Climate Change Negotiations', in I. Mintzer and J. A. Leonard (eds) *Negotiating Climate Change: The Inside Story of the Rio Convention* (Cambridge: Cambridge University Press).
De George, R. (1981) 'The Environment, Rights, and Future Generations', in E. Partridge (ed.) *Responsibilities to Future Generations. Environmental Ethics* (New York: Prometheus Books), 157–166.
den Elzen, M. G. J., M. Schaffer and P. L. Lucas (2005) 'Differentiating Future Commitments on the Basis of Countries Relative Historical Responsibility for Climate change: Uncertainties in the Brazil proposal in the context of a policy implementation', *Climatic Change* 71, 277–301.
Dobson, A. (1991) *The Green Reader* (London: André Deutsch).
Dobson, A. (1998) *Justice and the Environment: Conceptions of Environmental Sustainability and Dimensions of Social Justice* (Oxford: Oxford University Press).
Dobson, A. (2005) 'Globalisation, Cosmopolitanism and the Environment', *International Relations* 19, 3, 259–273.
Dove, M. R. (2006) 'Indigenous People and Environmental Politics', *Annual Review of Anthropology* 35, 191–208.
Elliot, L. (1998) *The Global Politics of the Environment* (London: Palgrave Macmillan).
Farber, D. A. (2007) 'Basic Compensation for Victims of Climate Change', *University of Pennsylvania Law Review* 155, 6, 1605–1656.
Friman, M. and B. Linner (2008) 'Technology Obscuring Equity: Historical Responsibility in UNFCCC Negotiations', *Climate Policy* 8, 339–354.
Gardiner, S. M. (2004) 'Ethics and Global Climate Change', *Ethics* 114, 555–600.
Geldart, C. and P. Lyon (1981) 'The Group 77: A Perspective View', *International Affairs* 57, 1, 79–101.
Gosseries, A. (2006) 'Constitutionalizing Future Rights?', *Intergenerational Justice Review* 3, 2, 10–11.
Gosseries, A. (2008) 'On Future Generations' Future Rights', *Journal of Political Philosophy* 16, 4, 446–474.
Grasso, M. (2007) 'A Normative Ethical Framework in Climate Change', *Climatic Change* 81, 223–246.
Grieco, J. M. (1988) 'Anarchy and the Limits of Co-operation: A Realist Critique of the Newest Liberal Institutionalism', *International Organization* 42, 3, 485–507.
Guha, R. and J. Martinez-Alier (1997) *Varieties of Environmentalism* (London: Earthscan).

Grubb, M., J. K. Sebenius and A. Magalhaes (1991) *Who Bears the Burden? Equity and Allocation in Greenhouse Gas Emissions and Abatement*. Paper for SEI project for UNCED.

Hamlin, T. B. (1989) 'Debt-for-Nature Swaps: A New Strategy for Protecting Environmental Interests in Developing Nations', *Ecology Law Quarterly* 16, 1065–1089.

Hayes, P. (1993) 'North–South Transfer', in P. Hayes and K. Smith (eds) *The Global Greenhouse Regime: Who Pays?* (London: Earthscan).

Hovi J., D. Sprinz and A. Underdal (2003) 'The Oslo-Potsdam Solution to Measuring Regime Effectiveness: Critique, Response, and the Road Ahead', *Global Environmental Politics* 3, 3, 74–96.

Hurrell, A. (2001) 'Global Inequality and International Institutions', *Metaphilosophy* 32, 1–2, 34–57.

Hume, D. (1975) 'An Enquiry Concerning Principles of Morals', in A. Ryan (ed.) *Justice* (Oxford: Oxford University Press).

Jordan, L. and P. Van Tuijl (2000) 'Political Responsibility in Transnational NGO Advocacy', *World Development* 28, 12, 2051–2065.

Keohane, R. O. (1984) *After Hegemony Co-operation and discord in the World Political Economy* (New Jersey: Princeton University Press).

Kiss, A. (1985) 'The Common Heritage of Mankind: Utopia or Reality?', *International Journal* 40, 423–441.

Knicley, J. E. (2012) 'Debt, Nature, and Indigenous Rights: Twenty-five Years of Debt-for-Nature Evolution', *Harvard Environmental Law Review*, 36, 79–122

Lane, M. B. and T. Corbett (2005) 'The Tyranny of Localism: Indigenous Participation in Community-Based Environmental Management', *Journal of Environmental Policy and Planning* 7.2, 141–159.

Leech, D. (2002) 'Designing the Voting System for the Council of the European Union', *Public Choice* 113, 3–4, 437–464.

Lehr-Lehnardt, R. (2005) 'NGO Legitimacy: Reassessing Democracy, Accountability and Transparency', In *Cornell Law School Inter-University Graduate Student Conference Papers*.

Leopold, A. (2001) *A Sand County Almanac: With Essays on Conservation* (New York: Oxford University Press).

Levy, D. and P. Newell (eds) (2005) *The Business of Global Environmental Governance* (Cambridge, MA: MIT Press).

Litfin, K. (1994) *Ozone Discourse: Science and Politics in Global Environmental Cooperation* (New York, NY: Columbia University Press).

Lohmann, L. (2009) 'Toward a Different Debate in Environmental Accounting: The Cases of Carbon and Cost–Benefit', *Accounting, Organizations and Society* 34, 3, 499–534.

Lohmann, L. (2010) 'Uncertainty Markets and Carbon Markets: Variations on Polanyian Themes.' *New Political Economy* 15, 2, 225–254.

Low, N. P. and B. Gleeson (1998) *Justice Society and Nature: An Exploration of Political Ecology* (New York: Routledge).

Lynas, M. (2009) 'How Do I know China Wrecked the Copenhagen Deal? I Was in the Room', *Guardian*, http://www.guardian.co.uk/environment/2009/dec/22/copenhagen-climate-change-mark-lynas

Masika, R. (ed.) (2002) *Gender, Development, and Climate Change* (Oxford: Oxfam).

Meadows, D. H., D. L. Meadows, J. Randers and W. W. Behrens III (1972) *Limits to Growth* (New York: New American Library).
Miller, D. (1999) *Principles of Social Justice* (Cambridge, MA: Harvard University Press).
Müller, B. (2001) 'Varieties of Distributive Justice in Climate Change', *Climatic Change* 48, 2, 273–288.
Myers, N. (2002) 'Environmental Refugees: A Growing Phenomenon of the 21st Century', *Phil. Trans. R. Soc. Lond. B* 357, 1420, 609–613.
Naess, A. (1973) 'The Shallow and the Deep, Long Range Ecology Movement: A Summary', *Inquiry* 16, 95–99.
Nagel, T. (2005) 'The Problem of Global Justice', *Philosophy & Public Affairs* 33, 2, 113–147.
Najam, A. (2005) 'Developing Countries and Global Environmental Governance: From Contestation to Participation to Engagement', *International Environmental Agreements: Politics, Law and Economics* 5,3, 303–321.
Nelson, W. N. (1974) 'Special Rights, General Rights, and Social Justice', *Philosophy & Public Affairs* 3, 4, 410–430.
Newell, P. and M. Paterson (2010). *Climate Capitalism: Global Warming and the Transformation of the Global Economy* (Cambridge: Cambridge University Press).
Newig, J. and O. Fritsch (2009) 'Environmental Governance: Participatory, Multi-level–and Effective?', *Environmental Policy and Governance* 19, 3, 197–214.
Nozick, R. (1974) *Anarchy State and Utopia* (New York: Basic Books).
Okereke, C. (2008a) *Global Justice and Neoliberal Environmental Governance: Ethics, Sustainable Development and International Co-Operation* (London: Routledge).
Okereke, C. (2008b) 'Equity Norms in Global Environmental Governance', *Global Environmental Politics* 8, 3, 25–50.
Okereke C. (2011) 'Moral Foundations of Global Environmental and Climate Justice', *Royal Institute of Philosophy Supplement* 69, 117–135.
Okereke, C. and D. McDaniels (2012) 'To What Extent are EU Steel Companies Susceptible to Competitive Loss Due to Climate Policy?', *Energy Policy* 46, 203–215.
Okereke C. and H. Schroeder (2009) 'How can the Objectives of Justice, Development and Climate Change Mitigation be Reconciled in the Treatment of Developing Countries in a Post-Kyoto Settlement?', *Climate and Development* 1, 10–15.
O'Riordan, T. (1977) 'Environmental Ideologies', *Environment and Planning* 9, 1, 3–14.
O'Riordan, T. (1981) *Environmentalism* (London: Pion Books).
Page, E. A. (2006) *Climate Change, Justice, and Future Generations* (Cheltenham: Edward Elgar).
Page, E. A. (2007) 'Fairness on the Day After Tomorrow: Justice, Reciprocity and Global Climate Change', *Political Studies* 55, 225–242.
Page, E. A. (2011) 'Cosmopolitanism, Climate Change, and Greenhouse Emissions Trading', *International Theory* 3, 1, 37–69.
Parks, B. C. and J. T. Roberts (2006) 'Environmental and Ecological Justice', in M. M. Betsill, K. Hochstetler and S. Demitris (eds) *Palgrave Advances in International Environmental Politics* (Basingstoke: Palgrave Macmillan).
Paterson, M. (2000) *Understanding Global Environmental Politics: Domination, Accumulation and Resistance* (Basingstoke: Macmillan).

Paterson, M. (2006) 'Theoretical Perspective of International Environmental Politics', in M. M. Betsill, K. Hochstetler and S. Demitris (eds) *Palgrave Advances in International Environmental Politics* (Basingstoke: Palgrave Macmillan).
Paavola, J. (2005) 'Seeking justice: International Environmental Governance and Climate Change', *Globalizations* 2, 3, 309–322.
Paavola, J. and W. N. Adger (2006) 'Fair Adaptation to Climate Change', *Ecological Economics* 56, 4, 594–609.
Pepper, D. (1984) *The Roots of Modern Environmentalism* (London: Routledge).
Pogge, T. (2001) 'Priorities of Global Justice', *Metaphilosophy* 32, 1/2, 6–24.
Potter, R., T. Binns, J. Elliott, and D. Smith (2008) *Geographies of Development: an Introduction to Development Studies* (3rd ed.) (New York: Prentice Hall).
Regan, T. (1987) 'The Case for Animal Rights', *Advances in Animal Welfare Science* 3, 179–189.
Ridely, M. and B. Low (1993) 'Can Selfishness Save the Environment?', *Human Ecology Review* 1, 1, 1–13.
Roberts, J. T. and B. Parks (2007) *A Climate of Injustice: Global Inequality, North-South Policies, and Climate Policy* (Cambridge MA: MIT Press).
Rawls, J. (1971) *A Theory of Justice* (Oxford: Oxford University Press).
Sachs, W. (ed.) (1993) *Global Ecology: A New Arena of Political Conflict* (London: Zed Books).
Sachs, W. (1999) *Planet Dialectics: Explorations in Environment and Development* (London: Zed Books).
Salter, R., B. Schutz and S. Dorr (eds) (1993) *Global Transformation and the Third World* (Boulder: Lynne Rienner Ltd).
Sandbach, F. (1980) *Environment, Ideology and Policy* (Oxford: Basil Blackwell).
Schlosberg, D. (2007) *Defining Environmental Justice: Theories, Movements and Nature* (Oxford: Oxford University Press).
Shrader-Frechette, Kristin (2002) *Environmental Justice: Creating Equity, Reclaiming Democracy* (New York: Oxford University Press).
Shue, H. (1992) 'The Unavoidability of Justice' in A. Hurrell and B. Kingsbury (eds) *International Politics of the Environment: Actors Interests and Institutions* (Oxford: Clarendon Press).
Shukla, P. R. (1999) 'Justice, Equity and Efficiency in Climate Change: A Developing Country Perspective', in F. L. Toth (ed.) *Fair Weather? Equity Concerns in Climate Change* (London: Earthscan).
Singer, P. (1975) *Animal Liberation: A New Ethics for our Treatment of Animals* (New York: Random House).
Skinner, E. (2011) *Gender and Climate Change: Overview Report* (Brighton: Institute of Development Studies).
Snidal, D. (1985) 'Coordination Versus Prisoners Dilemma: Implications for International Cooperation and Regimes', *American Political Science Review* 79, 923–942.
St Augustine (1467/2003) *The City of God*, Translated by Henry Bettenson (London: Penguin).
Strong, M. (1999) 'Hunger, Poverty, Population and Environment', *The Hunger Project Millennium Lecture* 7 April 1999. Madras (India), The Hunger Project http://www.thp.org/reports/strong499.htm [no page].
Sunstein, C. R. and C. M. Nussbaum (eds) (2004) *Animal Rights: Current Debates and New Directions* (New York, NY: Oxford University Press).

Thomas, D. S. and T. Chasca (2005) 'Equity and Justice in Climate Change Adaptation Amongst Natural-Resource-Dependent societies', *Global Environmental Change* 15, 2, 115–124.
Tol, R. S. and R. V. Roha (2004) 'State Responsibility and Compensation for Climate Change Damages – a Legal and Economic Assessment', *Energy Policy* 32, 9, 1109–1130.
UNFCCC (1992) United Nations Framework Convention on Climate Change, Available at www.unfccc.int – http://unfccc.int/resource/docs/convkp/conveng.pdf
Vanderheiden, S. (2008) *Atmospheric Justice: A Political Theory of Climate Change* (New York: Oxford University Press).
Viotti, P. R. and M. V. Kauppi (2011) *International Relations Theory: Realism, Pluralism and Globalism* (New York: Macmillan).
Walker, G. and H. Bulkeley (2006) 'Geographies of Environmental Justice', *Geoforum* 37, 655–659.
Wapner, P. (1995) 'Politics Beyond the State', *World Politics*, 47, 3, 311–340.
Wapner, P. (1996) *Environmental Activism and World Civic Politics* (Albany: State University of New York Press).
Wapner, P. (2006) 'A Meaningful Democratic Platform on Climate Change' (with Benjamin Goldstein) in *Tikkun* 12, 5, September/October.
Warpner, P. (2007) 'The UN and Civil Society', in T. Weiss and S. Daws (eds) *The Oxford Handbook on the United Nations* (New York: Oxford University Press).
WCED (1987) *Our Common Future: The World Commission on Environment and Development* (Oxford: Oxford University Press).
Willetts, P. (1978) *The Non-aligned Movement: The Origins of a Third World Alliance* (London: F. Pinter).
Wise, S. M. (2003) *Drawing the Line: Science and the Case for Animal Rights* (Cambridge, MA: Perseus Books).

14
Transparency and International Environmental Politics

Aarti Gupta and Michael Mason

Introduction

Transparency is increasingly the subject of scrutiny and controversy in the social sciences (Florini, 2007; Graham, 2002; Gupta, 2008; Langley, 2001; Mol, 2008). The concept is linked to the most politically charged debates of our times, relating to due process and good governance, human rights, global security, and the need for oversight of markets in an era of unfettered globalization (Grigorescu, 2007; Lord, 2006; Roberts, 2004). In this chapter, we analyze the rise and effects of what we posit to be a 'transparency turn' in international environmental politics (IEP) as well.[1] A call for transparency increasingly informs actor expectations in a diverse array of environmental issue areas. Both state and non-state actors embrace transparency as a necessary feature of decision-making and/or regulatory action to address transboundary environmental problems. Diverse actors are championing transparency as a means to enhance efficiency, accountability, and/or effectiveness of international environmental policy outcomes. Furthermore, a wide range of governance modalities (informal, formal, voluntary, mandatory) call for disclosure, suggesting a malleability between transparency norms and diverse institutional forms of environmental governance that deserves more sustained theoretical and empirical attention.

In much scholarship and policy practice, transparency is assumed to be a necessary precondition for more accountable, democratic, and effective policy and governance outcomes (Dingwerth, 2007; Dryzek, 1999; Held, 2004; Keohane, 2006). Yet, a growing number of transparency analysts are now revealing not only the promise but also the pathologies of relying on transparency to meet specific governance ends (Bannister and Connolly, 2011; Fung et al., 2007; Hood and

Heald, 2006; Lord, 2006). Such scholarship suggests that the ideal(s) of transparency might be contested or not attained in practice. There remains thus a compelling need to investigate the growing pervasiveness and implications of an embrace of transparency in IEP, one that extends beyond state-led international environmental regimes (Mitchell, 1998), to include private and market-based governance as well (Langley, 2001).

In doing so, a first step is to clarify how transparency is being conceptualized in diverse scholarly writings, and how we operationalize the concept in this chapter. Transparency is most often associated with openness and flows of information (Gilbert, 2007; Etzioni, 2011; Heald, 2006; Mitchell, 2011). Etymologically, it refers to *rendering visible* (Michener and Bersch, 2011). For purposes of this chapter, we conceptualize transparency as *disclosure of information intended to evaluate and/or steer behavior*. Our interest here is in the phenomenon of 'governance by disclosure' (Gupta, 2008), by which we understand public and private governance initiatives that employ targeted disclosure of information as a way to evaluate and/or steer the behavior of selected actors. We view the proliferation of governance by disclosure in IEP as clearly reflective of a transparency turn in this realm. Furthermore, such a transparency turn is 'multidirectional' (Michener and Bersch, 2011: 5), insofar as it is fueled by multiple architects and recipients of transparency, going beyond states to include corporations, civil society groups, international organizations, consumers, and citizens.

Governance by disclosure initiatives in the international environmental domain include, for example, the multilaterally negotiated Aarhus Convention on the right to environmental information, where disclosure of environmental information from states is linked to a right to know and enhanced citizen participation and access to justice (Krämer, 2012; Mason, 2014). A range of prior-informed consent-based global treaties governing trade in restricted pesticides, hazardous waste, or genetically modified organisms (GMOs) also seek to govern risk by furthering a right to know about global transfers of risky substances, and a right to choose whether to accept such transfers (Jansen and Dubois, 2014; Langlet, 2009). In addition, private eco-labeling schemes in forestry, fisheries, or the organic food sector rely on information disclosure to achieve a variety of governance ends, including empowering consumers to choose and improving sustainability performance (Auld and Gulbrandsen, 2010; Dingwerth, 2007; Dingwerth and Eichinger, 2010). Diverse corporate voluntary initiatives are also premised on the power of disclosure to achieve various ends, including the Global

Reporting Initiative (GRI) (calling for sustainability reporting by private corporations) or the NGO-led Publish What You Pay (calling for disclosure of revenue earnings by extractive industries operating in resource-rich developing countries) (Clapp, 2007; Pattberg and Okechukwu, 2009). Lastly, information disclosure is central to efforts by international organizations, such as the World Bank or the World Trade Organization, to improve accountability through disclosure (Roberts, 2002, 2004; Grigorescu, 2007).

All of the above initiatives vary greatly in terms of who is pushing for disclosure, from whom, and to what end. Notwithstanding these differences, underlying all of them is the fundamental presumption that information matters. If so, it becomes important to assess both the causes for a growing embrace of transparency and the effectiveness of governance by disclosure in a global context. We do so here by first reviewing diverse conceptual perspectives on transparency in international relations, distinguishing between institutionalist and critical theoretical perspectives, including relevant variants of each. We next advance an analytical framework for comparative assessment of the *uptake, institutionalization, and effects* of governance by disclosure in the international environmental realm. The following section briefly illustrates the evaluative prowess of this framework by applying it (selectively) to a diverse set of environmental governance by disclosure initiatives. In concluding, we put forward some general propositions about the prospects for transparency to improve the procedural quality and substantive outcomes of international environmental governance. We close by identifying areas for future research.

Theorizing transparency: Conceptual perspectives

Various strands of scholarly literature in comparative politics and international relations have analyzed the causes and consequences of an embrace of transparency in the international environmental realm. National-level legal and economic analyses of environmental policy were among the first to draw attention to what Ann Florini (1998) has labeled 'regulation by revelation' in specific developed country contexts (for example, Graham, 2002; Beierle, 2004; Fung et al., 2007; Gouldson, 2004; Konar and Cohen, 1997). Such scholarship has now expanded to generate insights about the diffusion and effects of right-to-know and access to information laws across the globe (for example, Florini, 2007). Such diffusion of transparency and disclosure norms and practices from industrialized to emerging economies – notably pollution release and rating programs – has stimulated further comparative research on transparency and its link to policy outcomes (for example, Blackman et al., 2004; Garcia et al., 2009; Gupta and Golder, 2005; Kathuria, 2009). This has been complemented by writings in international relations focusing on transparency's effects in improving oversight of international financial markets and enhancing (or reducing) the prospects for global security (Best, 2005; Lord, 2006; Soederberg, 2001; Stasavage, 2003).

Such analyses have been undertaken from both institutionalist and critical theoretical perspectives (see Paterson, this book). In international relations scholarship, *institutionalist* perspectives have emphasized the importance of openness, communication, and information exchange as central to more effective and accountable (global) governance. The basic premise is that institutions – conceived as organized clusters of rule-making and collective behavior – can increase cooperation among states by correcting for information asymmetries and rationalizing decision-making (see, for example, Mitchell, 1998, 2011; Keohane, 2006). In research on IEP, two strands of institutionalism – liberal institutionalism and rational institutionalism – have generated significant work on transparency.

Liberal institutionalism (underpinning 'international regime analysis') posits that transparency in transboundary environmental rule-making promotes inter-state cooperation by rendering more open and publicizing shared interests and actor commitments (Konar and Cohen, 1997; Stephan, 2002). The global institutionalization of information disclosure in the environmental realm ranges from multilateral obligations on notification and prior-informed consent in governing transfers of risky or hazardous substances, to the transnational diffusion of voluntary sustainability reporting standards. In analyzing such cases of transparency in international politics, liberal institutionalists are apt to attribute the lack of effective disclosure-based governance to shortfalls of design or capacity, and concurrent failure to embed transparency within the decision contexts of both disclosers and recipients (Bauhr and Nasiritousi, 2012; Florini, 2007; Fung et al., 2007). This promotes a functionalist concern with institutional design and capacity-building relating to governance by disclosure mechanisms.

Rational institutionalists view institutions as sets of incentives shaping the calculations of rational actors with regard to their environmental preferences. The nature and scope of institutionalized disclosure is argued to affect the payoff functions of polluters, whether these are firms (Garcia et al., 2009; Peck and Sinding, 2003) or states (Barrett, 2003; Bosetti et al., 2013). In this economic approach to the role of transparency in politics, the structured disclosure of environmental information assures relevant actors that free-riding polluters are easier to identify and thereby sanction. Dysfunctionalities in transparency arise or are explainable by actor preferences being distorted or skewed by the disclosure of incomplete or unreliable data, or the lack of comparability, comprehensibility, or accessibility of environmental information.

In contrast to institutionalism, *a critical theoretical perspective* on transparency emphasizes that its uptake, institutionalization, and effects need to be analyzed within broader, often contested, political-economic and normative contexts within which disclosure is being deployed. Critical perspectives thus stress the historicity and socio-political conditioning of transparency and disclosure practices, and acknowledge the unavoidable normativity (value-laden structure) of transparency. Two important strands of a critical perspective on transparency in international politics are constructivism and critical political economy.

Constructivist analyses of science, knowledge, and information have long highlighted the changing authority and accountability relationships around the generation and sharing of information in governing transboundary environmental challenges (see Lövbrand, this book). As writings in this vein suggest, current global environmental challenges, such as climate change or safe use of biotechnology, are characterized by fundamental normative and scientific uncertainties over what is valid knowledge and whose information counts. If so, agreeing on what is 'more and better' information, that is, on the scope and quality of transparency, is inevitably a matter of political conflict (Forsyth, 2003; Liftin, 1994; Jasanoff, 2004). Such insights are centrally relevant to the study of governance by disclosure. They suggest that the effects of disclosure depend *not* on reducing information asymmetries, and hence promoting more rational outcomes, but rather on whose information counts and is accorded primacy. From such a perspective, transparency itself is contested political terrain, whereby the very process of negotiating the scope and practices of disclosure serves to selectively frame and hence constitute the object of governance (Jasanoff, 2004; Gupta et al., 2014).

Critical political economy perspectives (for example, Clapp, 2007; Clapp and Helleiner, 2012; Levy and Newell, 2005; Newell, 2008a, 2008b; Stevis and Assetto, 2001) constitute another important lens into the uptake, institutionalization, and effects of transparency in international politics. These perspectives emphasize the current dominance in international environmental politics of what Steven Bernstein (2001) labels 'liberal environmentalism' – an authoritative complex of norms framing environmental governance challenges according to market liberal rights and values. From this perspective, transparency's uptake and effects in international environmental politics need to be understood within a broader (unequal) political economic context, one in which private actors have a major role both in shaping and deploying public modes of information disclosure. Given the global policy currency of liberal environmentalism, critical political economy posits that transparency, if adopted, will have minimal market-restricting effects.

Such a critical take acknowledges that transparency may *reproduce rather than disrupt* socially and ecologically harmful concentrations of public and private power. This is of particular relevance to an analysis of transparency in a global context, characterized by North–South disparities in the capacities to demand disclosure, and to access and use disclosed information. It points as well to a paradox of the transparency turn in global politics: that in certain instances, the desired quality and quantity of disclosed information (such as its breadth, comparability, comprehensibility, comprehensiveness, or accessibility) *follows upon* rather than *precipitates* changes in the broader normative and political context. Thus, greater levels of actionable transparency may only be obtainable *after* broader democratic, participatory, and environmental sustainability gains are secured in a given context, rather than being the cause for such gains (Florini and Saleem, 2011; Gupta, 2010b). This leaves open a fundamental question: is transparency epiphenomenal? Even as transparency becomes ubiquitous in global environmental governance, its transformative potential remains uncertain and contingent.

As one approach to assessing transparency's transformative potential systematically, we outline below an analytical framework with which to explain the *uptake and institutionalization* of governance by disclosure in specific instances, and evaluate its *effectiveness*, with effectiveness understood both as improved procedural quality and enhanced substantive outcomes. This framework is informed by constructivist and critical political economy approaches, combined in what we label a 'critical transparency studies' perspective.

Evaluating transparency: An analytical framework

In this section, we advance an analytical framework that allows for (comparative) analysis of the application and shifting representations of transparency, as shaped by political-economic and geopolitical constellations of power. We draw on current thinking and debates to advance a set of propositions about the drivers of transparency's uptake, its institutionalization, and its varied effects in international environmental politics.

Drivers of disclosure: Democratization and marketization

As the varied literature on transparency to date has documented, a growing embrace of transparency in global politics is being stimulated, partly, by a rights-based democratic push for individual liberty, choice, and participation (Graham, 2002; Gupta, 2008; Mason, 2008). We label this a *democratization* driver, insofar as democratic forms of environmental governance seem to call for open and inclusive forms of collective choice. A democratization driver of transparency is seen to underpin, for example, the spread of 'right to know' and freedom of information laws in multiple national contexts over the last quarter century (Florini, 2007).

This has now evolved, however, into a broader association of transparency with securing multidirectional accountability, and a more legitimate and democratic global polity (for example, Dingwerth, 2007; Keohane, 2006; Mason, 2005). Those positing such relationships assume that disclosure of relevant information is often a necessary step in holding actors to account for their (in-)actions according to set environmental standards. A reasonable expectation is that insofar as information is disclosed by those responsible for decisions that significantly affect the interests of others, such disclosure will facilitate individual and institutional answerability or even change. However, this involves assumptions about the capacity and responsiveness of particular actors, as well as the political systems within which they operate (Fox, 2007), including the assumption that democratic institutions foster greater accountability for environmental harm.

A key question then becomes whether a democratic rationale for disclosure is significant in the growing uptake of transparency in the environmental domain, and whether it is necessarily *liberal* democratic. While there is a substantial literature on the relationship between democratic decision-making and ecological sustainability, much of it informed by theories of deliberative democracy (for example, Baber and Bartlett, 2005; Bäckstrand, 2010; Smith, 2003), the relationship between transparency and democratization remains less studied.

Tensions arise from the fact that efforts to improve the democratic quality of environmental governance, through information disclosure, often go hand-in-hand with a neoliberal privileging of market-based solutions to global sustainability challenges and 'light touch' regulation of the private sector (Gupta, 2010b; Mason, 2008, 2014). This can stimulate an uptake of market-based and voluntary transparency as a default option to avoid more stringent or costly governance pathways (Lyon and Maxwell, 2011). This *marketization* driver of disclosure can result in an embrace of transparency with minimal market restricting effects, and one that exempts corporate actors from stringent disclosure (Florini, 2008; Haufler, 2010). Alternatively, however, disclosure of (certain types of) information might well be seen as essential to the establishment and functioning of *newly created markets* in environmental goods and services, such as those for carbon or genetic resources. In such cases, transparency might be promoted by powerful actors, such as corporations and policy elites, as desirable and necessary in order to create and facilitate markets, rather than being perceived as a threat in its potential to restrict markets (in, for example, environmentally harmful products).

A key element of a transparency research agenda thus consists of whether and how these two drivers of democratization and marketization shape uptake and institutionalization of transparency in the international environmental realm, and the extent to which they work in complementary or conflicting ways.

Institutionalization of disclosure: Shifting sites of sovereign authority

A democratization or marketization imperative for disclosure, and the ways in which these drivers intersect, necessarily shapes the institutionalization of transparency in specific instances. Distinctive regulatory possibilities and limits are created, depending upon whether disclosure is led by/solicited from public authorities or private actors. Following an extensive assessment of the institutionalization of information in governance, Mol (2008: 87–89) discerns, for example, a set of institutional tendencies in what he terms 'informational governance', notably *a decentering of state-based regulation and opening up of political space for other (non-state) actors*. Comparative in-depth case analysis is required to assess whether this indeed holds in the global environmental realm. In a given instance of governance by disclosure, the mapping out of institutionalization needs to identify the quantity (scope) and quality (attributes) of the disclosed information; the infrastructural designs steering the exchange of information between disclosers and recipients; and the role of information intermediaries in facilitating (or blocking) the usability and reach of environmental information.

Much scholarly and policy attention has focused, for example, on attributes of disclosed information as being central to the success of transparency-based governance. These include whether disclosed information is (perceived to be) accessible, comprehensible, comparable, accurate, or relevant (Dingwerth and Eichinger, 2010), and whether it is standardized or non-standardized. A related, increasingly important, development in transparency politics is the rise of information *intermediaries or infrastructures* that seek to validate or increase the utility of disclosed information for specific stakeholders (Etzioni, 2010; Fung et al., 2007; Graham, 2002; Gupta, 2008; Lord, 2006). These include auditors and verifiers of disclosed information, or civil society groups seeking to render disclosed information more user-friendly (see also Langley, 2001, for an early and detailed discussion of this). Such new transparency 'powerbrokers' may produce shifts in the loci of authority and expertise in environmental governance that shape how disclosure will be institutionalized and the concurrent effects that it will produce.

Effects of disclosure: Normative, procedural, and substantive

In addition to identifying the drivers of disclosure and dynamics of institutionalization, assessing the effectiveness of disclosure-based governance is the third pillar of the analytical framework outlined here (see also Young, this book). An evaluation of transparency's effectiveness requires, however, prior conceptual clarity on *distinct categories of effects*. The effectiveness of traditional environmental regulatory instruments is often primarily assessed in terms of reduced environmental harm (EEA, 2001; Mitchell, 1998; Young, 1999). Yet the ends sought to be furthered by governance by disclosure are broader, requiring an analytical openness to a variety of effects.

In line with this, we posit here three categories of effects: normative, procedural, and substantive. First, underlying most governance by disclosure initiatives is the normative rationale that recipients of information have a 'right to know' about damaging environmental behaviors or products (Beierle, 2004; Rowen-Robinson and Rothnie, 1996). Given the broad association of transparency with a right to know (for example, Florini, 2007), a goal to inform is thus a first step in embedding transparency entitlements within environmental governance. If so, the extent to which governance by disclosure furthers a right to know (and whose right to know) is an essential element in assessing its *normative effectiveness*. This requires analysis of whether a right to know is contested or not, and whether and how it is being institutionalized in practice in a given disclosure initiative.

Going beyond a right to know, transparency is also closely associated, in multiple political and legal analyses, with *procedural effects* related to the quality of governance processes (Graham, 2002; Gupta, 2008; Stasavage, 2003; van den Burg, 2004; Keohane, 2006). The procedural goals of disclosure can thus include holding disclosers accountable, as well as enhancing participation or the informed choice of information recipients. Assessing the procedural effectiveness of governance by disclosure thus requires assessing the extent to which these goals are accepted, institutionalized, and attained in practice.

Finally, disclosure also aims for improved *substantive* outcomes, such as reduced emissions, risk mitigation, and environmental improvements (for example, Fung et al., 2007; Gouldson, 2004; Stephan, 2002; Mitchell, 1998). A key example is the much-analyzed United States Toxic Release Inventory, where an ultimate goal of disclosure is reduced emissions of toxic pollutants (for example, Konar and Cohen, 1997). In global environmental governance, these substantive regulatory goals converge on the prevention or mitigation of significant transboundary environmental harm or harm to the global commons. Yet the link between transparency and environmental improvements remains both little examined and challenging to assess. This is related to long-acknowledged causality challenges inherent in such assessments. It is also perhaps linked to a more dominant association of transparency with a procedural turn in environmental governance, whereby its empowerment potential (and link to accountable, participatory and legitimate governance) may often be privileged over its role in securing substantive environmental gains (for example, Rose-Ackermann and Halpaap, 2002, see also Gupta, 2008). Yet the relationship between transparency and substantive environmental improvements is ever more important to assess, insofar as disclosure might be relied upon, more so than previously, as an innovative means by which to secure (transboundary) environmental improvements in a neoliberal, marketized global governance context.

In the next section, we analyze some of the issues briefly sketched above in an illustrative comparative empirical examination of a number of governance by disclosure initiatives in IEP.

Evaluating transparency: A comparative empirical assessment

A critical transparency studies approach is amenable to multiple research methodologies. In this section, we briefly illustrate the merits of a *comparative case study approach,* as one that is particularly well suited to explaining the unfolding of transparency initiatives across diverse modalities of international environmental governance. In doing so, we touch upon four significant examples of governance by disclosure in IEP – two state-led multilateral agreements (the Aarhus Convention and the Cartagena Protocol on Biosafety) and two private voluntary disclosure initiatives (the Carbon Disclosure Initiative and the GRI). We select these cases to illustrate the *multidirectional and hybrid (public–private)* nature of transparency's role and impact in IEP. We first provide a brief introduction to each below.

Aarhus Convention

The 1998 Convention on Access to Information, Public Participation in Decision-Making and Access to Justice in Environmental Matters (henceforth Aarhus Convention) is the most far-reaching expression in international environmental law of transparency and information disclosure norms. Negotiated under the auspices of the UN Economic Commission for Europe, it codifies citizen rights to information access, public participation, and access to justice in environmental matters. These are non-discriminatory entitlements for individual and civil society actors within and across the territories of all parties to the convention. The convention imposes mandatory information disclosure duties on public authorities, including both passive (request-based) and active measures. Private entities are excluded from these provisions, although a 2003 Protocol on Pollutant Releases and Transfer Registers indirectly creates disclosure obligations on private operators of relevant industrial facilities. An oft-noted characteristic of the Aarhus Convention is its innovative compliance mechanism, which includes rights of public actors to submit allegations of non-compliance to the convention compliance committee and launch liability claims against those alleged to be breaching *any* environmental law within the jurisdiction of an Aarhus member state. These rights strengthen the scope and depth of Aarhus obligations on information disclosure (Mason, 2014).

Cartagena Protocol on Biosafety

Negotiated under the auspices of the Convention on Biological Diversity, this multilateral agreement seeks to govern the safe transboundary transfer and use of GMOs. Ensuring safe trade in GMOs is a controversial global anticipatory governance challenge, one where the very existence of risk remains contested, but also one where information disclosure is central to current global governance efforts. The Cartagena Protocol was negotiated at the insistence of developing countries who were concerned about the spread of novel genetically modified seed and crop varieties into their territories without their knowledge. They demanded the right to know about and choose whether to permit or restrict such transfers. The Cartagena Protocol thus calls for the *advance informed agreement* of an importing country prior to trade in some GMOs. It thus seeks to further an importing country's normative right to know about entry into its borders of GMOs through labeling requirements, as well as the procedural and substantive goals of importer choice and regulatory oversight over GMOs (Clapp, 2007; Gupta, 2010a, 2010b).

Carbon Disclosure Project

In the realm of corporate social responsibility, the Carbon Disclosure Project (CDP) is arguably the most important voluntary mechanism advancing disclosure of corporate greenhouse gas emissions, energy use, and, more recently, exposure to deforestation risks. Founded in 2001 by a non-profit organization based in London, the CDP represents a consortium of institutional investors holding, by 2013, $87 trillion of assets (Carbon Disclosure Project, 2013). The CDP's core disclosure-based climate governance strategy revolves around an annual questionnaire to be completed by the world's largest companies that requests information on climate-related risks and opportunities, and their carbon management strategies. There is now a substantial level of reporting: according to the CDP, over 3,000 companies responded to the 2011 questionnaire, including 81 percent of the Global 500 (Carbon Disclosure Project, 2013). Individual public responses to the questionnaire are available on the CDP website, although CDP signatories, as investor members, can benefit from advanced comparative analysis of reported data. The CDP has been the focus of much international environmental governance research recently (for example, Kolk et al., 2008; Newell and Paterson, 2010; Knox Hayes and Levy, 2014).

Global Reporting Initiative

The GRI is a quintessential example of voluntary, private governance by disclosure in the international environmental realm. It consists of an extensive global multi-stakeholder effort to produce detailed guidelines for sustainability reporting by private companies. It is one of the longest established and most broad-ranging corporate sustainability disclosure initiatives now in place. It is also of growing interest to scholars of corporate social responsibility and private (transnational) governance (for example, Clapp, 2007; Dingwerth, 2007; Pattberg and Enechi, 2009). Emerging research on the GRI suggests that, notwithstanding elaborate, time-consuming and resource-intensive efforts to generate large amounts of sustainability data and disclose it, its utility to intended beneficiaries is often minimal (for example, Dingwerth and Eichinger, 2010).

Space constraints preclude a full application of our analytical framework to each of the four cases noted above. In what follows, we focus on two specific elements of our framework: first, whether and how the *institutionalization of transparency in each case decenters state-led regulation*; and second, the *normative, procedural, and substantive effects* of such governance by disclosure initiatives in the international environmental realm.

Institutionalizing governance by disclosure: Decentering state-led regulation?

For the multilateral disclosure regimes represented by the Aarhus Convention and the Cartagena Protocol, the institutionalization of disclosure reveals limits on usable disclosure by recipients in each case (for detailed analyses of these two cases, see Gupta, 2014; Mason, 2014). The manner of institutionalization in each case indicates, furthermore, a partial confirmation of the proposition advanced in the previous section, that transparency might decenter state-led regulation and open up political space for other actors. An analysis of institutionalized disclosure in these cases indicates that *transparency qualifies state sovereign authority but does not necessarily weaken it.* In one sense, this is no more than the familiar pooling of state sovereign powers well established in public international law. Transparency-based international regimes create state entitlements and duties based on the voluntary consent of parties to a treaty. This is reflected as well in the general access to information provisions in the Aarhus Convention, and the more specific disclosure rules on GMOs in the Cartagena Protocol.

This said, however, the propensity of states to adopt multilateral transparency norms and rules reflects their sensitivity to perceived domestic and external impacts on sovereign authority, constituting a geopolitics of information disclosure that reflects power differentials within and between developed and developing countries. This reveals, as well, a need to *differentiate between types of states* in assessing impacts of institutionalized disclosure on state-led regulation and authority. One conspicuous institutional logic running through the two examples evoked here is that of developed countries promoting global diffusion of transparency rules with *high policy currency in their domestic contexts.* This is evident from the diffusion of pollutant release and transfer registers (Aarhus Convention), and from the specific limits on global

GMO-related disclosure resulting from market liberal property rights norms and rules dominant in GMO exporting countries (Cartagena Protocol).

Furthermore, the institutionalization of transparency in such state-led disclosure arrangements is also *inflected by private authority* and is shaped by a dominant marketization impetus underpinning transparency's institutionalization in IEP. In the case of the Aarhus Convention, private enterprises are excluded from its mandatory information disclosure duties, which fall on public authorities, although pollution and release and transfer registers create indirect obligations on private operators. In the Cartagena Protocol, the major weight given to the *caveat emptor* (buyer beware) dictum in regulatory choices about GMOs puts poorer (GMO importing) countries at a disadvantage in their attempts to be informed about and mitigate risks relating to GMO imports. Similarly, the protocol's disclosure obligations are limited to information that is already known to exporters, ensuring that existing market practices do not have to change to comply with this global disclosure-based regime. Again, however, this effect varies by differential state power, as more powerful countries and trading blocs (such as the EU) can force GMO exporting countries to generate and disclose new information, typically by bypassing the Cartagena Protocol in favor of unilateral measures.

For the voluntary sustainability disclosure yielded by the GRI and the CDP, an original democratization impetus for transparency in both cases is again skewed in practice by a more dominant marketization imperative driving institutionalization of disclosure. There is ongoing bargaining between non-state rule-makers and corporate disclosers over the quantity and quality of disclosed information; for example, there are significant tensions between the corporate comparability goal of the GRI and the discretion allowed to companies to incentivize their self-reporting. There is also an important trend for sustainability reporting to become commodified: the CDP and GRI have, directly or indirectly (via commercial intermediaries), generated extensive paywalls behind which enhanced interpretive products are available, weakening their public transparency claims (Knox Hayes and Levy, 2014; Dingwerth and Eichinger, 2014).

In sum, the discussion above indicates a general finding across the cases with regard to drivers of disclosure and its institutionalization: that a marketization driver of transparency's uptake dominates, even as it contributes to a reconfiguring (rather than necessarily a weakening) of state sovereign authority. We turn next to the effects of disclosure-based governance.

Effectiveness of governance by disclosure: Normative, procedural, and substantive effects

To recall, we propose a broad typology of effectiveness in order to capture a range of (potential) effects attributable to transparency-based governance – normative, procedural, and substantive. We address below whether such effects are being realized in the cases discussed.

Normative effects: The normative aim of governance by disclosure is ensuring the 'right to know' about environmental performance and/or harm, geared mainly to civil society recipients but also to states and corporate actors. For individuals (as citizens or consumers), the moral authority infusing the right to know is closely affiliated with democracy and participation. Its strongest legal expression in IEP is in the access to information entitlement of the Aarhus Convention, where it attains the status of a cosmopolitan human right. While the right-to-know also features prominently in other governance by disclosure initiatives, it is either restricted to national settings (for example, domestic right-to-know laws); subsumed within problem-specific treaty entitlements (for example, Cartagena Protocol); or facilitated in actor- and sector-specific domains by civil society organizations (for example, CDP and GRI).

Across all these manifestations – including Aarhus rights – the public right to know is, however, restrained or diluted by countervailing norms, unsettling its governance legitimacy. The most potent are moral and legal norms underpinning the private authority of actors in market liberal systems of resource allocation; thus, we encounter environmental disclosure obligations constrained by corporate voluntarism (non-financial reporting systems), commercial confidentiality (Aarhus Convention), or the caveat emptor ('buyer beware') dictum (GMOs).

In summary, while right-to-know serves as a widely accepted normative justification of transparency in international environmental politics, its legal application tends to be compromised by the political deployment of market liberal or state sovereign norms.

Procedural effects: The procedural goals of transparency in environmental governance include empowering information recipients to perform meaningful governance roles, notably holding disclosers accountable and making choices that are more informed. Comparative analysis of the disclosure initiatives noted above suggests, however, that little sustained empowerment of intended information recipients is occurring to date in the cases examined.

The Aarhus Convention has arguably made the greatest legal progress with transnational public entitlements to environmental information, and non-compliance notifications, as a way to empower. Yet there have been repeated procedural blockages by convention parties to public information requests, and this opposition is often justified in relation to the discretion allowed to parties when implementing treaty obligations. The formal procedural rights for public actors created by the Aarhus Convention are, furthermore, not necessarily mirrored in the disclosure regime for global biosafety governance, whereby a holding to account by importing countries is rendered difficult by the limited disclosure obligations placed on those exporting potential risk. Procedural shortcomings concerning access to information by civil society actors are also apparent in voluntary environmental reporting, where civil society actors are intended to be either a primary or secondary recipient of information. The claim that these initiatives share – that disclosure is at least partly a means of public accountability – thus falls short in practice. For both the GRI and CDP, there are weaknesses in public participation, both at the system governance level and in terms of the usability of disclosed information for accountability claim making. According to Knox Hayes and Levy (2014), these voluntary corporate disclosure systems are limited by their focus on managerial processes and lack of comparability across forms, as well as by the lack of sanctions for non-compliance.

Substantive effects: Transparency and disclosure of environmental information can support substantive regulatory goals, such as reduced pollution emissions and the conservation of biodiversity. The direct substantive effect often attributed by proponents of disclosure is that the sharing of information will render producers of environmental damage or risk more responsive to regulatory pressures. Not surprisingly, the more process-oriented Aarhus Convention does not contain substantive environmental standards, assuming instead that its substantive aims of improved environmental health and safety will flow from procedural openness, access to participation, and justice. Other multilateral environmental agreements, including the Cartagena Protocol, are more centrally animated by harm prevention goals. Yet there are negligible treaty-based data sources on the environmental effects of the relevant disclosure measures, for reasons that include evaluative uncertainties, measurement difficulties, and a preoccupation with trade rather than environmental outcomes. It is instructive that, in the GMO governance case, various countries have bypassed the Cartagena Protocol's disclosure-based approach to governance by opting for unilateral moratoria or bans to achieve environmental and health protection goals.

In the voluntary realm of sustainability reporting systems that focus on managerial processes, the evidence on substantive environmental effectiveness is also slight. Dingwerth and Eichinger (2014) note the rising number of corporate reports registered under the GRI, but caution that the lack of data specificity and comparability prevents any meaningful assessment of environmental performance patterns. Knox Hayes and Levy (2014) reach the same conclusion in relation to corporate carbon disclosure systems, which, they claim, do not appear to be shifting core product or marketing strategies in a low-carbon direction. As with mandatory governance by disclosure, the monitoring and analysis of environmental outcomes by voluntary disclosure systems is still in its infancy.

Conclusions

This chapter began with the proposition that there was a 'transparency turn' in international environmental politics, suggesting a need to systematically assess its drivers and effects. The overview of disclosure-based governance undertaken here allows us to conclude that information disclosure is becoming a widely accepted norm and practice in multilateral and transnational governance of environmental harm and improved sustainability performance. At the same time, any claims about the 'rise and rise of transparency' in international politics (Raab, 2008: 600) need to be tempered by acknowledgment of competing trends that restrict the scope of actionable disclosure. These can be related to commercial confidentiality, the particularities of a neoliberal context within which disclosure is being promoted, scientific unknowabilities and uncertainties, or even to limits on the desire of society to know about (production and distribution of) environmental harm (Dingwerth and Eichinger, 2010). The transparency turn in global environmental governance is thus evident but partial.

This chapter has also highlighted the broader (contested) normative context that shapes the embrace of transparency by various actors, and its uneven institutionalization in the international environmental realm. As is evident from the empirical examples addressed here, struggles over the scope and practices of transparency are intensifying, and its effects are not necessarily positive in governance terms. The future research agenda of transparency studies then requires engaging, at the very least, with the rise of what Mol (2014) calls 'reflexive transparency', that is, secondary layers of disclosure necessary to render primary disclosure usable, as well as the infrastructures and powerbrokers of a transparency turn in governance. Such developments can only be understood by critically analyzing transparency's particular material and discursive contexts of application. Such an understanding is necessary to negotiate the multi-layered, contested realms of transparency-based IEP.

With regard to the legitimacy and effectiveness of such an approach to policy and governance, the relatively recent embrace of transparency as a regulatory tool cautions against a too quick dismissal of its potential to generate substantive environmental improvements. Furthermore, the transformative promise of governance by disclosure goes beyond substantive impacts to include normative and procedural effects as well. As indicated above, however, these effects are being circumscribed in practice by market liberal norms. This yields a proposition that requires further comparative assessment: that transparency-based governance may well face a *legitimation deficit*, insofar as it fails to fulfill its emancipatory and transformative potential.

The sources of legitimacy underpinning transparency-based governance derive, as we have indicated in this chapter, partly from a democratization impetus to governance, creating expectations among domestic and transnational publics that information disclosure will facilitate accountability claims against state and non-state actors responsible for producing significant environmental harm. Disclosure-based governance also fosters political legitimacy insofar as it enriches public understanding of what is proper in relation to the collective decisions of (potential) harm producers.

This notwithstanding, the inadvertent, indirect harm typically associated with transboundary environmental problems lends many governance by disclosure initiatives an air of experimentation concerning their intended substantive effects; and legitimacy becomes less feasible when expected from steering mechanisms coordinating dispersed decision-makers and affected publics. Furthermore, as the multidirectional character of governance by disclosure indicates, there remains a political struggle over legitimate arenas for disclosure rule-making and implementation, across diverse contexts and across hybrid configurations of state and non-state authority. There are, to be sure, cogent suggestions that increasing transparency in both state-led (vertical) and non-state (horizontal) multi-level governance can increase political legitimacy, if fed into more inclusive, deliberative systems of decision-making (Bernstein and Cashore, 2007; Dryzek and Stevenson, 2011). However, it may also be that increasing transparency and information disclosure will instead amplify the current legitimation deficits in global environmental governance, by locating the systemic sources of harm production in broader relations of political and economic power (Newell, 2008a). It is in this sense that governance by disclosure might itself face a legitimation deficit. A question for future research remains then the conditions under which transparency-based governance can overcome such potential legitimation deficits and thereby fulfill its transformative potential. This would also be fruitful to assess comparatively vis-à-vis other international policy domains.

Another aspect of such a future research agenda turns on the merits and attainability of 'full' versus 'partial' transparency and the implications for policy and governance outcomes. In the policy domains of international finance, security, and diplomacy, an oft-posed question turns on whether 'full disclosure' is necessary, feasible and/or desirable, in fulfilling the transformative potential of transparency. Most analyses conclude that complete disclosure in such areas is both unattainable and undesirable, given the merits of retaining varying degrees of secrecy, anonymity, or privacy in many instances (for example, Birchall, 2011; Lord, 2006). This raises, however, an intriguing question that we have only begun to touch upon in international environmental scholarship on transparency: how precisely is the (global) environmental realm distinct? The imperative to balance transparency with secrecy, privacy and anonymity arguably does not hold in this policy domain to the extent that it does in others, yet this has not been systematically analyzed. Future research could fruitfully shed light on the specific features of global environmental governance challenges that either impede full disclosure, or make its pursuit more or less desirable. What the limits of full disclosure, and the merits and demerits of partial transparency in international environmental politics, are remains an important empirical question, given the geopolitical and material contexts for such disclosure.

Note

1. This chapter draws upon and synthesizes the introductory chapter and main conclusions of a recently concluded comprehensive comparative assessment of 'Transparency in Global Environmental Governance', now forthcoming as an edited volume with MIT Press (Gupta and Mason, 2014).

Annotated bibliography

Fung, A., M. Graham and D. Weil (2007) *Full Disclosure: The Perils and Promise of Transparency* (New York: Cambridge University Press).

This important book is one of the first to examine whether transparency works as intended in a variety of policy fields, such as health, finance, education, national security, and environmental pollution control. It introduces the concept of 'targeted transparency' and outlines conditions under which targeted disclosure of information can achieve desired aims. It is primarily concerned, however, with the national context of the United States, and deals almost exclusively with *public* transparency, that is its scope is government-mandated, national-level US transparency policies.

Florini, A. (ed.) (2007) *The Right to Know: Transparency for an Open World* (New York: Columbia University Press).

This edited volume, compiled by a leading scholar of transparency studies, focuses on the diffusion of 'right-to-know' laws globally, and their functioning in diverse national contexts (for example, India, China, Nigeria) and across a range of institutional domains (for example, international financial institutions, the security sector, as well as environmental governance).

Gupta, A. and M. Mason (eds) (Forthcoming 2014). *Transparency in Global Environmental Governance: Critical Perspectives* (Cambridge, MA: MIT Press).

This is the first comprehensive examination of the nature and implications of a transparency turn in international *environmental* governance. It advances a 'critical transparency studies' lens to analyze whether transparency informs, empowers, and improves environmental outcomes in this policy domain. It does so by bringing together ten in-depth case analyses of 'governance by disclosure', including both public state-led and private voluntary disclosure initiatives in international environmental politics.

Lord, K. (2006) *The Perils and Promise of Global Transparency: Why the Information Revolution May not Lead to Security, Peace or Democracy* (New York: SUNY Press).

While not addressing environmental issues, this important contribution to transparency studies from political science and international relations offers a critical perspective on transparency, arguing that it is not the panacea that it is often assumed to be. Lord's book focuses on the role of transparency in conflict resolution and international security, and on the potential harmful uses of transparency in these policy domains by authoritarian governments and non-state actors.

Mol, A. (2008) *Environmental Reform in the Information Age: The Contours of Informational Governance* (New York: Cambridge University Press).

This comprehensive treatment in environmental sociology explores how (global) environmental governance is changing under the conditions of the Information Age. Mol advances the concept of 'informational governance' here in examining how information is increasingly becoming central to globalized and multilevel and multi-actor networked environmental reform processes.

Works cited

Auld, G. and L. H. Gulbrandsen (2010) 'Transparency in Nonstate Certification: Consequences for Accountability and Legitimacy', *Global Environmental Politics* 10, 3, 97–119.

Baber, W. F and R. V. Bartlett (2005) *Deliberative Environmental Politics: Democracy and Ecological Rationality* (Cambridge, MA: MIT Press).

Bäckstrand, K. (2010) *Environmental Politics and Deliberative Democracy: Examining the Promise of New Modes of Governance* (Cheltenham: Edward Elgar).

Bannister, F. and R. Connolly (2011) 'The Trouble with Transparency: A Critical View of Openness in e-Government', *Policy and Internet* 3, 1, article 8.

Barrett, S. (2003) *Environment and Statecraft: The Strategy of Environmental Treaty-Making* (Oxford: Oxford University Press).

Bauhr, M. and N. Nasiritousi (2012) 'Resisting Transparency: Corruption, Legitimacy and the Quality of Global Environmental Politics', *Global Environmental Politics* 12, 4, 9–29.

Beierle, T. C. (2004) 'The Benefits and Costs of Disclosing Information about Risks: What do we Know about Right-to-Know?', *Risk Analysis* 24, 2, 335–346.

Bernstein, S. (2001) *The Compromise of Liberal Environmentalism* (New York: Columbia University Press).

Bernstein, S. and B. Cashore (2007) 'Can Non-state Global Governance be Legitimate? An Analytical Framework', *Regulation and Governance* 1, 4, 347–371.

Best, J. (2005) *The Limits of Transparency: Ambiguity and the History of International Finance* (Ithaca, NY: Cornell University Press).

Bianchi, A. and A. Peters (eds) (2013) *Transparency in International Law* (Cambridge: Cambridge University Press).

Birchall, C. (2011) 'Introduction to "Secrecy and Transparency": The Politics of Opacity and Openness', *Theory, Culture and Society* 28, 7–8, 7–25.

Blackman, A., S. Afsah and D. Ratunda (2004) 'How Do Public Disclosure Pollution Control Programs Work? Evidence from Indonesia', *Human Ecology Review* 11, 3, 235–246.

Bosetti, V., C. Carraro, E. De Cian and E. Mosetti (2013) 'Incentives and Stability of International Climate Coalitions: An Integrated Assessment', *Energy Policy* 55, 44–56.

Carbon Disclosure Project (2013) 'Climate Change Programme', available at https://www.cdproject.net/en-US/Programmes/Pages/CDP-Investors.aspx

Clapp, J. (2007) 'Illegal GMO Releases and Corporate Responsibility: Questioning the Effectiveness of Voluntary Measures', *Ecological Economics* 66, 2–3, 348–358.

Clapp, J. and E. Helleiner (2012) 'International Political Economy and the Environment: Back to the Basics?', *International Affairs* 88, 3, 485–501.

Dingwerth, K. (2007) *The New Transnationalism: Transnational Governance and Democratic Legitimacy* (Basingstoke: Palgrave).

Dingwerth, K. and M. Eichinger (2010) 'Tamed Transparency:How Information Disclosure Under the Global Reporting Initiative Fails to Empower', *Global Environmental Politics* 10, 3, 74–96.

Dingwerth, K. and M. Eichinger (2014) 'Tamed Transparency and the Global Reporting Initiative: The Role of Information Infrastructures', In A. Gupta and M. Mason (eds) *Transparency and Global Environmental Governance: Critical Perspectives* (Cambridge, MA: MIT Press).

Dryzek, J. S. (1999) 'Transnational Democracy', *The Journal of Political Philosophy* 7, 1, 30–51.

Dryzek, J. S. and H. Stevenson (2011) 'Global Democracy and Earth System Governance', *Ecological Economics* 70, 11, 1865–1874.

Eckersley, R. (2004) *The Green State: Rethinking Democracy and Sovereignty* (Cambridge, MA: MIT Press).

EEA [European Environmental Agency] (2001) 'Reporting on Environmental Measures: Are we Being Effective?', *Environmental Issue Report* No. 25 (Copenhagen: European Environmental Agency).

Etzioni, A. (2010) 'Is Transparency the Best Disinfectant?', *The Journal of Political Philosophy* 18, 4, 389–404.

Fenster, M. (2010) 'Seeing the State: Transparency as Metaphor', *Administrative Law Review*, 62, 617.

Florini, A. (1998) 'The End of Secrecy', *Foreign Policy* No. 111, Summer, 50–63.

Florini, A. (ed.) (2007) *The Right to Know: Transparency for an Open World* (New York, NY: Columbia University Press).

Florini, A. (2008) 'Making Transparency Work', *Global Environmental Politics* 8, 2, 14–16.

Florini, A. and S. Saleem (2011) 'Information Disclosure in Global Energy Governance', *Global Policy* 2, SI, 144–54.

Forsyth, T. (2003) *Critical Political Ecology: The Politics of Environmental Science* (London: Routledge).

Fox, J. (2007) 'The Uncertain Relationship between Transparency and Accountability', *Development in Practice* 17, 4/5, 663–671.

Fung, A., M. Graham and D. Weil (2007) *Full Disclosure: The Perils and Promise of Transparency* (Cambridge: Cambridge University Press).

Garcia, J. H., S. Afsah and T. Sterner (2009) 'What Firms are More Sensitive to Public disclosure Schemes for Pollution Control? Evidence from Indonesia's PROPER Program', *Environmental and Resource Economics* 42, 2, 151–168.

George, A. L. and A. Bennett (2005) *Case Studies and Theory Development in the Social Sciences* (Cambridge, MA: MIT Press).

Gilbert, J. (2007) 'Public Secrets: "Being-with" in an Era of Perpetual Disclosure', *Cultural Studies* 21, 1, 22–44.

Gouldson, A. (2004) 'Risk, Regulation and the Right to Know: Exploring the Impacts of Access to Information on the Governance of Environmental Risk', *Sustainable Development* 12, 136–149.

Graham, M. (2002) *Democracy by Disclosure: The Rise of Technopopulism* (Washington, DC: Brookings Institution Press).

Grigorescu, A. (2007) 'Transparency of Intergovernmental Organizations: The Roles of Member States, International Bureaucracies and Nongovernmental Organizations', *International Studies Quarterly* 51, 3, 625–648.

Gupta, A. (2006) 'Problem Framing in Assessment Processes: The Case of Biosafety', In Mitchell, R. B., W. C. Clark, D. Cash and N. M. Dickson (eds) *Global Environmental Assessments: Information and Influence* (Cambridge, MA: MIT Press), 57–86.

Gupta, A. (2008) 'Transparency under Scrutiny: Information Disclosure in Global Environmental Governance', *Global Environmental Politics* 8, 2, 1–7.

Gupta, A. (2010a) 'Transparency to What End? Governing by Disclosure through the Biosafety Clearing House', *Environment and Planning C: Government and Policy* 28, 1, 128–144.

Gupta, A. (2010b) 'Transparency in Global Environmental Governance: A Coming of Age?', *Global Environmental Politics* 10, 3, 1–9.

Gupta, A. (2010c) 'Transparency as Contested Political Terrain: Who Knows What About the Global GMO Trade and Why Does it Matter?', *Global Environmental Politics* 10, 3, 32–52.

Gupta, A. (2014) 'Risk Governance Through Transparency: Information Disclosure and the Global Trade in Transgenic Crops', In A.Gupta and M. Mason (eds) *Transparency in Global Environmental Governance: Critical Perspectives* (Cambridge, MA: MIT Press).

Gupta, A., M. J. Vijge, E. Turnhout and T. Pistorius (2014) 'Making REDD+ Transparent: the Politics of Measuring, Reporting and Verification Systems', In A. Gupta and M. Mason (eds) *Transparency in Global Environmental Governance: Critical Perspectives* (Cambridge, MA: MIT Press).

Gupta, S. and B. Golder (2005) 'Do Stock Markets Penalize Environment-Unfriendly Behavior? Evidence from India', *Ecological Economics* 52, 1, 81–95.

Haas, P. (1989) 'Do Regimes Matter? Epistemic Communities and Mediterranean Pollution Control', *International Organization* 43, 3, 377–403.

Haufler, V. (2010) 'Disclosure as Governance: The Extractive Industries Transparency Initiative and Resource Management in the Developing World', *Global Environmental Politics* 10, 3, 53–73.

Heald, D. (2006) 'Varieties of Transparency', In C. Hood and D. Heald (eds) *Transparency: the Key to Better Governance?* (Oxford: Oxford University Press), 25–43.

Held, D. (2004) 'Democratic Accountability and Political Effectiveness from a Cosmopolitan Perspective', *Government and Opposition* 39, 2, 364–391.

Holmes, A. (2011) 'The Limits of Carbon Disclosure: Theorizing the Business Case for Investor Environmentalism', *Global Environmental Politics* 11, 2, 98–119.

Hood, C. and D. Heald (eds) (2006) *Transparency: The Key to Better Governance?* (Oxford: Oxford University Press).

Jansen, K. and M. Dubois (2014) 'Global Pesticide Governance by Disclosure: Prior Informed Consent and the Rotterdam Convention', In A. Gupta and M. Mason (eds) *Transparency and Global Environmental Governance: Critical Perspectives* (Cambridge, MA: MIT Press).

Jasanoff, S. (ed.) (2004) *States of Knowledge: The Co-Production of Science and Social Order* (London: Routledge).

Kathuria, V. (2009) 'Public Disclosures: Using Information to Reduce Pollution in Developing Countries', *Environment, Development and Sustainability* 11, 5, 955–970.

Keohane, R. (2006) 'Accountability in World Politics', *Scandinavian Political Studies* 29, 2, 75–87.

Kolk, A., D. Levy and J. Pinkse (2008) 'Corporate Responses in an Emerging Climate Regime: The Institutionalization and Commensuration of Carbon Disclosure', *European Accounting Review* 17, 4, 719–745.

Konar, S. and M. A. Cohen. (1997) 'Information as Regulation: The Effect of Community Right to Know Laws on Toxic Emissions', *Journal of Environmental Economics and Management* 32, 1, 109–124.

Knox Hayes, J. and D. Levy (2014) 'The Political Economy of Governance by Disclosure: Carbon Disclosure and Non-Financial Reporting as Contested Fields of Governance', In A. Gupta and M. Mason (eds) *Transparency and Global Environmental Governance: Critical Perspectives* (Cambridge, MA: MIT Press).

Krämer, L. (2012) 'Transnational Access to Environmental Information', *Transnational Environmental Law* 1, 1, 95–104.

Langlet, D. (2009) *Prior Informed Consent and Hazardous Trade* (Alphen aan den Rijn: Kluwer Law International).

Langley, P. (2001) 'Transparency in the Making of Global Environmental Governance', *Global Society* 15, 1, 73–92.

Levy, D. and P. Newell (2005) *The Business of Global Environmental Governance* (Cambridge, MA: MIT Press).

Litfin, K. T. (1994) *Ozone Discourses, Science and Politics in Global Environmental Cooperation* (New York: Columbia University Press).

Litfin, K. T. (ed.) (1998) *The Greening Sovereignty in World Politics* (Cambridge, MA: MIT Press).

Lord, K. M. (2006) *The Perils and Promise of Global Transparency* (New York: SUNY).

Lyon, T. P. and J. W. Maxwell (2011) 'Greenwash: Corporate Environmental Disclosure and the Threat of Audit', *Journal of Economics and Management Strategy* 20, 1, 3–41.

Mason, M. (2005) *The New Accountability: Environmental Responsibility Across Borders* (London: Earthscan).

Mason, M. (2008) 'Transparency for Whom? Information Disclosure and Power in Global Environmental Governance', *Global Environmental Politics* 8, 2, 8–13.

Mason, M. (2014) 'So Far but No Further? Transparency in the Aarhus Convention', In A. Gupta and M. Mason (eds) *Transparency and Global Environmental Governance: Critical Perspectives* (Cambridge, MA: MIT Press).

Michener, G. and K. Bersch (2011) *Conceptualizing the Quality of Transparency*. Paper prepared for the 1st Global Conference on Transparency, Rutgers University, Newark, 17–20 May, 2011. Available at http://gregmichener.com/Conceptualizing_the_Quality_of_Transparency–Michener_and_Bersch_for_Global_Conference_on_Transparency.pdf

Mitchell, R. B. (1998) 'Sources of Transparency: Information Systems in International Regimes', *International Studies Quarterly* 42, 1, 109–130.

Mitchell, R. B. (2011) 'Transparency for Governance: The Mechanisms and Effectiveness of Disclosure-based and Education-based Transparency Policies', *Ecological Economics* 70, 11, 1882–1890.

Mitchell, R. B., W. C. Clark, D. Cash and N. M. Dickson (eds) (2006) *Global Environmental Assessments: Information and Influence* (Cambridge, MA: MIT Press).

Mol, A. (2006) 'Environmental Governance in the Information Age: The Emergence of Informational Governance', *Environment and Planning C: Government and Policy* 24, 4, 497–514.

Mol, A. (2008) *Environmental Reform in the Information Age: The Contours of Informational Governance* (Cambridge: Cambridge University Press).

Mol, A. (2014) 'The Lost Innocence of Transparency in Environmental Politics', In A. Gupta and M. Mason (eds) *Transparency and Global Environmental Governance: Critical Perspectives* (Cambridge, MA: MIT Press).

Newell, P. (2008a) 'The Political Economy of Global Environmental Governance', *Review of International Studies* 34, 3, 5–7–529.

Newell, P. (2008b) 'The Marketization of Environmental Governance: Manifestations and Implications', In J. Park, K. Conca and M. Finger (eds) *The Crisis of Global Environmental Governance: Towards a New Political Economy of Sustainability* (London and New York: Routledge), 77–95.

Newell, P. and M. Paterson (2010) *Climate Capitalism: Global Warming and the Transformation of the Global Economy* (Cambridge: Cambridge University Press).

O'Neill, O. (2006) 'Transparency and the Ethics of Communication', In C. Hood and D. Heald (eds) *Transparency: the Key to Better Governance?* (Oxford: Oxford University Press), 75–90.

Pattberg, P. and E. Okechukwu (2009) 'The Business of Transnational Climate Governance: Legitimate, Accountable, and Transparent?', *St Anthony's International Review* 5, 1, 76–98.

Peck, P. and K. Sinding (2003) 'Environmental and Social Disclosure and data Richness in the Mining Industry', *Business Strategy and the Environment* 12, 3, 131–146.

Raab, C. (2008) 'Review: Transparency: the Key to Better Governance?', In C. Hood, D. Heald (eds) *Public Administration 86*, 2 (Oxford: Oxford University Press), 591–618.

Roberts, A. (2002) 'Multilateral Institutions and the Right to Information: Experience in the European Union', *European Public Law* 8, 2, 255–275.

Roberts, A. (2004) 'A Partial Revolution: The Diplomatic Ethos and Transparency in Intergovernmental Organizations', *Public Administration Review* 64, 4, 410–424.

Roberts, A. (2006) 'Dashed Expectations: Governmental Adaptation to Transparency Rules', In C. Hood and D. Heald (eds) *Transparency: The Key to Better Governance?* (Oxford: Oxford University Press), 107–125.

Rowan-Robinson, J., A. Ross, W. Walton and J. Rothnie (1996) 'Public Access to Environmental Information: A Means to What End?', *Journal of European Law* 8, 1, 19–42.

Smith, G. (2003) *Deliberative Democracy and the Environment* (London: Routledge).

Soederberg, S. (2001) 'Grafting Stability onto Globalization? Deconstructing the IMF's Recent Bid for Transparency', *Third World Quarterly* 22, 5, 849–864.

Stasavage, D. (2003) 'Transparency, Democratic Accountability and the Economic Consequences of Monetary Institutions', *American Journal of Political Science* 47, 3, 389–402.

Stephan, M. (2002) 'Environmental Information Disclosure Programmes: They Work But Why?', *Social Science Quarterly* 83, 1, 190–205.

Stevis, D. and V. J. Assetto (eds) (2001) *The International Political Economy of the Environment: Critical Perspectives* (Boulder, CO: Lynne Rienner).

van den Burg, S. (2004) 'Informing or Empowering? Disclosure in the United States and the Netherlands', *Local Environment* 9, 4, 367–381.

Weil, D., A. Fung, M. Graham and E. Fagotto (2006) 'The Effectiveness of Regulatory Disclosure Policies', *Journal of Policy Analysis and Management* 25, 1, 155–118.

Young, O. (ed.) (1999) *The Effectiveness of International Environmental Regimes: Causal Connections and Behavioral Mechanisms* (Cambridge, MA: MIT Press).

15
General Conclusion

Michele M. Betsill, Kathryn Hochstetler, and Dimitris Stevis

In this conclusion, we draw on the book's chapters to briefly reflect on the status of the field of international environmental politics (IEP) as a whole and to return to the cross-cutting themes and queries raised in the introduction – North–South relations, domestic–international linkages, and the role of the state. Although many of the field's most important developments can be seen only by examining a variety of substantive and normative issues together, some of the conclusions are especially well grounded in individual chapters of the book, indicated here with the chapter authors' names in parentheses.

This book clearly shows that the field of IEP has become broader and deeper over time in terms of research agendas, substantive concerns, theoretical approaches, and the geographical and disciplinary origins of researchers (Hochstetler and Laituri, chapter 4; Paterson chapter 3; Stevis, chapter 2), as demonstrated by the sheer number of publications and publishing venues. Equally importantly, however, there has been a broadening in terms of research agendas and areas. Since the early 1990s, for instance, methodological concerns, societal politics, and environmental justice have become more prominent in the literature, and today gender is emerging as a new area for inquiry. The IEP scholarship has also become deeper in two senses. Not only have additional theoretical views, particularly constructivist and structuralist, joined the fray, but the quality of the theoretical exchanges has also become more sophisticated. IEP research has also spread geographically well beyond the United States and Europe. A desirable project for the future would be to bring together reviews of IEP from a number of countries not only to ascertain convergences and divergences but, also, in order to move beyond the boundaries of the Euro-American literature. Another project, as becomes apparent below, would be to bring together reviews of how different disciplines study global environmental politics in order to ascertain complementarities, synergies and overlaps.

The diversification of approaches is reflected in a second overall conclusion, which is that the field of IEP continues to lack a single normative core idea of the kind that efficiency represents for the discipline of economics. The now-venerable concept of sustainable development has been useful in both academic and policy debates, but carries too many meanings to work well as a conclusive evaluative criterion (Happaerts and Bruyninckx, chapter 12). The search for alternatives is reflected in the recent development of normative criteria that capture only some parts of the sustainability concept, such as environmental justice (Okereke and Charlesworth, chapter 13), effectiveness (Young, chapter 11), or transparency (Gupta and Mason, chapter 14). Even in the more circumscribed domains of these alternative evaluative criteria, multiple definitions compete with each other. In addition, one can easily imagine environmental policies and projects that might be effective, but not just, or vice versa. Such observations suggest that the ambiguities and contradictions of the sustainable development criterion are not the result of inadequate analytical effort but may reflect inevitable tensions among the diverse concerns of IEP.

Cross-cutting themes

In the Introduction we identified three specific cross-cutting themes that we expected to be important across the book's chapters. The first, North–South relations, turned out to be every bit as central as anticipated and is a factor in nearly every chapter. The second theme, the interface between local and higher levels of politics, captures the changing nature of global environmental governance in theory and in practice. Finally, research on non-state actors and new forms of governance contributes to ongoing discussions about the role of the state in global environmental politics and in International Relations (IR) more broadly.

The North–South dimension has become increasingly important in the practice and study of global environmental politics over time (Stevis, chapter 2). For example, it is relevant to all of the different frameworks for normative evaluation discussed in Part Three. The sustainable development concept was forged to bridge the concerns and futures of the so-called developed and developing worlds (Happaerts and Bruyninckx, chapter 12). The North–South dimension is also fundamental in discussions of environmental and ecological justice and gender, especially in the context of globalization (Detraz, chapter 6; Okereke and Charleworth, chapter 13). We see a North–South dimension in the regime effectiveness literature as well, where differentiated commitments, technology/financial resource transfers, and capacity issues are central in regime development and implementation (Young, chapter 11). The North–South dimension also appears in virtually every case study in this book, from climate change (chapters by Betsill, Lövbrand, Okereke and Charlesworth) to transboundary environmental security concerns (Swatuk, chapter 9) to trade in toxic wastes (Clapp, chapter 5). These case studies show that North–South relations dominate inter-governmental negotiations on the environment. Non-governmental actors, whether based in the market (Clapp, chapter 5) or civil society (Betsill, chapter 8), also reflect the North–South dimension. Despite the centrality of North–South relations, it must be pointed out that there is no single Southern or Northern voice in global environmental politics or IEP scholarship. Instead, the North–South dimension is dynamic and constantly reconstructed, sometimes a divide and sometimes a bridge.

The significance of the North–South dimension does not mean that intra-North and intra-South issues are marginal. As the politics of climate change demonstrate, there are important differences within the North while Southern environmental politics is a growing and increasingly important and diverse phenomenon (Swatuk, chapter 9). Yet, as the various chapters collectively indicate, this geographic heuristic should not lead us to reify the country as a unit. Many of the issues discussed cross boundaries, whether as a result of natural or social processes, while important stakeholders may operate at various scales other than that of the nation-state, North or South.

With respect to our second proposed cross-cutting theme, the various chapters clearly show that the interface between local and higher levels of politics is more complicated than the 'international–domestic' dichotomy used by many IR scholars. Analytical concepts such as 'two-level games' and 'second image reversed' are no longer sufficient to capture the complex, cross-border dynamics of global environmental politics. With increasing global integration and environmental policies, concepts such as 'multilevel governance' are better for understanding the interplay of processes occurring at various levels (chapters by Betsill, Biermann, Detraz, Happaerts and Bruyninckx, Swatuk). To speak of *multilevel* governance is to note that there may be considerably more than two levels engaged in shaping IEP, cutting across and within the special analytical boundary of the nation-state.

Similarly, the conceptual move from government to *governance* stresses the growing number and variety of kinds of authority relations that are part of global environmental politics (chapters by Betsill, Biermann, Clapp, Gupta and Mason, Lövbrand). This does not imply that the state is withering away but, rather, that the meaning and roles of the state and interstate agencies are dynamic and contested and thus require renewed attention.

Where next?

The concluding section of each of the book's chapters outlines the most important emerging issues for the chapter's topic. Here we examine a set of overarching substantive themes and broader research areas or agendas that emerge from the chapters as a whole.

New substantive themes

With respect to substantive environmental issues, we see a great deal of debate emerging over the nature of specific environmental problems, such as climate change, biotechnology, and so on. These query the likely causes and impacts of these problems as well as their solutions. The book's chapters introduce actors and analytical constructs which must be understood to advance many of these debates. We think the debates within the field are more sophisticated than the dichotomous debates (population/no population, growth/no growth) that characterized previous eras. The debates are also more diverse among IEP scholars than they are in the general public and in most policy debates among governments. In the future, it would be interesting to consider to what extent such growing sophistication within IR scholarship on global environmental politics has been able to reconfigure policy debates and popular understandings of environmental issues.

We also see a resolute move toward a political economy of the environment with few if any environmental issues seen as separate from the overall political economy. Important research areas, such as sustainable development or ecological modernization, can hardly avoid taking a political economy approach. There is increasing attention to the challenge of simultaneously addressing poverty and environmental degradation in an era of globalization (chapters by Detraz, Happaerts and Bruyninckx, Okereke and Charlesworth). Moreover, IEP scholars increasingly recognize the importance of global financial and economic institutions in environmental governance (chapters by Betsill, Biermann, Clapp). Related to the move toward political economy, questions about production and consumption, the interface of the environment and health, commodity and policy chains and the built environment are likely to become more important in the future. The centrality of North–South relations is in large part tied to scholarly recognition of the importance of the political economy of the environment (Paterson, chapter 3).

With respect to broad empirical research agendas, we see a strong move toward considering normative issues, albeit increasingly based on solid empirical research. These include issues of equity, democracy, participatory processes, and so on. This trend complements the move to political economy and a focus on North–South relations. Questions of governance are likely to remain high on the agenda but increasingly will be affected by discussions over levels of governance and the changing nature and variety of actors involved in governance processes (chapters by Betsill, Biermann, Detraz, Lövbrand). Procedural issues and the ability to assert and claim authority will continue to be central dimensions of governance.

In our view IEP scholars ought to pay attention to the causes of environmental problems with the same empirical sensitivity that we have been using to understand their impacts and efforts at solving them (Paterson, chapter 3). Tracking global environmental problems even before they are viewed as such (for example, China's energy and transportation policies) will enrich our understanding of other aspects of the various issues and may also allow us to offer more proactive policy recommendations.

In all these new areas of study, the field is likely to be influenced by the substantive shifts of a number of actors in global environmental politics over the last decade. The United States' move toward unilateral international policies and frequent anti-environmentalism is especially notable. The European Union is often counterpoised to the United States as the world's new environmental leader, but those accomplishments could be exaggerated, especially with the challenges of enlargement and the state of its economy. On the other side of the global North–South divide, Southern countries now also sometimes find themselves as the drivers of environmental initiatives, such as the Johannesburg Conference in 2002 or the desertification negotiations (Happaerts and Bruyninckx, chapter 12). Corporate actors, once assumed to be universally opposed to environmental regulation, now develop policy tools and techniques for environmental protection (Betsill, chapter 8; Clapp, chapter 5). Complex issues like the global trade and management of transgenic organisms as well as the rise of emerging economies like

China and India have already rearranged global negotiating coalitions and positions altogether, in ways that seem likely to become more common.

Methodological issues

In general, methodological concerns have not been central in IR studies of global environmental politics, with the regime effectiveness and environmental security literatures being notable exceptions (chapters by Hochstetler and Laituri, Swatuk, Young). We detect signs that this is slowly changing and expect that the methodological sophistication of IEP scholars will grow in the future. For example, we anticipate developments in the use of quantitative techniques that complement the in-depth qualitative case studies that have characterized much of IR scholarship on global environmental politics. The field of IEP is also well situated to integrate important methodological insights and techniques from other disciplines, such as geography (for example, geospatial information technologies) and ecology. The prominence of global change models requires, in our view, a renewed debate on the use of models and forecasting, with particular emphasis on the integration of human practices and choices into what are largely naturalistic assumptions (Lövbrand, chapter 7). Debates in archaeology over the impact of climatic changes on past civilizations could provide us with a useful introduction as would a revisiting of the IEP literature from the 1970s (Stevis, chapter 2). In sum, we believe the field would greatly benefit from more explicit consideration of methodological issues and the development of a mixed methods approach and hope that trends in this direction continue.

Theoretical issues

All of the substantive debates and developments outlined above seem likely to contribute to theoretical developments as well. Such changes provoke theoretical debates as they point out the inconsistencies between theoretical assumptions and concrete unfolding events. When the global hegemon has clearly become an environmental laggard rather than leader, for example, certain theoretical explanations of global environmental change appear less tenable. Such changes often provoke rethinking that goes beyond the particular event.

With respect to broader theoretical perspectives, we expect that the substantive attention to questions of equity is also likely to offer a strong challenge to the theoretical perspectives of liberal institutionalism and liberal constructivism which have been largely inattentive to these questions. This will likely accelerate emergent debates within these perspectives on how to integrate questions of equity. Emphasis on the ecological foundations of global environmental politics is similarly likely to influence radical social perspectives and force IEP scholars to address ecological issues more seriously than they have done in the past. As a result we also expect to see an acceleration of the emergent debates within these perspectives.

In general, we believe that the theoretical diversity of IR research on global environmental politics is salutary and appropriate for such a young field with so many concerns and disciplinary foundations. We have been careful in this book to sample widely across different approaches to IEP scholarship in the English-speaking world. A next step would be to more consistently include theories and understandings from the global South, especially given the growing importance of North–South relations and normative concerns in global environmental politics. One drawback of such a pluralist orientation is the field's resulting inability to give clear and consistent guidance to those who govern the environment. Without this, the impact of IEP scholarship on the practice of global environmental politics may be limited.

The discrepancy between the study and practice of global environmental politics has been and ought to be a central theoretical and normative issue for IEP scholars. Why is there divergence during some periods or with respect to some issues, but not in others? For example, why was transnational politics, so seemingly appropriate for IEP, not taken up until almost 25 years after its introduction? Are there variations across countries and what explains them? While we think that IEP ought to address policy issues and offer specific proposals, this should not be at the expense of speaking truth to power. Tailoring IR research on global environmental politics to specific political or economic exigencies would be tantamount to taking theoretical pursuits out of physics or ecology or any other systematic field of study.

Linkage to other disciplines

Our focus in this book is on how the scholars in the field of IR approach the study of global environmental politics. The IR field is uniquely situated to take the lead in dealing with important aspects of global environmental politics, whether conflict or governance. At the same time, however, the study of global environmental politics clearly benefits from drawing on other disciplines in the social and natural/ecological sciences. The question, therefore, is how and when IR scholars should create bridges with other disciplines. In the chapters' presentations of the history of particular areas of study, we have noted that the IR research on global environmental politics has often followed the lead of the ecological and physical sciences while economics has also played a central role (Stevis, chapter 2). There is no doubt that IEP scholars have to pay close attention to these disciplines and even recognize their leading role in dealing with particular issues. It seems to us, however, that additional social sciences, and IR in particular, also have much to offer to the study of global environmental problems. Which topics are invisible or trivial from the perspective of IR and how might other disciplines make those more visible? To what extent does the study of global environmental politics require theories and approaches distinct from other areas of research within IR?

Several examples of contributions from other social science disciplines are evident in the chapters. Geography can offer some important methodological tools to enable IEP scholars to better link to the natural and physical sciences (Hochstetler and Laituri, chapter 4). On the issue of governance, geography, with its emphasis on scale, contributes to debates about multilevel governance in global environmental politics (Betsill, chapter 8; Biermann, chapter 10). Urbanists and regional planners have long been at the forefront of ecological thinking. The human habitat, for instance, was one of the major themes of the Stockholm Conference. The organization of space is central to the creation and solution of environmental problems and there is much that IEP scholars can learn from those disciplines. Sociological perspectives have informed IEP scholarship and often have been the source of important and now mainstream concepts, such as risk analysis and ecological modernization (Clapp, chapter 5). For example, studies of transnational actors rely on concepts of framing and social movements borrowed from sociology (Betsill, chapter 8). Finally, world systems approaches inform IR scholarship on global environmental politics, particularly related to issues of globalization and equity (Okereke and Charlesworth, chapter 13).

Archaeology, history, and cultural anthropology have much to offer us, as well. Archaeology and history can provide us with long-term perspectives as well as important cases that will help us better evaluate some of the assumptions behind key environmental issues of the day, such as climate change, biodiversity, deforestation, and desertification. Cultural anthropologists have produced some of the most intriguing studies of the impacts of human practices over time but, also, are particularly sensitive to practices and relations – for example, the importance of non-state actors and the 'powerless' – that IR scholars and others focusing on large systems and modernity may easily miss.

One important question that we must address is that of the economization of IR and IEP. The adoption of microeconomic thinking by important strands of IR has been the subject of serious debate going back a few decades. The move toward economic thinking and instruments in environmental affairs is certainly evident in IEP scholarship as well (Biermann, chapter 10; Clapp, chapter 5). While many economists and IR scholars may argue that economic instruments are just that, others have made a convincing argument that every social policy instrument carries with it key political and social assumptions (chapters by Clapp, Detraz, Happaerts and Bruyninckx, Lövbrand, Okereke and Charlesworth). Assigning prices to the environment or life is not a simple technical device. Rather, it indicates that everything can be subjected to the same common denominator and that alternative criteria, whether moral or political, cannot be used to guide policy choices.

How much and what kind of borrowing from other disciplines IR scholars need to do depends on what we wish to understand. If we are looking at what explains the formation of multilateral environmental policies, one could make the argument that IEP scholars do not need theories beyond those of IR. Other IR issue areas have contributed a great deal to emerging understandings of the multiple actors, processes, and levels of governance now important in IR generally and IEP in particular. IR already offers significant tools for understanding such governance issues. At the same time, however, the centrality of natural and physical sciences in this process does distinguish IEP from other areas of IR scholarship. As the literature on effectiveness has already noted, there may be large gaps between successful *political* outcomes and successful *environmental* ones (Young, chapter 11). Ecological and physical sciences will continue to be important for understanding the underlying causes of environmental degradation. Socially sustainable outcomes may also require the insights of disciplines like sociology and anthropology, as suggested above (Okereke and Charlesworth, chapter 13). In short, while IEP has found a hospitable and productive home in IR, IR has never been the only source of its insights and should not be expected to be in the future.

In any case, the contributions of IEP scholarship to IR should also be briefly noted. IEP scholars have exemplified certain new global developments – the rise of non-state actors and private governance, the changing meaning of security, the fungibility of nation-state boundaries, the importance of North–South and global equity concerns – in ways that have been clarifying and suggestive for IR as a whole. The willingness of IEP scholars to reach across disciplines for the necessary tools to understand these phenomena is one of its most important contributions – hence our suggestion that a project that brings together how different disciplines study the global environment is timely.

Index

Note: The letter 'n' following locators refers to notes.

Aarhus Convention, 357, 366–71
accountability, 173–4, 256–7, 373
  corporate, 129
  in global environmental governance, 356–80
  through disclosure, 356–62
activism and advocacy, 187, 191–3, 253
'actor network theory,' 56
adaptation, 28, 149–50, 230–3, 263, 308, 344
  adaptive capacity, 90, 221–2, 224
  and complexity, 90
  National Adaptation Programs of Action (NAPAs), 140
Africa, 122, 126–7, 140, 143, 224, 227, 231–4, 261, 343
  see also individual countries
African Union, 232, 261
Agenda 21, 172, 188, 302–3, 308, 315
  Local Agenda 21, 312
  see also United Nations Conference on Environment and Development
agenda-setting, 20, 196, 279
agent-based modeling, 90, 291
agriculture, 220, 230, 233, 310, 366
  farming, 153, 233–4
  see also trade
anarchy, 46–51, 58, 66, 163, 213, 216, 227, 330
Antarctica, 274, 279
  Antarctic Treaty System, 336
Anthropocene, 229, 231, 289
anthropocentrism, 64, 335
anti-globalization movement, 107, 314
apartheid, 224, 234, 330
Aristotle, 328–9

Article XX, 113–14
  see also General Agreement on Tariffs and Trade
Asia, 127, 214, 227, 230
  see also individual countries
Australia, 27
authority, 47–9, 59, 64, 167, 187–202, 250, 255, 360, 384–5
  private, 187, 195, 363–4, 368–70

balance of power, 213, 341
Bangladesh, 126
Basel Convention on the Transboundary Movement of Hazardous Wastes and their Disposal (1989), 114–15, 126–8, 339
behavioral change, 57, 60, 193, 195, 275–8, 320, 357
  see also effectiveness
Belgium, 305
Bhopal, India, 128, 331
  see also Union Carbide
bilateral agreements, 115, 119–20, 231
biodiversity, 54, 61, 120, 151–4, 166, 251, 285, 302, 336, 340, 371, 388
  and gender, 151–4
  species loss, 151, 215, 249
  see also biosafety; Cartagena Protocol; Convention on Biological Diversity
bioenvironmentalists, 15–22
biofuel, 232–4
biosafety, 114–15, 118, 120, 366–7, 371
  see also Cartagena Protocol
biosphere, 212, 227–8, 236
Biosphere Conference (1968), 18
Bookchin, Murray, 17, 19, 336–8
'boomerang' strategy, 191, 204

Botswana, 122
Brandt Commission Report, 301
Brazil, 112, 121, 152–3, 261, 302, 309, 342, 346
  biodiversity in the Amazon, 137, 151–4
  gender and population, 147–51
  industrialization, 112
  see also United Nations Conference on Environment and Development; World Social Forum
British Petroleum, 186
Brundtland Report (1987), 7, 141, 300–6, 318, 339
  see also World Commission on Environment and Development; global solutions; liberalism; North–South
Buddhism, 334
bureaucracy, 217
  administrative planning, 170–1, 307
business, 52–4, 118–19, 186–7, 191–2, 251–2, 257, 311, 314
  see also industry
Business Action for Sustainable Development (BASD), 118
business-as-usual, 232, 249
Business Council on Sustainable Development, 54, 118, 186, 314

C40 Cities Climate Leadership Group, 200
Canada, 18, 32, 115, 170
capitalism, 51–4, 95, 186, 214, 337
  capital accumulation, 46, 52–4, 66
  carboniferous capitalism, 229, 235, 236
  and critical theory, 65–7, 95, 149, 189, 217, 337
  modern, 217, 229
  and structuralist analysis, 47, 52, 95, 214–17
  transformation of, 53, 229, 248
Carbon Disclosure Project (CDP), 124, 366–7
carbon offsets, 232–3, 346
carbon tax, 288, 346

carrying capacity, 17
  see also scarcity
Carson, Rachel, 17, 19, 217, 331
Cartagena Protocol, 366–71
  see also transparency
causation, 235, 291
  see also research design
Centre for Science and the Environment, 348
certification schemes, 171, 194, 281
chemicals, 18, 113
  agriculture, 114, 128, 357
  trade, 61, 113, 126, 383
  see also Rotterdam Convention; Stockholm Convention
Chernobyl, 332
China, 22, 309, 346, 348, 375, 385
  and climate change, 346
  industrialization, 112, 348
chlorofluorocarbons (CFCs), 53, 293
Cities for Climate Protection Program (CPP), 194, 200
civil society, 29, 58, 122, 192–3, 204, 256–9, 281, 284, 311, 357, 364, 370–1
  and industry, 58, 192
  organizations, 370
  see also global civil society
class, 30, 51–3, 60–2, 149, 332–7, 345
Clean Development Mechanism (CDM), 125, 252, 346
Clean Technology Fund, 123
Climate Action Network (CAN), 207
Climate Action Report 2014, 222, 243
  see also U.S. State Department
Climate Alliance, 200
climate change, 27, 56, 90, 122, 169–70
  adaptation, 90, 140, 149, 230–1, 344
  climate justice, 54, 61, 140, 149–51, 343–8
  EU policy, 116
  and gender, 140, 150
  governance, 199–203
  industry response, 52, 118–20, 367
  Intergovernmental Panel on Climate Change (IPCC), 175–8
  and migration, 149–50
  mitigation, 200, 230–2, 344

negotiations, 96, 118, 176, 203, 315, 344
North–South, 339, 346
trade, 54, 114, 251, 255, 283
see also Kyoto Protocol; United Nations Framework Convention on Climate Change (UNFCCC)
Climategate, 173, 178
climate vulnerability, 222
Club of Rome, 147, 301
Limits to Growth, 301, 331
coal, 52, 56, 121, 232, 233
Cold War, 22, 30, 145, 164, 211, 218, 279, 317
Commission on Global Governance, 214, 247–8, 257
commodification, 54, 337, 369
commodity index funds, 123
common pool resources, 82, 86, 291–2
commons, 32, 67n1, 68, 86
see also global commons
comparative advantage, 109
compensatory justice, 329
complexity modeling, 90
conflict, 17, 21, 22, 31, 53, 60, 85, 145–7, 214, 218, 222–5, 231
over resources, 219–21
consensus, 179, 228, 318
constructivism, 32, 47, 168, 179, 360, 386–7
and anarchy, 47, 51
'constructivist turn,' 188
and knowledge, 55, 161–3, 169–70, 174, 179, 360–1
liberal, 32, 386
methods, 80, 162
social constructivism, 161–2, 167, 177
and transnational actors, 172, 188–9, 194, 197
and transparency, 173–4, 360–1
see also critical perspectives
consumerism, 217, 337
consumption, 52, 148–9, 199, 217, 220, 231, 303, 313
and agriculture, 220
and equity, 303
Northern, 149, 301, 303, 313, 320
structuralists, 309

transnational networks, 199, 313
see also Club of Rome
contested concept, 229, 304, 320
Convention on Biological Diversity, 54, 366
Convention on the Conservation of Migratory Species of Wild Animals, 336
cooperation, 162–8, 253
under anarchy, 49–51, 163
and ecoauthoritarianism, 48
environmental conflict, 222–5
game theory, 49
and liberal institutionalism, 331, 359
and realism, 48
scale, 65, 222
scientific information, 175–6
transnational networks, 189
Copenhagen Consensus, 17, 28
co-production, 169
see also knowledge
core/periphery, 60–1, 95–6, 214
corporate social responsibility, 129, 367
corporate accountability, 129, 371
counterfactual, 196, 275–6
credibility, 166–7, 173, 177, 257
critical environmental security studies, 90, 147–8, 152, 211–34
critical perspectives, 147, 174, 211, 359–60
and IPE, 360–1
and methodology, 57, 92, 216, 218
postmodernism, 161–3, 179, 212, 228
and science, 57, 88, 162, 168, 216, 263
and TNCs, 116–18
and trade, 111–13
and transparency, 360–1
see also critical environmental security studies; feminism

dams, 93, 121, 155n7, 233
Narmada Dam Scheme, 121
World Commission on Dams, 253
debt relief, 307
decentralization, 64, 309, 335, 337

deep ecology, 332–7
deforestation, 23, 61, 121, 155n6, 388
degrowth, 320
deliberation, 173, 174, 258
democracy, 64, 81, 173, 198, 235, 304, 337, 370, 385
  and environment, 29
  see also global governance
democratic peace, 236
democratization, 198, 362–3, 369, 373
  and accountability, 173, 198, 362
dependency theory, 47, 60
desertification, 62, 274, 309, 385, 388
  see also United Nations Convention to Combat Desertification
developing countries, 117, 128, 256–7, 260–1, 331, 366
  climate change, 62–3, 122, 177
  decentralization, 312
  digital divide, 91
  environmental security, 220
  finance, 120, 124
  in the global economy, 61, 112, 119
  hazardous waste trade, 126
  see also individual countries; North–South
developmentalists, 20, 28
diplomacy, 162, 168, 169, 178, 347, 374
  see also statecraft
discourse analysis, 197, 306
  see also critical approaches; feminism
distributive justice, 61, 141, 328, 336
Doha Round, 115
  see also General Agreement on Tariffs and Trade (GATT)
dynamism, 52

earth observation technologies, 88, 169
earth system governance, 249, 308
Earth System Governance Project, 264
Easter Island, 217
eco–authoritarianism, 21, 47, 47, 333
ecocentrism, 64
ecocolonialism, 62
ecodevelopment, 20, 25

ecofeminism, 139–40
  see also feminism
eco–labeling, 357, 367
ecological debt, 336
ecological economics, 17, 26
ecological footprint, 64, 228, 315
ecological justice, 32, 329, 347
ecological marginalization, 220, 234
ecological modernization, 26, 52, 67, 110, 129, 384, 388
ecological shadow, 228
Ecologist (The), 68
economic growth, 52–3, 64, 107, 109–11, 192, 233, 259, 305, 310, 317, 334
  and environmental justice, 20, 151
  neoclassical economists, 109–10, 130n1
  World Bank, 259
  see also capitalism; Club of Rome
economic recession, 67
  see also financial crisis
ecopolitics, 26, 229
ecosocialism, 328–55
effectiveness, 14, 21, 28, 50, 55, 82, 83–6, 260–3, 275–80, 382, 389
  methods, 386
  normative, 364
  procedural, 365
  substantive, 371
  see also international regimes
efficiency, 111, 113, 140, 195, 260, 276, 310, 346, 356, 382
employment, 233, 314–15
  job creation, 310, 318
'empty belly' environmentalism, 332, 338–46
Enlightenment, 216
environmental agreements, 29, 86, 113–15, 122, 260, 278, 283, 307, 336, 371
  see also international environmental agreements
environmental conservation, 140, 337
Environmental Defense Fund (Environmental Defense), 121
environmental degradation, 46, 48–9, 51–4, 61–2, 111, 137, 142, 144, 151, 187, 202, 221, 384, 389

economic growth, 310
globalization, 142, 146
health, 144–6
and science, 54, 57, 216, 389
and trade, 111
see also environmental security studies; tragedy of the commons
environmental equity, 15
see also environmental justice; inequality
environmental justice, 14–15, 26, 30, 62, 94, 150, 153, 231, 234, 329–49, 381–2
climate justice, 54, 140
and gender, 142–5
intergenerational, 329–30, 333, 339
Environmental Kuznets Curve (EKC), 109
environmental peacemaking, 225
environmental performance standards, 116, 119
Environmental Politics, 14
environmental racism, 143, 332
environmental security, 24, 49, 90, 137, 145, 227
and climate change, 230
critical, 211
feminist, 147
see also environmental security studies; scarcity; security
environmental security studies (ESS), 145, 211, 212, 219, 237
Environmental Studies Section (International Studies Association), 20, 26, 79
epiphenomena, 49, 285, 361
epistemic communities, 55, 163–8, 177, 315
see also knowledge
epistemology, 162, 173
Equator Principles, 124
Europe, 18, 27, 125, 216, 274, 286
climate change, 305
environmental policies, 305, 310, 366
in the study of IEP, 18, 385
transboundary air pollution, 277, 286
see also European Union; individual countries
European Emissions Trading Scheme, 125
European Union (EU), 115, 261, 385
European Commission, 310
sustainable development, 261, 310
e-waste, 127
expertise, 54, 80, 139, 141, 151, 154, 163, 192, 195, 227, 364
postpositivism, 166–8, 179
transnational actors, 172–5, 198
see also epistemic communities; knowledge
export subsidies, 111
Extractive Industries Review (EIR), 122

fairness, 263, 276, 280, 281, 287–8, 315, 329, 341
feminism, 47, 138, 141
ecofeminism, 139, 156
finance, 107–10, 120–5, 129, 374
see also investment
financial crisis, 67, 310, 317
see also economic recession
financial markets, 120–5, 359
fisheries, 217, 220, 274, 278, 357
food security, 144–5, 155n2
food insecurity, 144, 153
foreign direct investment (FDI), 116
see also investment
forestry, 58, 85, 117, 169, 189, 357
Forest Stewardship Council (FSC), 58, 194, 253
fossil fuels, 112, 121–3, 231, 236
Foucauldian analysis, 55, 170
Founex Report (1972), 20, 23
fragmentation, 50, 250, 254, 259, 283
see also international regimes
frames, 23, 63, 203n3, 217, 254
environmental justice, 150, 332
problem definition, 155n4
transnational actors, 59, 164, 196–7, 388
France, 305
Friends of the Earth International, 58, 69, 186, 204, 336
functionalism, 332, 359

game theory, 49, 86
  see also rationalism
Gandhian non–violence, 334
gas, 230, 331
gender, 60–2, 137–54, 214, 329, 341, 383
  culture, 139
  and labor, 62, 139, 150
  and social construction, 30, 137 9, 143, 152–3
genealogy, 13, 14, 211
General Agreement on Tariffs and Trade (GATT), 109, 115, 127
  Article XX, 113–14
General Assembly, 175, 260
genetically modified organisms, 115
  GMOs and transparency, 357, 366–7
geopolitics, 16–18, 20–1, 24–5, 368, 374
geospatial analysis, 79–80, 87, 97, 386
geospatial informational technologies (GIS), 87–9
Germany, 342
global change, 23, 28, 198, 249, 386
  see also global environmental change
global civil society, 29, 204, 257, 281, 284
  transnational actors, 193, 254
  see also civil society
global commons, 22, 114, 228, 261, 365
  unequal access, 338–9
Global Compact, 124, 252
Global Environment Facility (GEF), 122
global environmental assessment, 161–84
Global Environmental Change, 14
global environmental change, 23, 27, 145, 263
  Human Dimensions Programme on Global Environmental Change, 282, 292
  see also environmental degradation; global change
global environmental governance, 28, 69, 118–19, 186, 193, 198, 245–63, 312, 340, 343, 365, 372, 374, 382
Global Environmental Politics, 14, 79, 97
global governance, 24, 53, 113, 185, 194, 199, 224, 245–63, 308–9, 366
  and transnational actors, 188, 193
  see also international regimes
globalization, 22, 54, 94, 110, 245, 312
  ecological, 142, 247, 248, 251
  economic, 107, 214, 308
  political, 254, 311
  social, 255
  structural origins of, 128
  see also global governance
global political economy, 25, 53, 109–11, 125, 128, 309
  see also international political economy
Global Reporting Initiative, 367
global warming, 220, 302
  see also climate change
Global Warming Potential, 56
Gore, Al, 88
Governance by disclosure, 124, 356–74
governmentality, 170–1, 198
Gramsci, Antonio, 197
  see also hegemony
Greece, 116
Green Economy, 129
  and United Nations Environment Program (UNEP), 110
greenhouse gas emissions, 52, 175, 199, 200, 232, 283, 286, 312
Gross Domestic Product (GDP), 111, 233, 316, 318
Gross National Product (GNP), 96
Group of 8 (G8), 309
Group of 77 (G77), 331
Greenpeace, 58, 186
Green Revolution, 20
(The) Green State, 66

Halifax Initiative, 122–3
Hardin, Garrett, 48, 86, 147
hazardous waste, 114, 283, 339, 357
  trade, 125–8

see also Basel Convention on the Transboundary Movement of Hazardous Wastes and their Disposal (1989)
hegemony, 27, 57, 245
hierarchy, 226, 337
human development, 305
Human Dimensions of Global Environmental Change Programme (The), 282, 292
human rights, 142, 192, 356
human security, 25, 145, 147, 150–2, 227–8

identity, 92, 164, 220, 258, 340
implementation, 29, 50, 91, 122, 155n1, 161, 180, 196, 198, 252–3, 260, 279–80, 302–4, 308–9, 346, 373
 see also individual agreements
incremental costs, 122
India, 95, 121, 122, 126, 128
industrial agriculture, 121
inefficiency, 111, 346
 see also efficiency
inequality, 62–3, 138, 141, 143–4, 198
 see also environmental justice; North–South
informational governance, 356–80
Information disclosure, 356–80
ingenuity, 229
institutional interlinkages, 50, 255
institutional interplay, 50, 255, 258–9, 276, 282–3, 291
institutionalism, 45, 46, 49, 66
 liberal institutionalism, 17, 21, 26, 59, 331, 386
 rational institutionalism, 359–60
 see also liberalism
integration theory, 188
interdependence, 51, 58, 66, 163, 168, 247, 248, 331
interdisciplinary research, 90
intergenerational justice, 329, 330, 333, 339–40
intergenerational solidarity, 303

intergovernmental organizations (IGOs), 200, 256, 260–1
 see also global governance; individual organizations
Intergovernmental Panel on Climate Change (IPCC), 56, 149, 167, 173–6, 256–7, 315
 and representation, 176–8
 Working Group reports, 175
Intergovernmental Political Sphere, 187, 190, 195, 203
intermediaries, 363, 369
International Chamber of Commerce, 186
International Council for Local Environment Initiatives (ICLEI), 194
 see also Cities for Climate Protection
International Council of Scientific Unions (ICSU), 17
International Environmental Agreements, 14
international environmental agreements, 113, 115, 122, 348
 bilateral agreements, 115, 119, 120
 multilateral environmental agreements (MEAs), 114, 189–93, 260, 283, 307
 see also individual agreements
International Environmental Justice (IEJ), 332
International Human Dimensions Programme on Global Environmental Change (IHDP), 282, 292
International Labour Organization (ILO), 257, 260
International Monetary Fund (IMF), 259, 309
international negotiations, 24, 94, 252, 313
 North–South, 313
 ozone depletion, 118, 302
 two-level games, 383
 see also individual agreements
International Organization, 21

International Organisation for Standardization (ISO), 54, 119
international political economy (IPE), 107–8, 129, 216
international regimes, 163, 165, 199, 248, 274, 280–2, 287, 368
desertification, 309
effectiveness, 263, 280, 287
study of, 263, 281
see also global governance
International Solar Cities Initiative, 200
International Studies Association (ISA): Environmental Studies Section (ESS), 79
International Union for Conservation of Nature (IUCN), 17, 336
International Union of Forestry Research Organizations, 189
interpretivism, 92
interspecies justice, 328–55
investment, 22, 31, 91, 95, 107–30, 318, 346
see also finance
ISO 14000, 119
Italy, 116

Japan, 27, 90, 142, 214, 261, 342
Japanese tsunami, 90, 98n6
Johannesburg Conference, 155n1, 190, 195, 253, 303–4, 313, 385
see also World Summit on Sustainable Development
Journal of Conflict Resolution, 17
Journal of Peace Research, 21
justice, 7, 19, 54, 61–2, 142–5, 150, 263, 328–47, 357, 366, 381, 383
compensatory, 329
distributional, 329, 333, 339
procedural, 329, 330, 333, 340–3, 371

Kenya, 95, 237
knowledge, 14, 24, 32, 54–8, 161–3
gender, 139, 152
indigenous/local, 315
as a political category, 162, 170
Kyoto Protocol, 114, 345

labor, 51, 139, 150, 257, 259, 300, 311, 314–15
land degradation, 122
see also agriculture; soil degradation
Latin America, 54, 126
see also individual countries
Least Developed Countries (LDCs), 346
legitimacy, 28, 57, 166–7, 176, 198, 253–4, 256–7, 262–3, 280–1, 287, 303, 328, 343, 370, 373
and scientific expertise, 166–8, 195
Leopold, Aldo, 334
liberal environmentalism, 26, 51, 59, 171, 180, 337, 360–1
see also shallow ecology
liberalism, 32, 45
liberal institutionalism, 17, 21, 26, 47, 49, 66, 321n1, 331, 359, 360, 386
neoliberalism, 22, 149, 161, 165
pluralism, 47, 68n8
liberalization, 109, 112–13, 128, 189–90, 251
see also globalization; trade
Limits to Growth, 301, 331
limits to growth, 16, 23–4, 27, 51–2
Lisbon process, 310
London Convention on Dumping at Sea, 336

maldevelopment, 141–2, 221–2
Malthus, Thomas, 147
neo–Malthusian approaches, 218, 220, 236
maquiladora firms, 128
Marine Stewardship Council, 253
marketization, 362–3, 369
Marxism, 47, 51–2, 60–1, 66, 168, 248, 332
see also structuralism; neo–Marxism
methodology, 14, 57, 78–97, 216, 218
formal modeling, 86, 89, 263, 291
methodological pluralism, 60, 97, 173–4
see also qualitative methodology; quantitative methodology
Mexico, 128, 224

migration, 89, 146, 148–51, 153, 155n3, 155n7
International Organization for Migration, 149
military and environment, 25, 31, 152, 213, 215, 217, 226, 235
militarization, 139, 146, 152
see also security
mining, 117, 142, 186
modernity, 61, 216, 228, 389
modernization, 141, 221, 229, 248
monitoring, 23, 91, 170, 189, 198, 372
Montreal Protocol on Substances that Deplete the Ozone Layer (1987), 114, 115, 254
moral considerations, 341, 347
see also justice
multilateral environmental agreements (MEAs), 114, 189–93, 260, 283, 307
multi–level governance, 3, 282, 309, 373
multinational corporations (MNCs), 21, 25, 31, 186, 187, 188–9, 191, 193, 202
see also industry; transnational corporations (TNCs)

Naess, Arne, 332–5
Narmada Dam scheme, 121
National Association for the Advancement of Colored People (NAACP), 143
National Science Foundation, 88
natural resources, 63, 85, 141, 301, 310, 338
conflict, 31, 85, 145, 219–22, 226
curse, 219, 229, 236–7, 273
trade, 31, 85, 112, 114, 226, 301, 310
Natural Resources Forum, 31
neoclassical economics, 108
neocorporatism, 311
neo–Marxism, 248
Netherlands, 305, 321
networks, 25, 27, 32, 59, 89, 153, 163–7, 186–95, 199–202, 226, 249–54, 308, 312–15
see also private governance

Network of Regional Governments for Sustainable Development (nrg4SD), 312
New International Economic Order (NIEO), 19
New Zealand, 27
Nigeria, 342, 375
nongovernmental organizations (NGOs), 58–9, 62, 94, 120–1, 140, 204, 300, 314, 343
environmental (ENGOs), 21, 25, 29, 68n7, 122, 127, 142, 281, 311, 313
international nongovernmental organizations (INGOs), 186–93, 199, 202
non–state actors, 6, 22, 30, 82, 167, 185–8, 199, 203, 245, 250–2, 256–7, 281, 284, 356, 363, 369, 373, 382, 389–90
see also transnational actors
North American Agreement on Environmental Cooperation (NAAEC), 115
North American Commission for Environmental Cooperation (CEC), 115
North American Free Trade Agreement (NAFTA), 31, 108
North–South (issues of), 331, 338–9, 346
and sustainable development, 306–7
see also developing countries; inequality; environmental justice
nuclear power, 232, 332
nuclear weapons, 16, 18, 19, 215

oceans, 19, 338–9
see also water
oil, 122, 123, 230
extraction, 117, 303
ontology, 161–5, 219
openness, 8, 119, 174, 225, 357, 359, 364, 371
see also transparency
Ophuls, 21, 48, 63

400 *Index*

Organization for Economic Cooperation and Development (OECD), 127
Our Common Future, 23, 141, 301, 302, 303
  see also Brundtland Report
ozone depletion, 23, 53, 114, 115, 118, 165, 220, 254, 302
  see also Montreal Protocol on Substances that Deplete the Ozone Layer (1987)

Pachari, Rajendra, 178
Pakistan, 155n6, 224
participation in global environmental governance, 245–72
Participatory Action Research, 93
participatory democracy, 198, 337, 385
participatory policymaking, 303, 309, 313
Partnerships for Sustainable Development, 253
persistent organic pollutants (POPs), 114, 120, 122
pesticides, 114, 128, 357
  see also Rotterdam Convention
Philippines, 224, 236
planetary boundaries, 211
plurality, 45–77
policy formulation, 196
policy instruments, 255, 281, 288, 307, 318, 346, 389
political ecology, 215, 237
political economy, 25, 53, 107–8, 109–11, 125, 128, 129, 216, 309
  see also global political economy; international political economy
political geography, 21, 162–3
pollution, 112, 117, 147, 274, 277, 301, 331, 333, 358, 369, 371
  air pollution, 116, 286
  marine pollution, 278, 283
  'pollution halo,' 117
  'pollution havens,' 117, 128, 130
  water pollution, 279
Polonoreste road project, 121
Pongola River Basin, 231, 233

population growth, 16, 137, 140, 146–53, 171, 220, 384
population control, 148
  see also Club of Rome; limits to growth; Malthus, Thomas
Portugal, 116
positivism, 79–80, 87, 92, 95–6
  generalization, 83, 277, 278, 291
  positivist methodology, 81–92
postmodernism, 161–3, 168–9, 170–1, 174, 179, 212, 217, 228
  deconstruction, 163
  see also critical perspectives
postpositivism, 80
  see also critical perspectives
poststructuralism, 212
  see also postmodernism; postpositivism; structuralism
poverty eradication, 317–18
Principles for Responsible Investment (PRI), 124
prior informed consent, 114, 357, 359
private governance, 29, 53, 281, 357, 367, 389
process and production methods (PPMs), 113
public attitudes, 79
public goods, 280, 291–2
  public bads, 280

qualitative methodology, 81, 92
  case studies, 82
  critical, 92–5
  positivist, 79, 81–3
  process tracing, 81–2, 98, 196, 277
quantitative methodology, 79–88, 263, 290–3, 386
  see also statistical analysis
quotas, 111, 252

race to the bottom, 112, 117
racial inequality, 62
Ramsar Convention on Wetlands, 336
rationalism, 55, 59, 81, 161–5
  see also positivism
Rawls, John, 328–9
realism, 26, 32, 45, 47, 48, 213, 217, 248, 330, 333
recycling, 120, 126, 127

regime analysis, 24, 76, 163, 280, 285, 359
  see also international regimes
regime complex, 50, 284
regulatory chill, 117, 118
religion, 139
  Catholic Church, 250–1
relocalization, 110
research design, 78–9, 81, 98
  case selection, 93
  coding, 89
  construct validity, 84
  data collection, 81, 85, 218
  hypothesis testing, 162
  see also qualitative methodology; quantitative methodology
resource capture, 220, 230, 234
resource scarcity, 153, 218, 301
  see also scarcity; conflict
Responsible Care, 119
Ricardo, David, 109
  see also neoclassical economics
Rio Earth Summit (1992), 118, 145, 231
  see also United Nations Conference on Environment and Development
Roosevelt, Theodore, 334
Rotterdam Convention, 114
Russia, 22, 261
Rwanda, 224

salience, 166, 176
sanctions, 278, 284, 287, 371
  see also regime compliance
Scandinavia, 18, 21
scarcity, 23–4, 87
  absolute, 16, 17, 19
  conjunctural, 16, 17
  environmental scarcity, 220–2
  see also resource scarcity; conflict
science and technology, 17, 21, 161, 169
  see also expertise; knowledge
science and technology studies (STS), 161
securitization, 213

security studies, 31, 145, 211–13, 215–28, 237
  see also environmental security studies
self-regulation, 29, 246
shallow ecology, 332–4, 337
Shiva, Vandana, 141
Sierra Leone, 222
small island states, 243, 263
social ecology, 330, 333, 336–8, 345–7
social movements, 187, 194, 388
social network analysis, 89, 98
social science theory, 97, 216
soil degradation, 332
  see also land degradation
South Africa, 122, 213, 224, 231–4, 303
Southern African Development Community (SADC), 232
sovereignty, 48, 66, 189, 197, 203, 284
Soviet Union, 22, 217
  see also Russia
Spain, 116
Specialization, 111
  see also comparative advantage; trade
species loss, 215, 249
  see also biodiversity; Convention on Biological Diversity
Sprout, Harold, 21
Sprout, Margaret, 21
statecraft, 170
  see also diplomacy
statistical analysis, 81–2
  see also quantitative methodology
Stockholm Conference, 18, 20, 51, 163, 190, 246, 313, 388
  see also United Nations Conference on Human Environment (UNCHE) (1972)
structuralism, 45, 47, 69
  methodology, 92–5, 217
  poststructuralism, 45, 47, 92, 212, 217, 219
  see also Marxism; world–systems theory
structural inequality, 62, 138, 141
subsidies, 111, 113, 234
  see also export subsidies

sustainability, 27, 46–8, 63–6, 142, 147, 187, 200, 245, 249, 263, 273, 276, 305–7, 313, 339–40, 357–8
  alternatives, 382
  analysis, 109, 316
  reporting, 356–80
  sustainable development, 140, 307–12, 319–20
  unsustainability, 49, 64, 67
  see also Brundtland Report; limits to growth
Sweden, 156
Switzerland, 90
symbolic politics, 300–27

tariffs, 111
  see also trade
technology transfer, 114, 117, 346, 383
Tinbergen Report, 301
toxic substances, 23, 25, 52, 61–2, 112–13, 126–8, 332, 365, 383
  Basel Convention, 114–15, 126–8, 339
  Rotterdam Convention, 114
  see also pesticides
toxic waste trade, 61, 126–8, 383
trade, 31, 84–5, 107–30, 191, 233, 255, 259, 301, 309–10, 359
  comparative advantage, 109
  competitiveness, 112, 117, 310
  liberalization, 109, 112–13, 128, 189–90, 251
  restrictions, 111, 113–14, 283
  specialization, 111
tragedy of the commons, 19, 48, 147, 231, 288, 289, 291
  see also Hardin, Garrett
transboundary natural resource management (TBNRM), 4, 6, 85, 237
transboundary pollution, 18, 274, 277, 286
transition perspective, 319–21
transnational actors, 59, 164, 172–5, 188–9, 193–8, 254, 388
  see also nongovernmental organizations

transnational corporations (TNCs), 58, 68n7, 108, 116–18, 130
  see also multinational corporations (MNCs)
Transnational Municipal Networks (TMN), 199–202
transnational networks, 185–210
transparency, 14, 29, 124–5, 173–4, 262, 356–75
  right to know, 357, 362, 364, 367, 370
  'transparency turn,' 356, 357, 361, 372
  see also 'Governance by disclosure'
transportation, 112, 199
Tuna–Dolphin debate, 108, 114, 197

uncertainty, 54–5, 161, 165, 229–30, 252
Union Carbide, 128
  see also Bhopal, India
United Kingdom (UK), 14, 18, 22, 32
United Nations Commission on Global Governance, 214, 247–8, 257
United Nations Commission for Sustainable Development (UNCSD), 317
United Nations Conference on Environment and Development (UNCED) (1992), 172, 301, 304, 306
  Rio Declaration, 302
  see also Agenda 21; Rio Earth Summit; Partnerships for Sustainable Development
United Nations Conference on Human Environment (UNCHE) (1972), 301
  see also Stockholm Conference
United Nations Convention to Combat Desertification (UNCCD), 309
United Nations Development Programme (UNDP), 122, 155n2

United Nations Educational, Scientific and Cultural Organization (UNESCO), 18, 151
United Nations Environment Programme (UNEP), 110, 122, 124, 140, 175, 259–61, 317–18
  Finance Initiative (FI), 124
  Global Environment Facility (GEF), 122
United Nations Framework Convention on Climate Change (UNFCCC), 122, 140
United Nations Third Law of the Sea, 348
United States, 14, 18, 22, 30, 32, 62, 86, 115, 121, 143, 222, 230, 261, 332, 334, 338, 343, 365, 381, 385
United States Department of State, 222
  see also Climate Action Report 2014
urban politics, 228, 308, 388
utilitarianism, 334–5

veil of ignorance, 329

war on terror, 212, 217
waste, 19, 52, 61–2, 109–18, 120, 125–8, 199, 283, 303, 306, 318, 332, 339, 357, 383
  see also Basel Convention on the Transboundary Movement of Hazardous Wastes and their Disposal (1989); e-waste; hazardous waste; toxic waste trade
water, 52, 85, 90, 116, 122, 220, 225, 232, 233–4, 274, 277, 318, 339
  dams, 93, 121, 253
  see also oceans
women, 61, 93, 137–53, 216, 308
  see also gender

Women's Environment and Development Organization (WEDO), 140, 142
Woodrow Wilson Center's Environmental Change and Security Project (ECSP), 218, 225
World Bank, 120–2, 191, 259, 309, 358
World Business Council on Sustainable Development, 54, 118, 186, 314
  see also World Summit on Sustainable Development (WSSD)
World Commission on Dams, 253
World Commission on Environment and Development (WCED) (1987), 301–2, 338–9
  see also Brundtland Report
World Development, 23
World Economic Forum (WEF), 309
World Environment Organization (WEO), 4, 115, 252, 259–61
World Health Organization (WHO), 260
World Meteorological Organization (WMO), 175, 256
World Order Models Project (WOMP), 22
World Social Forum, 313
World Summit on Sustainable Development (WSSD) (2002), 115, 118, 155n1, 190, 303
  Johannesburg Declaration, 155n1
world-systems theory, 47, 60, 95
World Trade Organization (WTO), 31, 54, 191, 259, 261
World Wide Fund for Nature, 58, 69, 186

'zone of agreement,' 49

Printed and bound in the United States of America